THE STRATIGRAPHY OF
THE BRITISH ISLES

THE STRATIGRAPHY OF
THE BRITISH ISLES

SECOND EDITION

DOROTHY H. RAYNER, M.A., Ph.D.

CAMBRIDGE UNIVERSITY PRESS

CAMBRIDGE

LONDON   NEW YORK   NEW ROCHELLE
MELBOURNE   SYDNEY

Published by the Press Syndicate of the University of Cambridge
The Pitt Building, Trumpington Street, Cambridge CB2 1RP
32 East 57th Street, New York, NY 10022, USA
296 Beaconsfield Parade, Middle Park, Melbourne 3206, Australia

First published 1967
Reprinted with corrections 1972, 1976
First paperback edition 1980
Second edition 1981

Text set in 10/13 pt Linotron Plantin, printed and bound
in Great Britain at The Pitman Press, Bath

British Library cataloguing in publication data

Rayner, Dorothy Helen
  The stratigraphy of the British Isles. – 2nd ed.
  1. Geology – Great Britain
  2. Geology, Stratigraphic
  I. Title
551.7'00941   QE261   80-49804

  ISBN 0 521 23452 2 hard covers 2nd edition
  ISBN 0 521 29961 6 paperback 2nd edition
  (ISBN 0 521 06047 8 hard covers 1st edition)
  (ISBN 0 521 29820 2 paperback 1st edition)

CONTENTS

PREFACE TO THE SECOND EDITION

Since the first edition was written, in the mid-1960s, many changes have passed over the geological scene. This edition reflects some of those changes and though the arrangement of the chapters is broadly the same, the scope is slightly enlarged to touch on those countries whose histories have parallels to that of the British Isles. In particular much more is known of the neighbouring continental shelf, and a general introduction to 'British Seas' appears in Chapter 8; thereafter the geology of the sea floors is summarized in the relevant chapters. The second part of the Epilogue reviews briefly the major economic resources that are related to stratigraphy. In this revision nearly half the figures are redrawn or new and these owe much to the skill and patience of Mrs R. J. Boud.

As in the first edition I have been much helped by various people, with expert assistance, discussion and access to unpublished material, and am most sincerely grateful to them: Professor E. H. Francis, Professor M. R. House, Dr L. F. Penny, Dr D. B. Smith, Dr R. B. Rickards; in the Leeds office of the Institute of Geological Sciences many friends and especially Dr M. A. Calver and Dr W. H. C. Ramsbottom; in the Edinburgh office Dr D. Evans and his colleagues for information on offshore geology; in Dublin Professor C. H. Holland, Dr C. J. Stillman and Dr J. H. Morris.

Dr Peigi Wallace and Professor D. Curry kindly lent me copies of *Special Reports* before their publication by the Geological Society of London (*Reports* are listed on p. 430) and I am much indebted to Dr J. C. W. Cope for the opportunity to look over the correlation charts of the Jurassic Report in Swansea. At the time of writing most, but not all, of the *Reports* have been published and though much of the recommended nomenclature has been incorporated some minor variations in treatment have been unavoidable.

Sir Kingsley Dunham gave generous assistance on economic resources and read much of that section in draft. Mr R. G. W. Brunstrom undertook the pages on hydrocarbons; Professor P. McL. D. Duff and Dr D. Naylor supplied information on Scotland and Ireland. I am also particularly grateful to Dr M. H. Dodson for guidance in geochronology and for recalculating a number of isotopic dates (cf. p. 3). In the late stages of the work I have been much indebted to the care and guidance of the staff at the Cambridge University Press.

It is a pleasure to thank all the foregoing but the usual proviso should also be emphasized – the faults and omissions, all too possible in a work of this length, are mine alone. Moreover while personal views should not obtrude in a textbook a little self-indulgence has crept into the last chapter.

As in the earlier edition I thank all holders of copyright for permission to reproduce copyright material; the sources of the figures are given in the legends. The revision was completed during the tenure of a Leverhulme Fellowship and that award is gratefully acknowledged.

1979–1980   D. H. R.

PREFACE TO THE FIRST EDITION

British stratigraphy has been studied and discussed for some 150 years and the writer who presents yet another synthesis should set out its scope and purpose. In this book I am largely concerned with surface and near-surface processes and the rocks that resulted from them. The outcome is stratigraphy in a somewhat restricted sense and not 'historical geology'. Igneous, metamorphic and structural aspects appear only cursorily, when bound up with stratigraphical problems.

It also happens that the present time is opportune for a review of the subject. In the past two decades much has been discovered and assembled about Ireland, and Irish stratigraphy has become more closely integrated with that of Great Britain, to the benefit of understanding on both sides. Advances in allied sciences, especially sedimentology and isotopic dating, have altered subtly both the factual syntheses and deductions made from them. The permanence of ocean basins and continents has once more been called into question and the stratigraphy and structure of these islands, on the edge of a continent, have a significance that is global as well as European. Lastly by historical accident Britain is the origin of many stratigraphical terms, especially of several systems and their larger sub-divisions, and definitions, redefinitions and much nomenclatural turmoil are in the air.

Thus some fresh breezes have found their way through the classic edifice of British stratigraphy, but to the writer of a textbook this also brings drawbacks and the student reader should perhaps be warned that it is precisely in the most active fields that a general work becomes dated, and that he may have to search further after a few years. For this I am not greatly concerned: any writer can only set down, as clearly as may be, the facts and interpretations of his age, make plain the areas of doubt and leave future research to confirm or reject his generalizations.

For whom is this book written? I have aimed at making the subject intelligible to those with little previous knowledge of it, but some acquaintance with the broad principles of geology. Such an acquaintance

can be gained from Read and Watson's *Introduction to Geology*, vol. I
(1962) or Holmes' *Principles of Physical Geology* (2nd edn., 1965) – to
name but two recently published works. In particular I have not explained
many terms available in the former, but some topics of especial importance to stratigraphy are taken a little further in the Introduction.

Selected references appear at the end of each chapter, and within the
last – which is some form of concluding (but scarcely 'conclusive')
epilogue, touching on a few broader or more controversial problems. All
references are collected together in the list at the end. They are probably
beyond the needs of the elementary student, but it has been my hope that
the more advanced, or perhaps the worker in more distant fields, while
familiar with the bulk of the text, may find the list useful as a first source
for more detailed study. Moreover, if one principle more than another has
imbued the work it is that while generalizations and oversimplified
deductions are inevitably the substance of a textbook, yet they stand
weakly alone, and that the evidence should be accessible as far as space
allows.

At the time of writing there is no formal British Code of Stratigraphical
Nomenclature, though the American Code, which is often taken as a
guide, may be found in several standard works. In these circumstances I
have thought it best to retain traditional or current British terms even
though a few may not be strictly correct, or there may be minor
inconsistencies between one system and another. At least this retention
should not introduce any new complexities.

Most of the chapters have been read in draft by one or more persons
more deeply versed than I in the various geological systems and to them I
am most sincerely grateful: Dr H. W. Ball, Mr M. A. Calver, Mr D.
Curry, Dr W. T. Dean, Professor D. T. Donovan, Dr K. C. Dunham,
Dr J. C. Harper, Professor J. E. Hemingway, Dr R. W. Hey, Professor
M. R. House, Professor W. Q. Kennedy, Dr J. W. Neale, Mr L. F. Penny,
Dr W. H. C. Ramsbottom, Professor R. M. Shackleton, Professor
F. W. Shotton and Dr J. V. Watson. Their friendly advice has added
detail and perspective to the descriptive sections and saved me from many
errors and questionable deductions. They are in no way responsible for the
shortcomings that remain. In addition I would like to record my thanks to
very many friends, in Leeds and elsewhere, who have provided information on many topics and borne patiently with my enquiries.

The illustrations reproduce or incorporate much copyright material; the
sources of the figures appear in their legends. I am glad to thank all
holders of copyright for permission to reproduce this material.

1966   D. H. R.

Table 1. *The Phanerozoic time-scale* xiii

The age of the base of each stratigraphical unit is estimated in millions of years. The sources are: **1,** Harland *et al.* (1964, pp. 260–2); **2,** Lambert (1971, p. 25); **3,** Armstrong (1978, p. 90). The first is still a useful guide for the smaller divisions, when compared with the other two; the third incorporates the revised constants and some new data. All indicate some imprecision, especially among the older systems, and it should be emphasized that the time-scale will continue to evolve as new techniques afford greater precision and reliable rocks or minerals are found more firmly related to stratigraphical zones.

The Appendix (pp. 418–29) includes those stratigraphical systems where comprehensive zonal sequences are available in the British Isles; the tables given there, however, incorporate a few revisions in terminology (e.g. among the Jurassic divisions) and thus differ a little from those in this time-scale which is transcribed from Harland *et al.* (*loc. cit.*).

	1	2	3
CAINOZOIC			
QUATERNARY			
PLEISTOCENE	1.5–2		
TERTIARY			
PLIOCENE	*c.* 7	7	
MIOCENE			
Upper	*c.* 12		
Middle	*c.* 18–19		
Lower	26	26	
OLIGOCENE			
Upper	?		
Middle	31–32		
Lower	37–38	38	
EOCENE			
Upper	*c.* 45		
Middle	*c.* 49		
Lower	53–54	54	
PALAEOCENE			
Upper	58.5		
Lower	65	65	65
MESOZOIC			
CRETACEOUS			
UPPER			
Maestrichtian	70		
Campanian	76		
Santonian	82		
Coniacian	88		
Turonian	94		
Cenomanian	100	95	⩾96
LOWER			
Albian	106		
Aptian	112		
Barremian	118		
Hauterivian	124		
Valanginian	130		
Ryazanian	136	135?	143

	1	2	3
JURASSIC			
UPPER			
Purbeckian	141		
Portlandian	146		
Kimmeridgian	151		
Oxfordian	157		
Callovian	162		
MIDDLE			
Bathonian	167		
Bajocian	172		
LOWER			
Toarcian	178		
Pliensbachian	183		
Sinemurian	188		
Hettangian	190–195	200?	212
TRIASSIC			
UPPER			
Rhaetian			
Norian			
Karnian	205		
MIDDLE			
Ladinian			
Anisian	215		
LOWER			
Olenekian			
Induan	225	240?	≥247
PALAEOZOIC			
PERMIAN			
UPPER			
Tatarian	230		
Kazanian	240		
LOWER			
Kungurian	255–258		
Artinskian	265–268		
Sakmarian	280	280	289
CARBONIFEROUS			
SILESIAN			
Stephanian	290–295		
Westphalian	310–315		
Namurian	325		341
DINANTIAN			
Viséan	335–340		
Tournaisian	345	370	367
DEVONIAN			
UPPER			
Famennian	353		
Frasnian	359		
MIDDLE			
Givetian			
Eifelian	370		

	1	2	3
LOWER			
Emsian	374		
Siegenian	390		
Gedinnian	395	415?	416
SILURIAN			
Ludlow			
Wenlock			
Llandovery	430–440	445?	⩾446
ORDOVICIAN			
UPPER			
Ashgill			
Caradoc	445		
LOWER			
Llandeilo			
Llanvirn			
Arenig	c. 500	515?	⩾509
CAMBRIAN			
UPPER	515		
MIDDLE	540		
LOWER	570	590?	~575

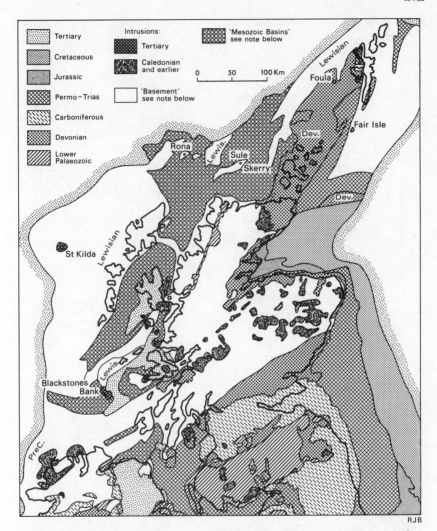

Fig. 1A. The geology of the British Isles and adjoining parts of the continental shelf, northern section. 'Mesozoic Basins' are deduced or known to contain sediments from Permo-Trias to Cretaceous or parts of that range. The 'Basement' regions range from Precambrian to pre-Permian or are undetermined; some deduced or probable systems are indicated by lettering. Reliability of offshore data varies from reasonably secure to decidedly speculative. All three maps are adapted from *Sub-Pleistocene geology of the British Isles and adjacent continental shelf*, 1:2 500 000, 2nd Edn (1979), by permission of the Director, Institute of Geological Sciences; N.E.R.C. copyright.

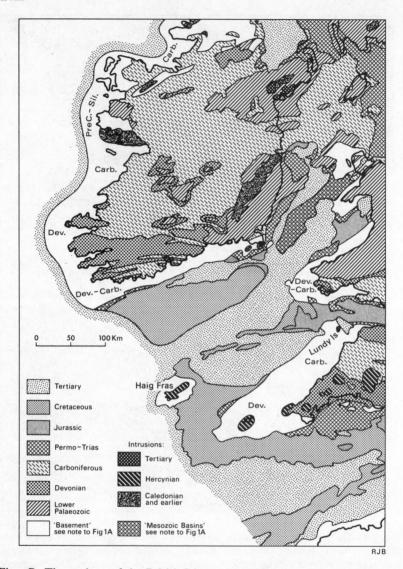

Fig. 1B. The geology of the British Isles and adjoining parts of the continental shelf, western section. Much detail on the south side of the English Channel is omitted; see also notes to Fig. 1A.

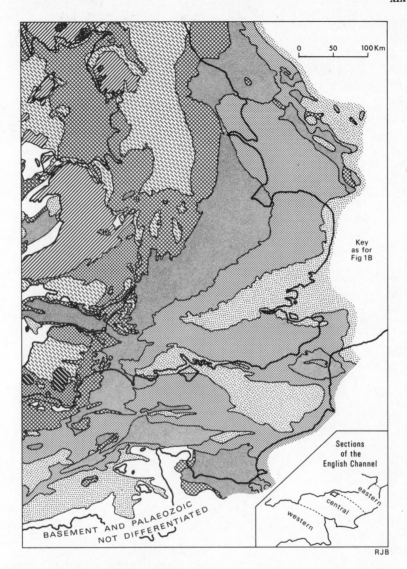

Fig. 1C. The geology of the British Isles and adjoining parts of the continental shelf, eastern section; see also notes to Fig. 1A.

I

INTRODUCTION

There have been many definitions of stratigraphy, differing slightly in scope and emphasis, but its core is the study of the earth's history, in so far as it can be interpreted from the outermost layers of the crust and particularly from the succession of stratified rocks. As a formal term it is a latecomer (appearing in 1865, according to the *Oxford English Dictionary*); but the description of stratified rocks had been placed on a scientific basis long before, in the late eighteenth and early nineteenth centuries – a phase of geological history that in this country will always be associated with William Smith. Since that time, when investigations began on the more accessible fossiliferous strata, with emphasis on the essential geological map, there have been far-reaching extensions in methods and scope. Even the last few decades have seen major developments.

Much more has been discovered about rocks at depth, both on land and beneath the sea, especially by geophysical methods and deep boreholes. Economic incentives are influential here and with problems of energy and material resources much to the fore these incentives are likely to continue. A range of structural and geochronological techniques has been applied to highly metamorphosed and deformed rocks, so that Precambrian history in particular is much better known; even the traces of Precambrian life are beginning to present a logical sequence. From a greater knowledge and analysis of modern sediments our interpretations of sedimentary rocks and their origins are on a more secure basis. With more international collaboration some standardization in stratigraphical nomenclature has been initiated. Finally the plate tectonic hypothesis has been an outstanding catalyst in promoting new syntheses, new collaborations between the various branches of earth sciences and the questioning of old assumptions. Not only have maps of the world's surface at various periods been radically altered but also the thinking of geologists in more subtle ways.

For the most part these developments are complementary to, and not at variance with, the older classic views of stratigraphy and earth history.

One major principle, however, has come into question, that of uniformitarianism. This is not, of course, a complete overthrow. The simplified version, that the present is key to the past, is still largely relevant to that part of earth history where it was first formulated – the Phanerozoic systems (i.e. Cambrian and later periods). It is the much greater understanding of Precambrian history, more than five-sixths of the total span, that has caused this revision. As outlined in the next chapter this enormous stretch of time saw fundamental changes in the composition of oceans and atmosphere, with concurrent changes in the chemistry of sediments and the slow evolution of the primitive forms of life. Even within Phanerozoic history there have been some unidirectional changes, most obvious in the biological sphere. For instance, the clothing of the land surface with an advanced rooted type of vegetation, which developed in late Silurian times, must have altered regimes of denudation, transport and deposition.

Climatic comparisons of past and present are less pronounced. In one sense the Quaternary glaciation is still with us, and the present time may well represent only the modest opening of an interglacial regime. In earlier periods when there were no polar ice-caps (possibly from the Trias to the Eocene) it is likely that the climatic belts were slightly different, which would influence the distribution of faunas and floras and their relations to latitude.

GEOCHRONOLOGY

Time, enormous stretches of time which are measured in millions of years, forms a background to most aspects of geology, particularly to historical geology and stratigraphy. In the last few decades the measurement of this basic dimension has become a major subject in its own right, geochronology, in which the age of a rock group or a geological event is measured in years.

These age determinations are based on the analysis of one or more pairs of isotopes (radioactive parent and daughter), the most commonly employed being given in Table 2. The first of each pair is transformed into the second at a known rate and from the relative amounts present an estimate can be made of the time since the mineral was formed, when it contained none of the daughter isotope. Analyses may be made on individual mineral grains collected from a rock outcrop or from whole rock samples. In modern work experimental errors are small, normally less than 5% and sometimes as low as 1%. However, there remain numerous uncertainties in the interpretation of isotopic ages, mainly because of the possibility that the daughter isotope may be lost through diffusion out of the mineral in which it was formed.

There are various aims in geochronology. One is to calibrate the Phanerozoic time-scale so that the boundaries of the systems and their major divisions can be expressed in years. Ideally the measurement would be made on a sedimentary rock that also contains reliable stratigraphical

Table 2. *Some isotopes used in age determination*

Method	Parent	Half-life	Daughter	Minerals
Uranium–lead	^{238}U	4 500	$^{206}Pb (+ 8^4He)$ ⎫	⎰ Zircon, uraninite
Uranium–lead	^{235}U	700	$^{207}Pb (+ 7^4He)$ ⎭	⎱ and pitchblende
Rubidium–strontium	^{87}Rb	50 000	^{87}Sr ⎫⎬⎭	⎰ Micas, potassium ⎱ feldspars
Potassium–argon	^{40}K	1 300	^{40}Ar ⎫⎬⎭	⎧ Micas, sanidine, ⎨ hornblende and ⎩ glauconite

The half-life (approximately, in million years) is that in which the quantity of the parent isotope is reduced by half. The decay constant (a factor quoted in the Rb–Sr method) is the proportion of a certain number of atoms that decay in a certain time. The isochron Rb–Sr method depends on the ratios ^{87}Rb to ^{86}Sr and ^{87}Sr to ^{86}Sr in several rock or mineral samples and is valuable in deciphering the history of igneous and metamorphic complexes.

Individual age determinations quoted in later chapters have been recalculated where necessary in accordance with internationally agreed decay constants (Steiger and Jäger, 1977).

fossils but this is only rarely possible. There are few sedimentary minerals that are formed at or near the time of sedimentation and can be separated in a pure state for analysis; they include some clay minerals in shales but here the separation problems are formidable. Glauconite is more generally useful and is not uncommon in marine rocks, but loses argon easily, particularly if it has been deeply buried. Lavas and tuffs containing zircon, sanidine and micas may be very useful if interbedded with suitable sediments. In general, however, these direct methods have been most successfully applied to late Mesozoic and Tertiary rocks, for instance in North America, and not so much in this country. The use of intrusive rocks is more indirect. To be a reliable gauge of stratigraphic position the intrusion should cut fossiliferous strata of one age and be overlain by, or contribute recognizable detritus into, a later group with only a small time-gap between the two. In this country many Palaeozoic dates are based on intrusions, including Caledonian and Hercynian granites, with varying degrees of precision. The methods quoted so far apply primarily to regions little affected by later metamorphism or intrusion – that is, once the radioactive clock has been set the rock has remained a closed chemical system. In other situations, such as a major orogenic belt, the clock may be re-set by later tectonic or thermal events, sometimes millions of years later than the formation of the rock. There are two main causes for these late dates, or apparent ages. Deep-seated igneous or metamorphic complexes may cool very slowly before reaching the crucial 'blocking' temperature at which isotopic diffusion ceases and the system becomes closed. These late effects usually follow uplift and erosion, and have been called cooling ages. They show up as systematic variations in age related to structural level, or even altitude, with certain minerals always older than the others; for example, muscovite usually appears older than co-genetic biotite because its blocking temperature is higher. Then in regions of multiple orogenic phases late dates often reflect the last tectonic or thermal event which has extensively 'overprinted' the earlier intrusive or metamorphic phases. Both situations commonly necessitate the cautious ascription of minimum ages only, and both are well known from the history of the Caledonian orogeny in Britain. One of the virtues of whole rock methods is that they help to distinguish early effects from later overprinting.

A third, and probably the most influential, outcome of geochronology has been to establish a time-sequence and broad correlations within the great Precambrian shield areas of the world. Here isotopic data form the essential framework within which the tectonic and metamorphic sequence can be set; together with this sequence the data establish and define the major provinces or ancient fold belts of the shield. Although British Precambrian rocks are small in area they have been investigated by the same methods and year by year a greater understanding of their grouping and history has been gained.

A final isotopic method, on a very different scale, is based on the radioactive isotope of carbon, ^{14}C, and it is often called radiocarbon dating. Very small amounts of ^{14}C are produced in the upper atmosphere and as carbon dioxide it is incorporated in the tissues of plants and animals. While these are alive the proportion of ^{14}C is in equilibrium with that of the atmosphere, but after the organism dies no more is absorbed and the isotope is gradually lost, the half-life being 5730 years. Organic substances that originated as much as 35 000–40 000 years ago can be dated by radiocarbon analysis, and by special techniques the dating can sometimes be carried back to about 50 000 years. The method is thus valuable in late Quaternary stratigraphy and for dating archaeological objects.

SEDIMENTARY ROCKS, STRUCTURES AND ASSOCIATIONS

The raw materials of stratigraphy are rocks (sedimentary, volcanic, metamorphic), their structures and relationships and any fossils they may contain. Within the scope of this book the sedimentary rocks are the most voluminous, and some of their characters are reviewed here.

Sedimentary structures, cross-bedding and graded bedding

These two types of bedding have a double significance in stratigraphy; they are important as environmental indicators and may also be used as guides to the orientation of strata (Fig. 2). The foreset beds of cross-bedding normally dip in the direction of the current immediately responsible for them. In certain conditions, such as a meandering stream, these

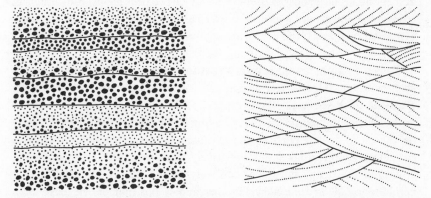

Fig. 2. Graded bedding and cross-bedding. On the left four beds are graded and show an upward decrease in grain size; on the right the majority of inclined foreset beds are concave upwards. (Adapted from several sources.)

directions may swing widely on either side of the valley trend as a whole, but in general when many readings are taken and analysed, in conjunction with larger structures such as channels, they constitute valuable data on the current directions of, say, a delta or a river system. In graded beds there is a gradual upward decrease in grain size, for instance from coarse sand at the base to silt or even mudstone at the top.

Graded bedding, from greywackes and quartzites in particular, has been used to determine the orientation (i.e. the original top and bottom) of formations in many folded belts' of the world, including the Scottish Highlands, Southern Uplands and north-west Ireland. The use of cross-bedding depends on the truncation of the foreset beds. In certain types these tend to curve tangentially into the normal bedding planes at the base but to be truncated at the top, having been eroded by the next influx of sediment. Nevertheless if such structures are examined in a succession whose superposition is indisputable they commonly include some indeterminate examples. It is therefore imprudent to rely on isolated items of either type of bedding, but when repeated many times and combined with structural evidence they are valuable stratigraphical guides.

Cross-bedding may develop in many situations, given the necessary current speed and sediment supply, the latter being usually of sand grade. It tends to characterize the shallow waters of shores, deltas or rivers. Aeolian cross-bedding may occur in rather large units and sometimes with slightly steeper foresets. A variety of ripple marks is formed by currents in shallow water but there are also deep water variants and an aeolian type. Rootlet beds result from plant growth both above and below water-level and abundant large desiccation cracks are one of the clearest indications of temporary emergence. Intertidal and subtidal muds and silts are often reworked by many types of marine organism (bioturbation), leaving tracks and burrows or largely destroying the original bedding.

Thus a combination of several structures may suggest sedimentation in shallow water. 'Deep water' is a more problematic ascription. Turbidites, resulting from sediment-charged gravity flows, are known to accumulate at the foot of the continental slope while fine pelagic muds spread widely on the more distant parts of the sea floor. Nevertheless some turbidites have been found in lakes or at relatively shallow depths. Among ancient environments deep waters are probably most securely deduced from a combination of sedimentary features and a position at the edge of a continental plate margin.

Disturbed bedding, slumps and gravity slides

In most rocks that are unaffected by folding the bedding planes are horizontal or nearly so, excluding such obvious cases as current-formed foresets or accumulations on the flanks of a reef. Occasionally however, a normally bedded sequence contains units which are strongly convoluted, or altogether disrupted, resulting in balled-up pillow-like structures. These are slumps, or slump units, and are most common in sandstones though occasionally found in siltstones and carbonate rocks. Some of the less disrupted types result from the curving over of the tops of foreset beds. It seems that most slumping took place on a subaqueous slope while the sediment still held enough water to be unstable, and the dislodged mass crumpled up as it slid. The slope, however, need not have been more than a few degrees and depth of water does not appear to be crucial. Some kind of triggering mechanism seems likely and earthquakes have been suggested. Slumping in this sense is largely a sedimentary structure, but tectonically induced gravity sliding may occur on a larger scale, as is evident in the Alpine chain. Here there are large transported blocks (making up mélanges or wildflysch) or great rafts of strata (olistostromes). A modest British example is mentioned on page 35.

Unconformities and other breaks in sedimentation

The most obvious result of a period of uplift and erosion is the major unconformity – an angular unconformity where there is a discordance of dip above and below, and a disconformity without that discordance. There may also be smaller breaks, variously known as diastems, lacunae or

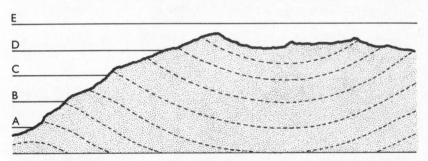

Fig. 3. Overlap and overstep. A sequence of beds, A–E, is laid down against an older landmass. Beds A–C show overlap against that landmass and the successive positions of their shores can be determined. Bed D, however, oversteps several of the older gently folded strata, and the shore positions for upper D and bed E cannot be determined.

hiatuses, in which evidence of erosion is inconspicuous or may only result from a prolonged dispersal of sediment. A non-sequence is a gap that is only detectable by palaeontological means, usually the absence of a zone or zones. It has also been suggested that sedimentation itself is a somewhat spasmodic process, and that the common bedding plane represents a lull in sedimentary influx.

Unconformity with overlap (Fig. 3) results from uplift and erosion, followed by a marine transgression farther and farther over the submergent land surface. Overlap is valuable in palaeogeographic reconstruction because it presents evidence of this retreating shoreline. No such inference can be drawn from unconformity with overstep, where a single formation oversteps or transgresses more than one underlying one, because there is no indication of how much farther the sea extended before the shoreline was reached. In both situations later folding or faulting may complicate the final structure. As a result the distinction between overlap and overstep may be debatable, and the palaeogeography similarly uncertain. An example affecting Ordovician geography is given on page 91.

Cyclic sedimentation

This refers to the vertical repetition, several times over, of a few sedimentary types, each cyclic unit being known as a cyclothem. An 'ideal cyclothem' may be either of the following:

d	coal	⎫	*a*	limestone	⎫
c	sandstone	⎬ cyclothem	*b*	sandstone	⎪
b	shale	⎪	*c*	shale	⎬
a	limestone	⎭	*d*	coal	⎪ cyclothem
d	coal	⎫	*c*	shale	⎪
c	sandstone	⎬ cyclothem	*b*	sandstone	⎬
b	shale	⎪	*a*	limestone	⎭
a	limestone	⎭			

Cyclic sedimentation is most conspicuous among various types of shallow water rocks, some of the best examples resulting from fluctuations between the sharply distinct environments just above and below sea-level – such as the simplified versions set out above. However other cycles have been recorded from evaporite, turbidite and fluviatile associations and overall there is much variation in scale and complexity. In practice the ideal type is nearly always complicated by minor omissions or small-scale oscillations.

The causes of cyclothems have been much debated and probably there are several. Some seem to reflect largely sedimentary events within the basin or catchment area, such as changes in river position or regime. For others a tectonic cause has been invoked or, particularly in recent years, a eustatic one. Eustatic changes are strictly those of ocean-level (not uplift or depression of the land) and the clearest mechanism is glacio-eustatic, or the withdrawal of water during major glaciations to form ice-caps. Outside periods of known glaciation – and several cyclic sequences belong to such periods – the most plausible cause is variation in ocean-floor topography related to changes in the rate of sea-floor spreading, so that with an enhanced rate there is overspill of oceanic water onto the edges of the continents. This view is supported by some degree of correlation between spreading rates and transgressions or regressions in late Cretaceous and Tertiary times. Unfortunately such comparisons cannot be directly sought in earlier systems, because no sea floor earlier than the Jurassic survives.

Stratal thickness, isopachs and rates of deposition

Thickness of strata is a basic component in stratigraphic syntheses, linked with lateral variations in thickness and changes in sedimentary type. Sedimentation is commonly accompanied by subsidence, and if there is evidence throughout that the formation was deposited near sea-level, then the thickness becomes also a measure of the subsidence. It is often convenient to summarize changes in thickness in the form of an isopach map (cf. Figs. 46, 67); the isopachs, being lines denoting equal thickness, form a system of contours based on points where the total thickness of the unit can be measured.

Thickness measurements normally refer to the rock unit as it exists at present, but there may have been changes during its history, for various reasons. Compaction is a diagenetic process affecting clay rocks. In their earliest stages these contain a large amount of water, which is gradually expelled by the weight of superincumbent layers so that the thickness is much reduced, perhaps to a quarter or less. Tectonic stress may also alter rock dimensions, either by lateral extension, with consequent thinning, or by compression to give 'tectonic thickening'. It has been estimated that in parts of North Wales the Lower Palaeozoic rocks have been about doubled in thickness owing to deformation during the Caledonian orogeny.

In favourable circumstances rock thickness may give an indication of subsidence but it is far more difficult to relate it to duration or the time taken to deposit a certain number of metres. Rates of sedimentation are elusive and the best attempts at measurement are either on a very small scale or a very large one. Certain laminated beds are formed by annual variations in the sedimentary regime, the best known being the periglacial varves; these may be counted in years. Then since isotopic determinations supply the duration of several of the later systems and of their major subdivisions, it is possible to deduce, for instance, that 300 metres of Lower Jurassic shales were laid down in about 10 million years. But to extrapolate within this span and assume that a certain 3 metres represent

100 000 years is hazardous, in that it presupposes a regularity of deposition that is hard to justify. Nevertheless, with some assistance from an exceptionally detailed zonal scheme, it has been suggested that a typical marine band of Namurian age (Upper Carboniferous), six to nine centimetres thick, represents no more than 10 000 to 12 000 years. This would be a very unusual degree of precision and probably should be taken only to indicate the *order* of time involved.

Lateral variation, facies and environments

In the same way that there are variations in thickness there are lateral changes in rock types with their associated structures and fossils. Such associations introduce the concept of facies – one essential to stratigraphy but not easy to define concisely. 'Facies' was first used in 1838 in a description of the Swiss Jura mountains, more or less in its modern sense, to describe lateral changes in lithology within a stratigraphical unit. Bound up with this is the concept of analogous lateral changes in the environments that gave rise to the sediments and which have their counterparts at the present day.

It is thus postulated that during a certain period the conditions of deposition varied from place to place – sand banks and channels here, mud-flats there, and shell banks a little way offshore; or perhaps a mangrove swamp, a lagoon, a barrier reef and a steep outer slope to the ocean floor. Each of these environments produces different bottom sediments, characterized by different faunal and floral assemblages; when 'fossilized' they become different facies – sandy, muddy, or calcareous, all of the same age and passing laterally into one another. The names given to the facies may pick out one aspect of the whole – the lithology (carbonate facies), fauna (shelly and graptolitic facies), tectonic or regional setting (trough and shelf facies), environment (lagoon facies). The lithological version is the most factual, universal and probably the most satisfactory, while the last two contain a greater element of deduction, but all have been used.

A detailed analysis of facies is an essential component in the assessment of past environments, and in favourable circumstances these may be set in a larger frame, in relation to contemporary seas, lands and mountain ranges, climatic belts and latitude, continental and oceanic margins, island arcs and vulcanism. A review of past and present environments shows that the major types are relatively few, though naturally with enormous variation in detail. Small though the British Isles be, their history has been diverse enough to exemplify most of these types:

fluviatile, of river valleys and lowlands
deltas and coastal swamps
deserts and semi-arid regions
shallow saline seas and salt-flats
reefs, with back-reef and fore-reef facies
glacial and periglacial environments

Two complex associations are more difficult to summarize in environmental terms. There were long periods (e.g. Jurassic and Cretaceous) when marine sediments accumulated under conditions that at times were undoubtedly shallow; these may perhaps be called shelf or epicontinental seas. The sediments were varied but only rarely included coarse clastics. Then in Lower Palaeozoic times a range of thick sedimentary sequences were formed which in earlier terms might have been called a geosynclinal facies and were primarily related to sedimentation at or near the margin of continental lithospheric plates (p. 14). The products of erosion were swept down into the neighbouring seas, typically by turbidity currents bearing much detritus, some of it coarse. The accumulated sediments include greywackes, siltstones and mudstones, with graded bedding and lamination common. When, as in the British Ordovician, these are accompanied by vigorous vulcanism, this is of a type characteristic of destructive convergent plate margins, or in more geographical terms, forming part of a volcanic island arc or volcanoes along the edge of a continent. Behind the plate margin these facies may pass into a shelf version that is normally thinner, finer, occasionally including carbonates and where vulcanism is reduced or absent.

STRATIGRAPHICAL NOMENCLATURE AND CORRELATION

The past two decades have seen considerable progress in making stratigraphical nomenclature more precise and more international. Various stratigraphic codes have been formulated in different countries but they have much in common and there is now an International Stratigraphical Commission. In turn correlation between different parts of the world has become rather easier. The main methods and terms that come within the scope of this book are lithostratigraphic, chronostratigraphic and biostratigraphic. In the formal codes and recommendations there are many more terms than are given below, but they are not essential in the present context and references to the technical publications are given at the end of the chapter. Geochronology, or the time-scale recorded in years, is not directly part of stratigraphical nomenclature but can be used in parallel.

Lithostratigraphy

This is a relatively simple concept whereby rock units or divisions are defined by, and named after, their dominant lithology. Sometimes a measure of correlation is implied, in that strata of the same name in different areas are thought to be of approximately the same age, but this is not necessarily so. Some conspicuously diachronous units (p. 14) can be accommodated in a lithostratigraphic nomenclature. The categories in order of decreasing size are: Supergroup, Group, Formation, Member, Bed. These terms are coming to be used in British stratigraphy, though this is an innovation. Their scope is not necessarily comparable from one system to another. The first (e.g. Dalradian Supergroup) may cover very large thick stratal assemblages whose history is longer than that of a major Phanerozoic system. The term bed may also be informal, in its common geological usage.

Chronostratigraphy

This system of nomenclature is primarily concerned with divisions of rocks based on time, and there are two sets of terms – those based solely on time, and those based on rock sequences:

Era	
Period	System
Epoch	Series
Age	Stage

Thus the strata of the Silurian System were formed in the Silurian Period. The boundaries between the divisions are defined by marker points that are considered to be synchronous – or at least as far as that can be determined. They may be faunal or floral (the appearance of a new species or assemblage), climatic, eustatic (reflecting changes in ocean-level), volcanic, or magnetic. The beginning, or base, of each chronostratigraphic division is defined by a marker point in an agreed type locality, or stratotype. The terminal or upper limit coincides with the base of the succeeding division.

The full succession of stratal or chronostratigraphic terms makes up the Standard Stratigraphic Scale – a scale which in theory is world-wide and is the most important foundation of stratigraphical correlation. It is these terms – especially system, series and stage – that are used in correlation charts and tables to show which rock sequences are contemporaneous. Those of the British Isles are given in the Appendix (pp. 418–29). Since the compilation of an International Standard Scale, with its host of agreed terms and stratotypes, is a lengthy process, regional (or non-standard) scales are useful within a country or larger geological province.

Correlation methods, including biostratigraphy

Correlation methods by which chronostratigraphic divisions can be recognized are varied, but in countries rich in fossiliferous strata, such as the British Isles, the use of fossils – or biostratigraphy – is the traditional method and still the most common. A small but fundamental unit here is the zone, or more precisely the biozone. In practice this is usually based on a small number of species (an assemblage biozone), though sometimes the abundance or acme of a single species will suffice. Commonly, but not always, a stage is recognized by more than one biozone. Moreover in many systems several groups of fossils are available as indices so that the zonal sequence employs, say, brachiopods, corals, goniatites, foraminifera, conodonts and plant spores; all these are used in Carboniferous stratigraphy. The limits of the biozones based on different groups frequently do not coincide, but all can contribute to the chronostratigraphic scheme.

In the British Isles the non-palaeontological markers in chronostratigraphy and correlation include the evaporites of the Permian cycles; widespread tuffs, tuff-groups and other volcanic products (Ordovician); climatic changes (glacial and interglacial, Quaternary); and possibly eustatic transgressions and regressions (Carboniferous). Given suitable rocks some ancillary dating methods can be derived from comparison with established palaeomagnetic data. These may relate to periods of normal or reversed polarity in the more recent systems, or to the magnetic pole positions which are known to have fluctuated over long periods of time to give an apparent polar wander path, unique to each plate of the earth's crust. Neither is a primary chronological method but may add refinement or supporting evidence on the dating of a rock or formation whose age is already known within certain limits. On a larger scale the thick groups of tillites found among late Precambrian sequences in many lands around the North Atlantic seem to be due to broadly contemporaneous glaciations.

Diachronism

This term refers to a lithological unit, or boundary between units, that transgresses the time-planes and is of different ages in different parts of its outcrop (Fig. 4). As such it reflects lateral facies change and the encroachment of one sedimentary environment over another. Thus a sand bank gradually overwhelms the mud-flats, or with increased run-off river

North   South

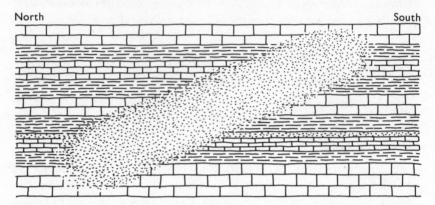

Fig. 4. Diachronism. A sequence of shales and limestones is interrupted by a diachronous sandy unit which is laid down first in the north of the area and later shifts gradually southwards.

muds spread farther and farther into a clear water lagoon. In lithostrati-graphic nomenclature such a diachronous unit can be given a name, while at the same time its transgressive nature is acknowledged. Even though we may suspect that diachronism is fairly common, especially among shallow water sediments, demonstrable examples are not very many. The diffi-culty lies in the establishment of a detailed reliable time-sequence that is itself securely independent of the shifting conditions. This normally demands an exceptionally fine series of zones and subzones; a classic Jurassic example is described on page 284, and other less well defined ones are mentioned elsewhere.

THE TECTONIC SETTING

Over the earth's surface as a whole there is a distinction between regions of relative stability – the major lithospheric plates, whether of continental or oceanic crust – and more mobile belts. Even in the former, however, some earth movements occur so that over wide spans, both of space and time, tectonics and stratigraphy are most intricately related. The scale also varies enormously. At its largest there is the global pattern of plate tectonics which ultimately affects all the processes and products in the history of the crust. More immediately obvious on the continents are the tectonic relations of the orogenic belts, with their marked compressive effects, thrusting and overfolding, the structures visible at outcrop being much influenced by the depth of erosion in the mountain range. As a third instance the gentler warping movements may also have considerable effects on sedimentation, through the amount of uplift in the source areas and of depression in the basins. Even climates must have been profoundly influenced in the past, as the continental masses migrated slowly from, say, tropical latitudes towards the poles, carrying with them changing sedimentary conditions, floras and faunas.

From late Precambrian times onwards the relations of the area of the British Isles to the nearby plates can be assessed in varying degree – more clearly of course in the later periods. Phenomena at plate margins are most cogent during Caledonian history and hence appear principally in Chapters 3 and 4. In this introduction we only need a brief synopsis, illustrated in Fig. 5, in which the major orogenic belts result from the collision of converging plates (i.e. destructive margins), whether one or both carry continental crust. The intervening oceanic crust is carried down and consumed along an inclined subduction zone. The surface indications of this process may include a deep ocean trench, an arc of volcanic islands or range of volcanoes along the continental margin, with extrusives typically calc-alkaline, and sometimes small marginal seas or basins behind the arc. When finally welded together by orogenic compression – plate collision and its multiple effects – traces of the now-vanished ocean crust may remain near the plate suture, but uplifted, displaced and fragmentary; these remnants comprise the classic ophiolites of the Alpine chain and are much sought after in other mountain belts.

The characters of divergent or rifted margins are less dramatic. In the early stages there may be a narrow oceanic rift (a Red Sea condition) with the extrusion of oceanic basalts, and as the plates diverge the continued formation of new ocean floor at a central spreading ridge. By then the trailing edges of the continents are passive with little of the rifting process reflected in their tectonics. The sedimentation and faunal distribution of the epicontinental seas, however, may reflect the increasing area of ocean nearby. A glance at the geological map will show how much western Europe is governed by orogenic belts; conversely in the middle of a major continental plate there is much more uniformity. Here the term epeirogenic has been used for gentle up and down warping movements.

In this summary one item has been left outstanding – the geosyncline, a term and concept left somewhat in limbo by the advance of plate tectonics.

When continents were still considered to be stationary, orogenic belts were commonly thought to be preceded by an elongate trough, or troughs, between two borderlands – the classic geosyncline. As far as modern orogenic belts and their histories are concerned the term has largely faded

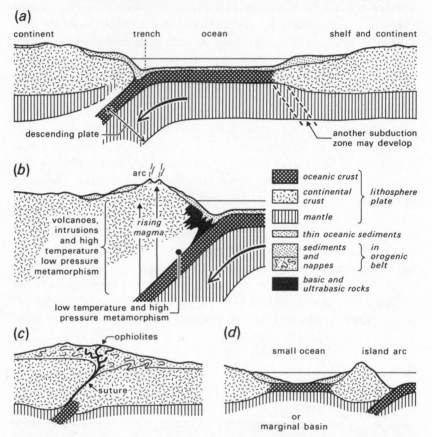

Fig. 5. Interpretative examples of an ocean, continents and subduction zones along convergent, or destructive, plate margins, and of the resulting orogenic belt. (Adapted from several sources.) (*a*) Two converging continents, that on the left overriding the ocean floor on a subduction zone, that on the right bordered by a shelf. (*b*) Magmatic and metamorphic effects in the orogenic belt above an active subduction zone. Volcanoes develop along the continental margin, with associated detrital sediments; wedges of basic and ultrabasic rocks may be derived from the ocean floor; magma rises from the hotter levels of the subduction zone at depth. (*c*) Final stage in continental collision and formation of a mountain range where the sediments are folded and thrust into nappe structures. Almost all the oceanic crust has been destroyed but occasionally small fragments are upthrust and distorted to form the 'alpine' ophiolites. (*d*) Situation similar to (*b*) but subduction takes place beneath an island arc which is separated from the continent by a back-arc basin or a minor ocean.

away, though in the late stage of continental convergence a relatively narrow sedimentary downwarp is a logical deduction. It is sometimes used however in accounts of older orogens where the elongate sedimentary and volcanic facies belts are appropriate but the surrounding tectonic elements are obscure.

Plate movements and palaeolatitudes

Much of the evidence for the movements of lithospheric plates and their relative positions has been derived from the ocean floor; however no floor is known older than Jurassic and the remnants of that age are relatively small. On the continents a much longer history of movement has gradually been compiled, principally through palaeomagnetism, though the reliability of the data decreases with increasing age; so far, pre-Carboniferous results are seriously sparse or questionable.

The position of the contemporary pole from rocks of known age can give the approximate latitude of the block of which those rocks form a part, but never the longitude. However in the later systems this inherent shortcoming is partly overcome through the assembled history of the ocean floor and in any sequence of postulated movements the crustal plates must ultimately arrive at their present longitudinal relations. In practice the deduced plate patterns from Triassic times onwards are not grossly controversial. Since that time North-West Europe, with the British Isles, has moved slowly and a little irregularly northwards.

Generalized latitudes for the later Mesozoic and Tertiary periods can thus be given (in Chapters 8 to 11), and some much more tentative suggestions made for the Upper Palaeozoic positions. Moreover as far back as this some checks are available from floras, faunas and climatic evidence. In earlier times, say late Precambrian to Silurian, there are many problems to overcome; in this country they include the complications of plate movements during the Caledonian orogeny. Accordingly no more than very general or uncertain estimates of palaeolatitudes appear in Chapters 3 and 4 and these are liable to major revisions. It may be that at certain times in earth history there were unusually large plate movements entailing relatively rapid major shifts in latitude. One such period possibly occurred in the late Precambrian but the data are too sparse for a firm pronouncement.

North-West Europe and the British Isles

The one large stable region of Europe is the Baltic or Fennoscandian

Shield where Precambrian rocks are exposed for many hundreds of kilometres and on which later formations lie almost undeformed. The British Isles are part of western Europe which, broadly, is dominated by orogenic belts, truncated by the relatively late Atlantic rifting. The strip of Precambrian rocks in north-west Scotland is a very small fragment of an 'American plate' left attached to the European side during the rifting process.

It will be seen from Fig. 6 that the Caledonian belt, or Caledonides, in Europe is approximately linear. There is no major interruption between Ireland and Scotland, but the deep depression of the northern North Sea separates the latter and Scandinavia, with little structural continuity between the two. Across the Atlantic there are ranges of similar age in maritime Canada, particularly Newfoundland, and also in Spitsbergen and East Greenland (Figs. 19, 32, insets).

In their great extent and influence the Caledonian movements in Britain take pride of place; they affected most of the uplands from Scotland and Ireland to mid-Wales and Leinster. 'Caledonian' strictly refers to structures of the dominant trend, north-east to south-west, and of Caledonian age, from late Precambrian to mid-Devonian; 'Caledonoid' refers to structures of any age with the same trend. This system of suffixes is applied, though rather less often, to other orogenies. Folding and metamorphism were more vigorous in Scotland and north-west Ireland than farther south and four great faults, or fault belts, divide the former country into five geological units whose distinction is evident even on a small-scale map (Fig. 7). The Moine Thrust belt, which extends from the north Sutherland coast to Skye, forms the western margin of Caledonian folding. It consists of a number of piled-up slices carried westwards on easterly dipping planes, the uppermost being the Moine Thrust itself. The whole is probably a late Caledonian structure. The Great Glen Fault is responsible for the through valley between the North-West Highlands and the Grampian or Central Highlands and is essentially a transcurrent fracture. The amount and direction of dislocation has been much debated, but a sinistral movement (i.e. the western block being moved southwards) of over 100 kilometres has been commonly held.

The Midland Valley of Scotland is a rifted structure, or graben, between the Highland Border Fault and the Southern Upland Border Fault. The first continues across into Ireland and has usually been deduced to emerge on the south side of Clew Bay in Mayo (Fig. 29); the existence of the Southern Upland Fault on this side of the Irish Sea is more doubtful, since there is a much more extensive late Palaeozoic cover in Ireland. Although

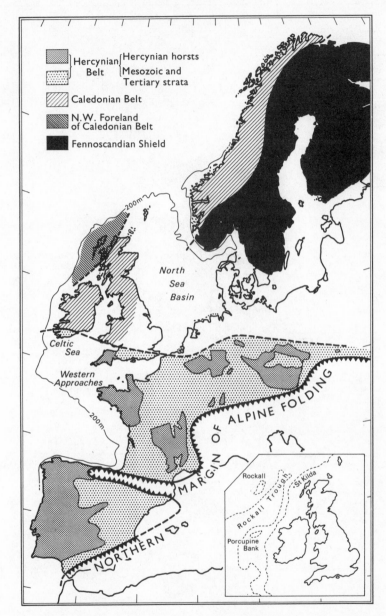

Fig. 6. A sketch map showing the principal tectonic units in north-western Europe, north of the main Alpine folds; the 200-metre contour west of Britain and France approximates to the edge of the continental shelf. Inset: topographic features of the sea floor west of the British Isles.

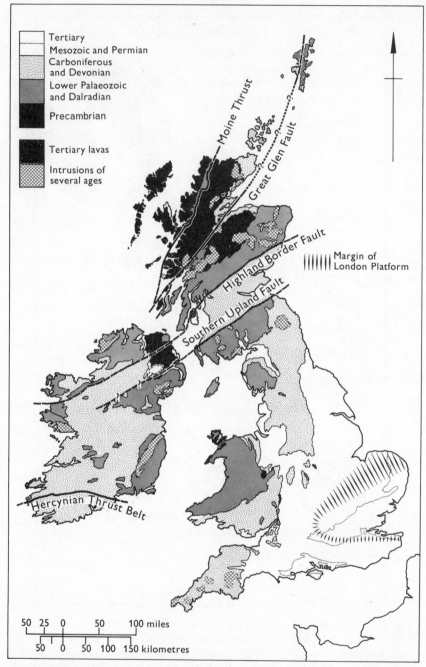

Tertiary

Mesozoic and Permian

Carboniferous
and Devonian

Lower Palaeozoic
and Dalradian

Precambrian

Tertiary lavas

Intrusions of
several ages

Moine Thrust

Great Glen Fault

Highland Border Fault

Margin of
London Platform

Southern Upland Fault

Hercynian Thrust Belt

50 25 0   50   100 miles

50   0   50 100 150 kilometres

Fig. 7. The principal stratigraphical regions of the British Isles with the more important faults in Scotland and Ireland. The northern continuation of the Great Glen Fault is highly problematic, as is the Southern Upland Fault in Ireland.

Caledonian structures dominate the Lower Palaeozoic outcrops of Wales, the Lake District and south-east Ireland they are not so complex as farther north and there is only low grade metamorphism. The term orthotectonic has been applied to the Caledonides north of the Highland Border Fault and to their continuations in Ireland, and paratectonic to the remainder (Fig. 12, inset).

The Hercynian, or Variscan, belt is not so clearly linear and a wide range of structures and strikes are found. It stretches across Middle Europe from southern Ireland through South-West England and Brittany to Belgium, West Germany and Czechoslovakia; the name is taken from the Harz Mountains. To the south there are further Hercynian massifs in central France, the Vosges and Black Forest, Spain and the Atlas Mountains. The Alpine folds overrun the southern margin of the belt and include some Hercynian cores or massifs. It is logical to suppose that plate tectonic processes operated along the Hercynian chain as in other orogenic belts, but the data are less clear (some frankly controversial) and only a minor section lies within the British Isles.

In Britain the strike varies a little on either side of east–west and the effects, structural and metamorphic, increase southwards to give strong deformation in south Cornwall. A partial northern boundary can be traced in southern Ireland as a belt of thrusts and aligned structures from Dingle Bay to Dungarvan in Waterford. Similarly the southern margin of the South Wales Coalfield is more strongly deformed than the northern. Eastwards from Avon and Somerset the Hercynian structures disappear under the Mesozoic cover of southern England but are effective in the Kent Coalfield and its continuation in the Boulonnais. The same probably applies to the London and East Anglian Platform, which is the western end of an important subsurface stable element, the London–Brabant Massif. Palaeozoic rocks are known from several boreholes and lie not very far below the surface in the centre, north of London.

In the Midlands and north of England much of the folding and faulting was imposed during the Hercynian earth movements but the structures commonly follow older trends. These include Caledonoid in the West Midlands and Charnoid in Leicestershire. The term Charnian refers to the Precambrian north-westerly structures of Charnwood Forest, but a comparable trend is widespread in the later rocks of the East Midlands and also in the southern North Sea basin. Across the latter, from southern Britain to the northern part of the Hercynian belt on the continent, Upper Palaeozoic history and Hercynian earth movements present a fairly coherent pattern. Slightly later tensional structures cut across in various places and at various times. In the north they include the Oslo Graben (with Permian lavas), those of the North Sea (p. 270), some comparable structures on the British mainland, and farther south the Middle Rhine Graben, which is principally Tertiary but was possibly initiated earlier. Some major structures of the sea floors around the British Isles are summarized towards the end of Chapter 8.

After Permian times north-western Europe became a relatively stable region, part of the great northern continent of Laurasia. Around Britain the main tectonic effects were faulting and rifted structures, especially in the North Sea where vertical movements were dominant. The east-west structures in south-east England are principally mid-Tertiary in age. They were once termed 'Alpine' but any direct relation with one or another Alpine phase seems improbable. That great orogenic belt lay far to the south, where colliding African and European plates, with minor elements jammed in between, produced complex folds and thrust sheets. In the more stable regions of the north the last great tectonic event was the opening of the North Atlantic rift; this was probably developing in Cretaceous times but was marked, in the British area especially, by extensive early Tertiary vulcanism.

REFERENCES

References appear in lists A or B in the Bibliography, with an explanatory note (p. 430).

General: Ager, 1973; Anderton *et al.* 1979; Donovan, 1966; Evans and Stubble-field, 1929; Gass *et al.* 1972; Owen, 1976; Read and Watson, 1975; Wills, 1951; Windley, 1977.

Geochronology: Cohee *et al.* 1978; Faure, 1977; Harland *et al.* 1964; Harland and Francis, 1971; Steiger and Jäger, 1977; York and Farquhar, 1972.

Sedimentary facies, etc. Reading, 1978; Selley, 1978.

Stratigraphical nomenclature: Harland *et al.* 1972; Hedberg, 1976; Holland *et al.* 1978.

Tectonics, general: Ager, 1975; Bullard *et al.*, 1965; Hallam, 1973a; Kay, 1969; Le Pichon *et al.* 1973; Oxburgh, 1974; Press and Siever, 1974, ch. 19; Smith and Briden, 1977; Tarling, 1978; Tarling and Tarling, 1972. (References relating to the Caledonides are given at the end of Chapter 4.)

Regional works: British Regional Geology (p. 431). Bowes and Leake, 1978; Charlesworth, 1963; Coe, 1962; Craig, 1965; Moseley, 1978a; Naylor *et al.*, 1980; Owen, 1974a,b; Rayner and Hemingway, 1974; Sylvester-Bradley and Ford, 1968; Wills, 1951, 1973, 1978; Wood, A., 1969.

PRECAMBRIAN ROCKS:
THE CALEDONIAN FORELANDS AND
OTHER ANCIENT GROUPS

Most of this book is concerned with the stratigraphy of fossiliferous Phanerozoic strata – or of strata which, if lacking fossils in one area, may ultimately be correlated with a fossiliferous sequence elsewhere. Many of them have relatively simple structures and have undergone little or no metamorphism. In these circumstances it is usually possible to apply the principles set out in the preceding chapter, especially the correlation of rocks by fossils, and to interpret their original environment by means of sedimentary composition and the fossil content. Age determinations and correlation by isotopic methods have added important new data on the older systems and major intrusions, and also some refinements to palaeontological methods and the Phanerozoic time-scale.

When we turn to the study of Precambrian rocks the picture is rather different. It is not only that fossils are very rare, and their stratigraphical value thus only applicable sporadically in late Precambrian sediments, but that the whole scale of events is so much vaster. Phanerozoic and recent rocks cover more than three quarters of the earth's land surface and floor all the oceans, but in time they represent only some 570 million years, or less than one fifth of the known Precambrian time-span. During this enormous stretch of time there were many periods of vulcanism, orogeny, intrusion and metamorphism so that, with a few exceptions, the rocks are much deformed and altered. They are most fully seen in the great Precambrian shield areas of the world, such as the Canadian or Baltic shields, or in the several stable cratons and mobile belts that make up the Precambrian basement of the African continent.

In some of the shields, where very ancient rocks are exposed for thousands of kilometres, it has been possible to distinguish a number of Precambrian provinces. In a province certain structural trends tend to be dominant; there are cognate characters in rock type, and in degree and type of metamorphism and history of intrusion; boundaries with neighbouring provinces are sometimes major dislocations. Such areal distinctions have long been recognized in the better known shields, and the common concentration of isotopic dates within a province has further substantiated them. So valuable has been the great advance in these dating methods in the last two decades that a broad history of Precambrian events is now established for several major shield areas.

Some of the ancient mobile belts resemble the Phanerozoic examples and presumably reflect similar orogenies; they also appear to have been previously the site of sedimentary and volcanic (supracrustal) accumulations; elsewhere such accumulations are not recognizable and it is possible that the rocks deformed along the belt were polycyclic, in that they were those of an earlier orogeny reconstituted in a later one.

Now that more is known of Precambrian history it is seen that although some processes were cyclic and resembled those of later periods, others were probably unidirectional, incorporating permanent changes. Important among these was the composition of oceans and atmosphere. From theoretical considerations of the elements available early in the earth's history, and from the development of living organisms, it seems probable that the early atmosphere lacked oxygen and that this was gradually built up through the photosynthesis of the first primitive chlorophyll-bearing plants. Such a change, it is deduced, would have affected surface processes such as weathering and in particular the deposition of ferric oxides. The formation of red-bed facies is thought to be linked with an oxygenated atmosphere, which developed somewhere around 1400 million years ago.

Precambrian fossils, or traces of Precambrian life, are known from several shield areas, notably in Russia and Australia. Although exceedingly rare compared with Phanerozoic fossils, and often problematic in biological ascription, they nevertheless have been employed as ancillary time-markers or tentative stratigraphical guides in Riphean and Vendian rocks (from about 1600 m.y. onwards) where these are largely unaltered.

The most abundant signs of Precambrian life are algal growth-forms (stromatolites) and microfossils (single cells or strings of cells); there are also trace-fossils and impressions of various kinds of plants and animals. These biological links, and the conformable or near-conformable sequences across the Precambrian–Cambrian boundary in many regions, suggest that the latter does not reflect any fundamental or global transformation about 570 million years ago but chiefly the appearance of invertebrate fossils with a durable shell or skeleton. Whether this was a relatively 'explosive' stage in evolution, possibly related to changing ocean composition, has been much debated; but the advanced complex structure of some of the earliest Cambrian groups, notably the trilobites, does at

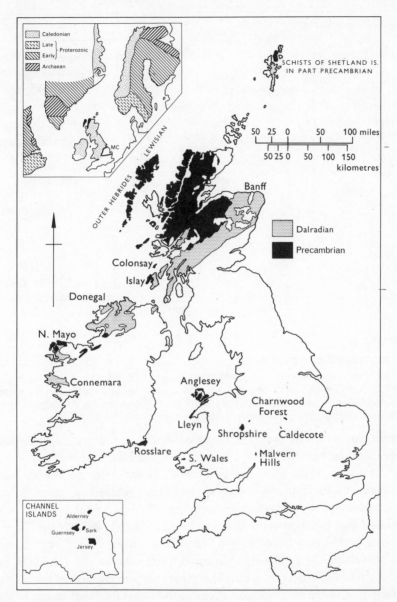

Fig. 8. Precambrian and Dalradian outcrops, the former including the Moine rocks in the central and northern Scottish Highlands and in north-west Ireland. Inset above: Caledonian belt and neighbouring Precambrian provinces, including South Greenland and East Canada; black, North-West Foreland; MC, Midland Craton. Inset below: Channel Islands.

Table 3. *Precambrian classification and some British assemblages*

CAMBRIAN base at		570 ± 20		
	Vendian	650 ± 20	....	Dalradian
	Upper Riphean		....	Several late Precambrian groups in various parts of the British Isles
		950 ± 50		
UPPER PROTEROZOIC	Middle Riphean			(Possible gap with rocks of this age not much represented or not well known.)
		1350 ± 50		
	Lower Riphean			
		1600 ± 50		
LOWER PROTEROZOIC			Laxfordian	Lewisian Rosslare Complex complex of S.E. Ireland here or later?
		2500–2600	Scourian	
ARCHAEAN				

The ages in million years are given for the base of each division; they are highly generalized in the older assemblages and all are liable to revision. The term Katarchaean is also used for rocks older than 3000–3300 m.y. Adapted from Watson (1975*a*, Fig. 1) and van Eysinger (1975).

least suggest a long period of metazoan evolution well back into Precambrian times, but one which left no trace in durable hard parts.

These universal questions have taken us far from the British Isles, where only the fringes of such rocks and their problems appear, largely in Scotland and north-western Ireland (Fig. 8). In Chapter 1 it was mentioned that the deposition in the Caledonian mobile belt, or the Caledonides, had already begun in late Precambrian times. Accordingly the two great groups related to early Caledonian history, Moine and Dalradian, are deferred to the next chapter. The remainder are considered here; they form part of the ancient foundations of the stable forelands on either side of the Caledonian belt, and earlier also made up small parts of the major continental plates – American and Baltic (or European) – that were separated by the early Caledonian ocean.

Table 3 shows that in age the Precambrian of the British Isles can be broadly divided into relatively late groups and a small assemblage of much earlier ones, and that there seems to be a major gap in between which is unrepresented. In North America rocks of this age, stabilized around 1100–900 million years ago, are widespread, for instance in the eastern parts of the Canadian Shield, and form the Grenville Province. A few samples dredged from the Rockall Bank, 400 kilometres west of the Hebrides, have given Grenville dates; to the east the nearest province of the Baltic Shield, in south Norway and south Sweden, has similar ages. In terms of major continental shields it has been suggested that the Midland block or craton represents a westerly tip of the Baltic Shield. However the fragmentary Precambrian groups of England are later and as yet no major extension of the Grenville Province is recognizable here, though a few dates of comparable age have been recorded from some Moine rocks.

THE LEWISIAN COMPLEX: THE FOUNDATIONS OF THE NORTH-WEST FORELAND

The highly deformed and metamorphosed rocks of this complex span a very long period and include the oldest known in this country. They form the foundations of the North-West Foreland of the Caledonian orogenic belt, lying west of the Moine Thrust (p. 18 and Fig. 9), and make up almost all the Outer Hebrides. Farther south there are smaller outcrops on the Hebridean islands of Tiree, Coll and Iona, and on the small islet of Inishtrahull off the north coast of Ireland. Slices of Lewisian basement are also found within the neighbouring Caledonian belt.

Isotopic age data are of prime importance in unravelling the long complex history of the group, which is summarized in Table 4; but before these were available it was realized that the Lewisian gneisses were

Table 4. *Summary of Lewisian Complex*

LAXFORDIAN	2200–1600 m.y.	
	Position of Scourie dyke swarm *c.* 2200 m.y.	
	——— (or a more extended time range?)———	
SCOURIAN	2800–2200 m.y.	{ Inverian episode { Badcallian episode 2800 m.y.
? PRE-SCOURIAN?		

Adapted from Watson (1975*b*, p. 18).

polycyclic – that they included rocks of diverse ages and origins which had been affected by several episodes of metamorphism, deformation and migmatization. A widespread rock type is a banded quartzo-feldspathic

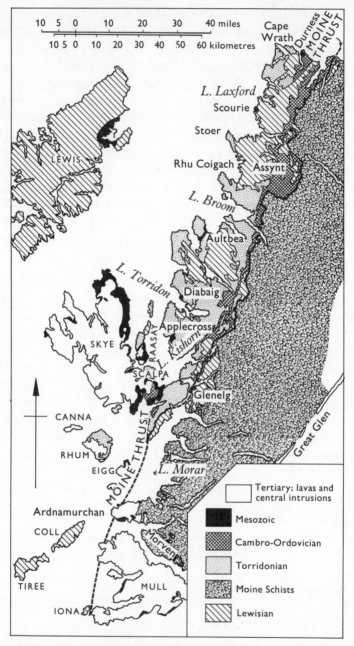

Fig. 9. Geological sketch map of the North-West Highlands of Scotland, with part of the Outer and Inner Hebrides. Intrusions and some Lewisian inliers in the Moine Schists omitted. (Redrawn from Phemister, 1960, pl. II.)

gneiss, with lesser altered sedimentary and igneous components. Some groups of surface accumulations, sedimentary or volcanic, can be recognized, but for the most part we are seeing the results of deep-seated crustal regeneration or reactivation. Much of the rock material, in origin, was Scourian, some possibly pre-Scourian, but in wide areas this was profoundly altered by Laxfordian events, so that the whole could also be called a province of Laxfordian mobility. It was finally stabilized about 1600 million years ago and at a later period became part of the Caledonian foreland. As a stable or cratonic element it is really only a very minor fragment, now adhering to the European continent, of the great Canadian Shield, of which the nearest section is the Precambrian basement of Greenland (Fig. 8, inset). From this our North-West Foreland was separated only very recently, in terms of the full geological time-scale, during the early Tertiary North Atlantic rifting. Southern and eastern Greenland have several similarities, Caledonian and pre-Caledonian, with Scottish geology and there are extensive provinces comparable in age to Scourian and Laxfordian. The areal distinction in the small Scottish section is not so marked because Scourian structures and dates are dominant only where they escaped most of the Laxfordian effects. This is principally in the central section of the mainland outcrop, from Scourie to Loch Maree and in lesser areas of the Outer Hebrides.

Many of the early formed gneisses are acid to intermediate in composition and of no clear origin. Others are intrusions, granitic, basic and ultrabasic; yet others supracrustal, probably including some sedimentary rocks and basic lavas. There are also those deduced to be basic intrusions showing modified layering. The oldest known gneiss-forming metamorphic event (the Badcallian, Table 4) apparently affected all rock units and has given dates around 2800–2700 million years. High grade metamorphism resulted, ranging from granulite to amphibolite facies; in some regions a north-easterly grain has survived. No dates earlier than the major event of 2800 million years are as yet recorded but a 'pre-Scourian' basement, already gneissose, has been postulated, which would take Lewisian history back to about 3000 million years or even earlier.

The largest supracrustal assemblage, the Loch Maree Group, with altered sediments and lavas, presents no direct data but may belong to later Scourian history. Here also there were many small intrusions, especially in the Outer Hebrides, and widespread pegmatites. There was a further late Scourian period of deformation and amphibolite-facies metamorphism (*c.* 2700–2200 m.y.). The Inverian structures are particularly characterized by north-westerly trending monoclinal belts with marked deformation and recrystallization – a heterogeneous pattern superimposed on the ubiquitous earlier gneiss-forming metamorphism.

The next stage is commonly thought to be marked by the intrusion of the Scourie dyke suite, a major swarm trending west-north-west, which was largely emplaced as tholeiitic dolerites with some more basic types. They are best seen on the mainland but in all cover a belt some 300 kilometres wide. A central date of about 2200 million years is suggested, though with some earlier and later members – the earlier for instance carrying traces of Inverian structures. On the view incorporated in Table 4 the dykes are seen broadly as a unified igneous episode and serving as a time-marker; an opposing opinion is given later.

The Laxfordian stages in the Lewisian history of the complex were also polycyclic, with metamorphic assemblages mainly in amphibolite facies and with heterogeneous deformation. The principal effects are seen in the northern and southern sections of the mainland outcrops and much of the Outer Hebrides; structural belts commonly trend north-west and in between them are kernels or larger relict masses where Scourian rocks are largely unmodified. Few if any signs of supracrustal rocks remain and Laxfordian history is essentially one of repeated regeneration of Scourian rocks with much granitic and pegmatitic intrusion. Dates from these intrusions range down to 1700 million years to give the main stabilization of the craton, with a scatter of later dates, possibly reflecting uplift and cooling.

So far this summary has concentrated on what might be called a majority or orthodox view and perhaps others should be mentioned. These offer some alternative interpretations of the early gneisses, with more emphasis on altered supracrustal rocks, especially volcanic, and less on a history of reworking; also the alternative view that the history of episodic deformation and metamorphism was more complex, with more phases than suggested in Table 4; and, more important, that the relation of the Scourie dykes to these phases suggests a long period of basic intrusion, so that the dykes do not form a reliable time-marker. Some divergences in opinion possibly reflect the interpretations and weight placed on items of field evidence; others may be resolved by further isotopic dating.

Lewisian outcrops within the Caledonian belt

Most of the inliers of Lewisian basement within this belt lie in the Moine country, east of the Moine Thrust, but minor outcrops occur also in western Ireland and the Shetland Islands. In the Scottish examples, despite the heavy overprinting by Caledonian activity, some Lewisian ages have been recorded even as far back as 2700 million years. These results, the rock types and metamorphism suggest that the basement more resembles the Scourian than the Laxfordian assemblages of the foreland. There are also some distorted basic dykes that apparently do not penetrate the Moine rocks above and these have been compared to the Scourie swarm.

In the Glenelg region, near the Moine Thrust opposite Skye, where the Lewisian basement is clearest it is overlain by an almost autochthonous Moine cover; farther to the north-east, in central Ross-shire and Suther-land, the allochthonous slices of Lewisian rocks are less easy to distin-guish. In north Mayo, Ireland, affinities have been noted between the rocks of the Lewisian inliers and the Laxfordian complex of the Scottish foreland.

THE COVER OF THE FORELAND: TORRIDONIAN ROCKS

The Lewisian basement is not overlain by a later cover throughout but only in parts of the Scottish mainland and certain islands of the Inner Hebrides. The two rock groups are in striking contrast, the younger being very largely unmetamorphosed sediments with simple structures, forming upstanding outliers or larger mountainous masses. Rather coarse red beds are characteristic. Despite this modern appearance a late Precambrian age has long been known from the overlying unconformable Lower Cambrian quartzites on the east (Fig. 15), and has been substantiated by isotopic dating. The term 'Torridonian' has traditionally been applied to these cover rocks, but it is rather a loose title for groups of distinct ages and the relation of some of the outlying sequences is questionable.

The main outcrop extends, irregularly, from Cape Wrath in the north to the Applecross region east of Skye (Fig. 9), and similar beds crop out on Rhum. The underlying Lewisian surface may be only undulating, but in places the earliest detritus gradually filled in a hilly landscape, with a relief up to 400 metres. In addition to the coarse red arkoses and sandstones with pebbly layers, there are breccias (especially near the base) and a range of finer sediments – sandstones, siltstones and interbedded shales – commonly grey. Various types of cross-bedding, channelling and other sedimentary structures suggest shallow water conditions; some were probably fluviatile, others possibly marine. The occasional microfossils collected from the shales have been interpreted as marine phytoplankton.

The oldest beds, the Stoer Group (Fig. 10), occur in relatively restricted outcrops from the Stoer Peninsula southwards to Loch Ewe; they are variably developed on a highly irregular floor with a maximum thickness of some 2000 metres. There is a typical range of sandy and pebbly red beds together with a layer of volcanic debris. Shales have provided an isotopic age, which is approximately that of deposition, of 974 ± 24 million years.* Poorly preserved microfossils are comparable to those of the Middle Riphean Russian sequence (Table 3).

The Torridon Group follows with an angular unconformity and although the rock types and sedimentary regime seem to have been similar

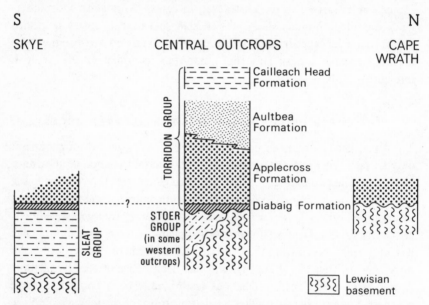

Fig. 10. Torridonian successions from Skye to Cape Wrath; relative thicknesses indicated but not to scale. (Data from Stewart, 1969, 1975.)

there was a substantial gap between the two. An isotopic age of 793 ± 17 million years is recorded from the shales and Upper Riphean microfossils have been collected from several horizons, a much more varied assemblage than the earlier one. The Torridon Group is by far the most extensive and typical on the Scottish mainland, amounting to 6000–7000 metres. A superposed succession has been recognized in the central section, for instance in parts of Applecross and around Loch Broom, although in such a facies some degree of diachronism is likely at the boundaries:

Caillach Head Formation, a grey and red cyclic sequence, local in distribution

* The note to Table 2 (p. 3) on recalculated dates applies here and in later chapters.

Aultbea Formation, fine to medium red and brown sandstones; for example, near Loch Ewe and Loch Broom

Applecross Formation, red and brown feldspathic sandstones, arkoses and conglomerates

Diabaig Formation, chiefly fine grey sandstones, siltstones and rare thin limestones, with local basal breccias

These formations are very variably developed. Thus the Diabaig thickens somewhat to the south, becomes thinner northwards and disappears in Sutherland; even the Applecross, by far the most persistent, is under 1000 metres thick at Cape Wrath and there is the only recognizable division. Some of the Applecross pebbles can be matched among Lewisian rocks but many types cannot and are deduced to come from a more distant source; there are resemblances to the rocks of the Precambrian basement in Greenland. Although a considerable range of current directions has been deduced in the Torridonian rocks as a whole, as might be expected in these diverse facies, ages and outcrops, those of the Applecross Formation agree well with such a westerly to north-westerly derivation.

A cyclic sequence is beautifully displayed around Applecross and Loch Torridon, passing upwards from the finer silts and sandstones through the cross-bedded sandstone and culminating in a massive convolute unit that may be as much as 10 metres thick. These and other examples of disturbed and disrupted bedding appear to have been produced by abrupt shocks (possibly caused by earthquakes) while the sands were uncemented and unstable.

The southerly extensions of the Torridonian cover include part of south-eastern Skye but here there are both tectonic and correlative problems. The beds are part of the Kishorn Nappe, within the Moine Thrust belt – a wedge between the Kishorn Thrust below and the Moine Thrust itself above. They have also been affected by Caledonian metamorphism. Within the 2000 metres' thickness an upper sequence is not unlike a thinner representative of the Applecross Formation, and that below has been called Diabaig, though it is considerably thicker than on the mainland. It comprises grey poorly sorted feldspathic sandstones with interbedded shales and lies unconformably on Lewisian gneiss. An alternative view is that this facies is too different from the type Diabaig for that name to be applicable and a separate term, the Sleat Group, is better (Fig. 10).

The question of lithological correspondence or correlation becomes even more acute with the cover outcrops of the southern Hebridean islands of Colonsay and Islay, which are only included here briefly for convenience. Moreover they lie south-east of the deduced course of the

Great Glen Fault, and thus may well have been some 100 kilometres farther south at this time. The Colonsay Group, including sediments of western Islay, contains little that resembles the northern outcrops. A lower facies comprises some 2000 metres of turbidite greywackes and these are overlain by a similar thickness of sandstones and mudstones with deltaic characters. The Bowmore Sandstones of central Islay are grey, locally pebbly and rather more like the Applecross of Skye. Although Lewisian rocks crop out in western Islay the junctions with these cover groups appear to be faulted.

Omitting these last items as too uncertain in age and relationship, the Torridonian cover can be seen as a long-continued phase of detrital accumulation on an irregular gneissose floor, near the edge of the major American continental plate. Sedimentation was broken at least once, with tilting, and may well not have been continuous elsewhere. Far-derived pebbles (perhaps from a section of the Canadian Shield) were part of the coarse detritus. Possibly, shallow marine muds invaded the coastal lands occasionally and brought in a typical Riphean phytoplankton. How these foreland conditions and their detrital sediments might be related to the early Caledonian sedimentation must be deferred till that history has been reviewed (Chapter 3).

ENGLAND, WALES, SOUTH-EAST IRELAND AND THE CHANNEL ISLANDS

South of the Scottish Highlands the Precambrian outcrops are few, mostly small and provide little coherent view of any South-East Foreland to the Caledonian belt. Some rocks are highly metamorphosed but in many areas they are not much altered; this, together with the overlying Cambrian of some Welsh and English outcrops, suggested long since a late Precambrian age. The majority of isotopic dates support this ascription but in a few regions, notably Anglesey, there is some additional evidence for a Lower Cambrian age as well. A simple distinction can be drawn, partly geographical and partly deformational, between those outcrops that lie within the Caledonian belt and those on its margin or beyond. The closest analogue overseas is with the south-eastern (or Avalon) platform of Newfoundland. Here there are several late Precambrian groups of little-altered sediments and volcanics; they are succeeded by Cambrian with trilobites of the Acado-Baltic Province, similar to those of the Welsh Borders (p. 68). The Ingleton Group, or Ingletonian, of north-west Yorkshire, remote in situation and questionable in age, is here grouped with its Lower Palaeozoic neighbours (p. 107).

Anglesey, Lleyn and Wexford

Rather more than half Anglesey (Fig. 16) is made up of the Mona Complex or Monian Supergroup. Although the oldest fossiliferous beds found above are Ordovician (Arenig), the Monian rocks are in general more highly metamorphosed, folded and locally intruded, and a late Precambrian age has commonly been accepted. There are three main components: the bedded succession, various gneisses and a granite. The bedded rocks were sediments with repeated volcanic intercalations; metamorphism ranges from greenschist to amphibolite facies. Their striking folds are particularly well exposed on the cliffs of Holy Island, where graded bedding and other criteria have demonstrated the orientation of the various units, as set out in Table 5.

Table 5. *Succession of Monian Complex, Anglesey*

FYDLYN FORMATION: acid tuffs and ? lavas
GWNA GROUP: varied pelitic sediments with subordinate limestones and stromatolites; spilitic pillow lavas; quartzites; the mélange (see text)
SKERRIES GROUP: massive sandstones with pyroclastic debris
NEW HARBOUR GROUP: pelites and semipelites with pillow lavas
RHOSCOLYN FORMATION: turbidites
HOLYHEAD QUARTZITE FORMATION: orthoquartzites
SOUTH STACK FORMATION: turbidites

Thicknesses are difficult to estimate but the thickest, the Gwna and New Harbour groups, perhaps amount to 2000–3000 metres. Adapted from Shackleton (1975, p. 77).

The mélange belongs to a type of disrupted accumulation better known in Alpine history than in this country – a great jumbled mass (olistostrome) containing huge blocks of limestone and quartzite, formed as a kind of tectonic slide breccia. The volcanic rocks are dominantly basaltic with affinities to oceanic basalts. All these features, together with turbidites and sporadic areas of glaucophane schists, may be interpreted as being related to a nearby subduction zone, but the Monian outcrops are so limited that its direction and extent are problematic.

There are also further questions on Anglesey stemming partly from debated field interpretations, especially the contacts between different rock units, and from the age data – problems which are only touched on here. Thus the gneisses have been considered as a 'basement', on which the bedded rocks were deposited, or as metamorphosed equivalents of the latter, the second being perhaps the majority view. The intrusion of the granite and the metamorphism of certain older parts of the bedded succession (i.e., gneisses perhaps equivalent to the New Harbour Group) have provided very similar dates (603 ± 34 and 595 ± 12 m.y.), so that these are taken to be related late Precambrian events. However some poorly preserved microfossils (palynomorphs) from the Gwna Group higher up have been compared to Lower Cambrian assemblages, so that the whole bedded succession, which is very thick, may have a long time-range or be an interrupted sequence. More firm age data would help to resolve these and other uncertainties.

Several of the Precambrian groups of Anglesey can be found on the Lleyn Peninsula of the Welsh mainland. South and east of the Menai Straits in the Bangor and Padarn ridges a different type appears, the Arvonian, consisting principally of acid ignimbrites. They are much less altered than most of the Monian rocks and on the Padarn ridge are overlain by the Lower Cambrian of the Llanberis outcrops. The Arvonian is thus generally taken to be a very late Precambrian group. However a major unconformity is improbable since tuff and ignimbrite layers continue a little way above the 'basal' conglomerate. A borehole through the lowest grits of the Harlech Dome (p. 64) also encountered pyroclastic rocks below a conglomerate. On balance, although the gap is probably small and the distinction in nomenclature largely formal, the Arvonian is here retained as Precambrian.

South-westwards, across the southern Irish Sea in Wexford, there are some resemblances to Anglesey. The younger of the two Precambrian groups, the Cullenstown Formation, consists of greywackes and quartzites and is very similar to certain of the non-volcanic Monian constituents and the two are regarded as equivalents. The Rosslare Complex is much more metamorphosed and deduced to be much older. It has been compared to Lewisian assemblages with banded polycyclic gneisses affected by successive phases of metamorphism and intrusion, the latter including basic dykes. It is logical to suppose that Wexford and Anglesey thus expose the ends of a Precambrian ridge but with a more ancient core towards the south-west. Such a feature comes into prominence again as the postulated Irish Sea Landmass in Lower Palaeozoic times (p. 88). Even at present a positive element has been detected, separating two subsiding basins, below the sea floor.

South Wales

All the Precambrian outcrops in this region (Fig. 21) are confined to

Dyfed (Pembrokeshire). The most important are found north of St Brides Bay, especially around St David's. There are two related rock groups, part of a single complex but distinguished as the volcanic Pebidian and intrusive Dimetian; the Lower Cambrian conglomerates contain pebbles of the former and can be seen overlying both. The Pebidian rocks are somewhat similar to the Uriconian of the Welsh Borders, rhyolitic and trachytic lavas and tuffs with bands of flow breccias and lava conglomerates. They are cut by the Dimetian intrusions, chiefly of granite, granophyre and quartz diorite.

A smaller group of rather similar volcanic and intrusive rocks occurs in south Dyfed where it is brought up on the Benton Thrust, a major Hercynian structure, and is overlain by Silurian beds. Here is found the Benton Volcanic Group of acid lavas, tuffs and breccias and nearby (though with faulted boundaries) the quartz diorites of the Johnston Group.

From various rock samples in both northern and southern outcrops a range of isotopic dates has been obtained, which concur with the field evidence of late Precambrian age and magmatism between 650 and 570 million years. In rock type and their position near the margin of the Lower Palaeozoic trough the Dyfed groups are more like those of Salop than of North Wales.

South Salop (Shropshire)

South-eastwards from Anglesey and Bangor, Precambrian rocks appear from under the thick Lower Palaeozoic cover of North Wales in south Salop, near the edge of the Caledonian belt. Here they are associated with small Cambrian and Ordovician outcrops and rather more extensive Silurian ones – the whole being a classic geological region (Fig. 11). The Precambrian forms upstanding masses such as the dissected plateau of the Longmynd or the conspicuous hills of the Wrekin and the Caradoc range; not only do these uplands dominate the present scenery but they were important influences in Lower Palaeozoic sedimentation, as described in later chapters.

The Church Stretton Fault is really a Caledonian fault-complex with a long history, at least from Ordovician to Mesozoic times. At its name-place it follows a through valley which separates the two largest Precambrian masses, the Eastern Uriconian on the east from the Longmynd on the west. There are other major faults that take a parallel course, important among them being the Pontesford–Linley belt, west of the Longmynd. The Western Uriconian is a series of faulted slivers along that

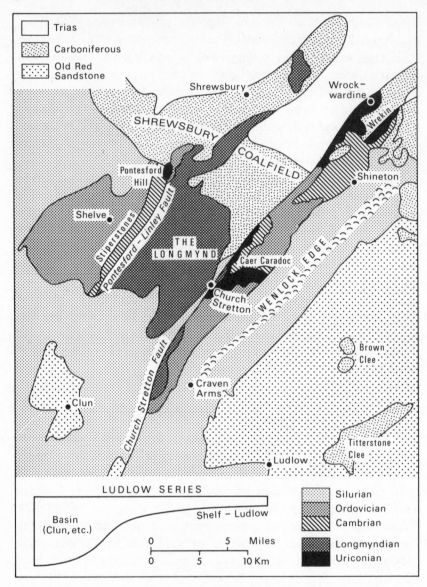

Fig. 11. Geological sketch map of south Salop (Shropshire) and adjoining regions; the belt of 'ballstones' is shown along Wenlock Edge (p. 93). Below, a section to show generalized thickness relations between basin and shelf in the Ludlow Series.

belt, the largest being Pontesford Hill. The Eastern Uriconian makes up Caer Caradoc and other uplands stretching north-eastwards to the Wrekin and Wrockwardine hills. Lower Cambrian quartzites overlie the Uriconian unconformably on the Wrekin and among the Caradoc hills. The few dates obtained from Uriconian rocks range from 687 to 642 million years; this is presumably a minimum age and before Cambrian times the Longmynd sediments were deposited on the Uriconian and then folded and faulted.

Since the various outcrops in south Salop are commonly separated by tracts of later strata their mutual ages depend on the relation to these dated rocks, field contacts and lithological similarities. The commonly accepted scheme is set out in Table 6. The structures of the Longmyndian

Table 6. *Precambrian rocks of Salop (Shropshire)*

	Succession and lithology	Thickness (metres)
LONGMYNDIAN:		
WENTNOR GROUP or 'Western Longmyndian'	purple sandstones and conglomerates with some grey-green siltstones and shales	3700
	~~ unconformity ~~~~	
STRETTON GROUP or 'Eastern Longmyndian'	purple and green shales, silt-stones and sandstones with some conglomerates and tuff beds	4300
	~~ unconformity ~~~~	
WESTERN and EASTERN URICONIAN:	acid to basic lavas, tuffs, agglomerates and minor intrusions (base not seen)	

Adapted from Dunning (1975, pp. 91–2).

sediments, especially graded bedding, demonstrate that the western parts of the Wentnor outcrops are inverted, and that the major structure of the upland is a slightly overturned syncline (Fig. 23), the Minton Group being below the Wentnor and separated from it by an unconformity. The strong petrological similarity between Western and Eastern Uriconian suggests that they are of similar origin and broadly contemporaneous. Fragments are incorporated in the basal Stretton beds and in the Wentnor conglomerates, so that the Uriconian is older than the sedimentary group. The latter was presumably laid down on an eroded surface of tuffs and lavas and incorporated debris from them. The Longmyndian sandstones approach sub-greywackes and show shallow water characters of a flood plain or delta-top environment. These rocks are not widespread in the Welsh Borders but small inliers crop out along the line of the Church Stretton

Fault farther to the south-west. Uriconian rocks, or volcanics of similar type, have a wider distribution in the Midlands.

The Malvern Hills

This last example south of the Welsh Borders is geographically and petrologically rather different. The north–south ridge of the Malvern Hills, rising up on the west side of the Severn valley (Fig. 24), is the most impressive of the English outcrops. Two rock groups are present. Most of the ridge is formed of the Malvernian gneisses, a heterogeneous group largely or entirely composed of altered plutonic rocks, much faulted and sheared. On the eastern side a small volcanic outcrop, the Warren House Group, is rather like the Uriconian and is sometimes correlated with it. The eastern boundary is a major fault against the Trias of the Severn valley; on the west the unconformable Cambrian and Silurian are strongly folded and locally inverted. Isotopic dates on the metamorphic rocks average around 600 million years but this can only reflect late stages in their deformation. Opinions have varied in comparing the main effects to those of Laxfordian age or to the Monian, or possibly an intermediate event between the two.

In its regional setting the Malvern range is something of a puzzle. The north–south trend remained influential for a long time; the Coal Measures are strongly affected farther north and similar trends are found on either side of the Severn to the south (e.g. Usk Anticline and Bristol Coalfield); major faulting affected the Trias and even nearby Jurassic rocks. The Malvern ridge is thus not only part of a more ancient Precambrian basement than is seen elsewhere in England, but it also forms an eastern margin to the dominant Caledonian trends of South and Central Wales.

Charnwood Forest and the foundations of the Midlands

Charnwood Forest is an irregular hilly region north-west of Leicester, where the rocks protrude through a Triassic cover (Fig. 47) in the form of a much faulted anticline, trending north-westwards and plunging to the south-east. This, the Charnian trend, affects much of the East Midlands, especially the Carboniferous rocks; south-easterly structures are also influential in the southern North Sea.

The Charnwood rocks are divided in upward succession into Black-brook, Maplewell and Brand groups, and are intruded by a number of porphyritic and dioritic masses. In the south-east there is a great variety of bedded pyroclastic rocks, agglomerates and tuffs, with lesser sedimentary components from conglomerates to slates. The rocks are unique in the British Isles for the impressions of the fossil, *Charnia*, collected from fine-grained tuffaceous siltstones in the uppermost Maplewell beds. These organisms consisted of some kind of frond-like structure attached to a disc; although their biological position is uncertain (both primitive coelenterates and algae have been considered) they are part of a widespread very late Precambrian assemblage, well known for instance in Australia.

Charnwood has long been considered as a late Precambrian outcrop – part of the foundations of the Midlands – but the age is not wholly clear. The immediate contacts are with Trias and Carboniferous, and while the finer rocks are more cleaved than the nearby Cambrian shales a good deal depends on the age of the intrusions in the south. These diorites have supplied dates up to 540 ± 57 million years, which could just allow the lower and middle groups, those intruded, to be lowest Cambrian or Precambrian. Further small masses of Lower and Middle Cambrian diorites are found nearby in south Leicestershire. On balance, Charnwood Forest is retained in this chapter.

Small Precambrian outcrops are also associated with various Midland coalfields and may form part of a varied Midland basement. Thus Uriconian-like rocks crop out on the west side of the Coalbrookdale Coalfield and in the Lickey Hills of south Staffordshire. Moreover a much greater expanse of such rocks than the present inliers is suggested during Permian times by the large proportion of similar blocks and pebbles that are found in some of the Midland breccias.

The Caldecote Volcanic Formation of Nuneaton is rather different and more like the Charnwood rocks. It forms a small much faulted strip of volcanic and intrusive rocks alongside, and locally below, the larger Cambrian outcrops. The intrusions include a variety of diorite (markfieldite) very like certain Charnian types, and also supplied large boulders in the basal Cambrian quartzite above. These exposures thus not only prove the Precambrian age of the local group but have been taken as a reinforcement for that of Charnwood.

Farther to the east several borings in Leicestershire, Lincolnshire, near Peterborough and in north Norfolk have met with tuffs and sediments. In the last this basement lies at a depth of 733 metres below Trias; nearer the Pennines the cover is more usually Carboniferous. This group is usually taken to be Precambrian, though Caledonian is a possibility, and forms part of the deep-seated foundations of the Midlands. Geophysical evidence suggests that an even deeper floor may be present in East Anglia (p. 98).

Some authorities have seen this 'Midland Foreland' of the Caledonian belt as the westerly tip of the Baltic Shield. Nevertheless there are substantial differences in age and rock composition between the late little-altered rocks of the Midlands and the older ones that compose the nearest outcrops of the shield; and between lies the deep faulted depression of the North Sea.

The Channel Islands and Hercynian belt

Although territorially part of Great Britain, the Channel Islands (Fig. 8, inset) have far more in common with the geology of Brittany and Normandy and will be only briefly considered. They consist largely of late Precambrian granitic and associated igneous rocks. A late Precambrian sedimentary succession (the Brioverian), little altered, has been found on Guernsey and Jersey together with lavas on the latter. More widespread are the effects of the Cadomian orogeny, with a range of metamorphic effects and late orogenic granitic intrusions from about 650 to 540 million years. Minor outcrops of post-orogenic detritus have been compared to Cambrian sediments on the French mainland. On Guernsey and Alderney a much older metamorphic basement is found, the various gneisses giving dates of about 2590 to 2170 million years. Though there are some similarities to other British outcrops, the Channel Islands like the rest of the Brittany Massif differ fundamentally in being unaffected by Caledonian events. Even in Hercynian times the massif remained a stable block, though cut by numerous granitic intrusions.

Across the western Channel, the small metamorphic complexes of South Devon and Cornwall (e.g. Start Point and the Lizard) are in an equivocal position. None of the isotopic results so far is clearly Precambrian, but they range back to 500 million years and are minimum dates only. A Precambrian basement at depth presumably underlies southern England, and might resemble the rather widespread late Precambrian of the Midlands or the Brioverian of the Channel Islands, but so far it is purely conjectural.

REFERENCES

Harris *et al.* 1975 (*Special Report* No. 6, Precambrian, with sections by Bishop *et al.*, Downie, Dunning, Max, Moorbath, Shackleton, Stewart, Watson).
General: Moorbath, 1975*a*; Read and Watson, 1975 (part I); van Eysinger, 1975; Windley, 1977.

Regional: Baker, 1973; Beckinsale and Thorpe, 1979; Bowes and Leake, 1978; Cribb, 1975; Evans *et al.* 1968; Ford, 1958; James, 1956; Patchett and Jocelyn, 1979; Shackleton, 1969; Stewart, 1969; Thorpe, 1974.
British Regional Geology: Central England, The Northern Highlands, North Wales, South Wales, The Welsh Borderland.

3

EARLIER CALEDONIAN HISTORY: MOINE, DALRADIAN, CAMBRIAN

In this context Caledonian history is taken to include those systems or rock groups that lie within or border the Caledonian belt and were largely affected by the Caledonian orogeny. In terms of sedimentation this history certainly goes well back into late Precambrian times. In tectonic and palaeogeographic setting the orogenic belt (or orogen) is also related to the processes at or near the margins of two opposed lithospheric continental plates.* Early in the history these were separated by an ocean of uncertain width (sometimes called Iapetus or Proto-Atlantic), but not a vast major ocean; in the somewhat clearer later history these plates converged and finally met, while the oceanic crust between them was carried down and destroyed by opposing subduction zones. The orogen thus conforms broadly to the pattern of plate tectonics outlined in Chapter 1, together with the appropriate tectonic and volcanic processes. The interpretation of the earlier history is however still rather problematic; the rock groups often occur in isolation and the chronological data are sporadic and of variable quality.

This chapter begins by reviewing the large dominantly Precambrian rock groups of Scotland and Ireland, and then in Cambrian times expands to take in the forelands as well. The small amount of Lower Ordovician beds of north-west Scotland (the Cambro-Ordovician) are included for convenience.

THE MOINE SCHISTS

The Moine rocks or Moine Succession, vary in grade but much of them are schistose and the traditional title is retained here. In the Scottish

---

* These are often named after the modern continents or parts thereof, for example, American on one side and Baltic or European on the other. At present their components are disposed in the north-west and south-east sectors of the British Isles and this is also a useful shorthand notation. This orientation should not be assumed, however, for Caledonian times, though in fact north and south do seem to apply in rather a general way.

Highlands they occupy most of the country between the Moine Thrust and the Great Glen Fault (Fig. 9), where an unconformable base above Lewisian inliers in the west has been detected, and are also found in the northern part of the Central or Grampian Highlands. The great bulk of the rocks are altered clastic sediments, now psammites and pelites, often poorly sorted and varying in metamorphic grade from greenschist to amphibolite facies. In eastern Sutherland there are extensive gneissose outcrops which have been thoroughly migmatized; similar rocks occur in parts of the Shetland Islands. The Moine inliers in Ireland are distant and only partly similar, and their affinities have been questioned. The main outcrops are around Loch Derg in Donegal, the core of the Ox Mountains in eastern Mayo and several areas in north-western Mayo. Among the psammitic rocks the Scottish resemblances are strong but there are also other types such as the altered limestones and volcanics, partly migmatized and associated with the Ox Mountains igneous complex.

In the central Scottish outcrops the structures are complex and multiple, incorporating several superimposed fold systems, but there appears to be a tendency for the larger folds to follow a dominantly Caledonian trend (around north-north-east) and, at least in the west, the axial planes dip south-eastwards. This attitude agrees with the suggestions that the principal movement during folding was westerly, which is also that on the Moine Thrust itself.

Local stratigraphical successions have been recognized in many regions, usually as alternating formations that are dominantly psammitic or pelitic. They are clearest in western Inverness-shire, where the metamorphic grade is relatively low, and especially in the regions around Morar and Glenelg. Here a good deal can be made out of the original sedimentary rocks and structures. Among the coarser psammitic types there are traces of lamination, ripple-marks, cross-bedding, slump units and small-scale erosion channels. The whole resembles a shallow water assemblage (possibly marine), and the directions of the foreset beds suggest currents dominantly from the southerly quadrant. The coarse sandstone and pebble contents are much more pronounced in the extreme westerly outcrops, which might also indicate a detrital source, perhaps a Lewisian upland, in that direction.

The degree of deformation makes any thickness estimate of even the clearest sequence hazardous, though earlier some 7000 metres was suggested in Morar. In all assessments of Moine stratigraphy much depends on the interpretation of the formation junctions. Some are known to be major dislocations; if, as has been postulated, these are a prevalent

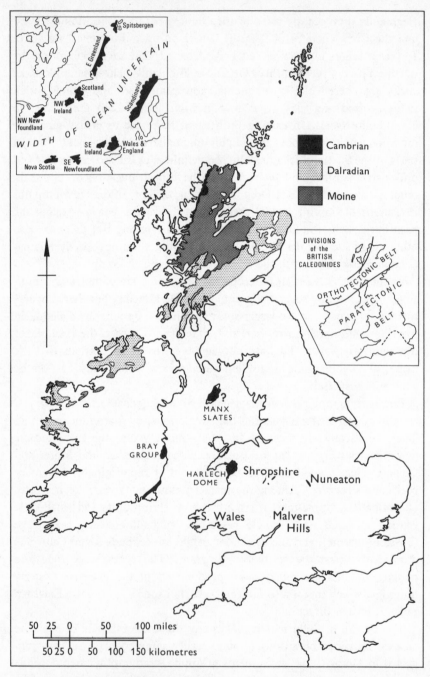

Fig. 12. Moine, Dalradian and Cambrian outcrops. Inset top left: sketch of suggested distribution of Caledonian outcrops in Cambrian times.

character then attempts to compile and name extensive stratigraphical successions would be premature.

Attempts to determine the age of the Moine Schists depend on the interpretation of the isotopic data and the stratigraphical relations with the Dalradian sediments. In the Grampian Highlands the Moine Schists underlie the lowest Dalradian with presumed conformity; they may therefore be accepted as late Precambrian and both groups were affected by early Caledonian (post-Cambrian) deformation – a relatively straight-forward situation. The majority of isotopic dates (around 450–420 m.y.) in Moine and Dalradian rocks reflect the late stages of uplift and cooling in the orogen. It is the larger outcrop west of the Great Glen that presents additional problems. Here some early established results from pegmatites suggested that a major deformation occurred at about 730 million years and the date of sedimentation was hence earlier than that; more recently a considerably older age (1028 ± 45) has been obtained from migmatites in the south, with a deduced sedimentary age of perhaps 1250–1050 million years. Thus there seem to be two Moine groups, 'Older' and 'Younger', and this division has wider implications. It renders any direct equivalence between Moine and Torridonian sediments distinctly improbable. This was originally based mainly on sedimentary resemblances, especially relating to the Moine rocks of Morar in the south-west where the sediments were least altered; it was also possible that the Stoer Group corresponded to the Older Moines and the Torridon Group to the Younger Moines. The latter equivalence is not entirely ruled out but the depositional age of the older group does seem to be substantially more than the 974 ± 24 million years of the Stoer sediments.

In external relations and chronology the Older Moines are nearer to the Grenville Province of Canada, but with the major distinction that they have later been involved in the Caledonian belt, whereas the very much more extensive Grenville Province was largely stabilized earlier, around 1100–900 million years, and has been little affected since. For the moment the western Moine outcrops remain in a doubtful position, both spatially in relation to the Caledonian foreland facies and in their chronology.

THE DALRADIAN SUPERGROUP

Dalradian rocks comprise the largest section of early Caledonian history. They are immensely thick and in time correspond perhaps to two or more later systems, but their definition is primarily lithological; hence the term supergroup. The lowest sediments are late Precambrian, the upper part

Cambrian, and Ordovician beds are probably present at the top. In Scotland the outcrops include much of the highlands between the Great Glen Fault and the Highland Border Fault (Fig. 13), from Banff on the

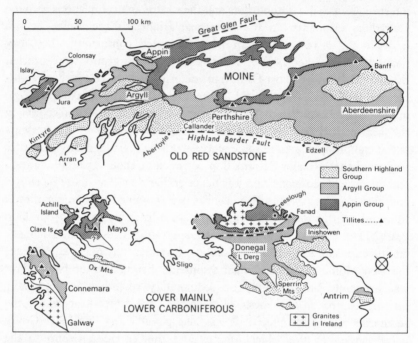

Fig. 13. Sketch map of the three Dalradian groups in Scotland and Ireland; intrusions in Scotland not shown. (Redrawn from Harris and Pitcher, 1975, Fig. 13.)

north-east coast to the Kintyre Peninsula in Argyll. The Highland Border rocks, faulted slivers along the fault and very poorly fossiliferous, probably belong to Upper Dalradian horizons. Parts of the Shetland Islands are Dalradian, though with some rather different lithological characters. The Irish outcrops, south-westwards along the Caledonian strike, are far more extensive than Moine or Lewisian in that country. They make up the highlands of Donegal, much of north-west Mayo and the mountains of Connemara, the last region being unique in lying south of the Highland Border Fault.

Along the 700-kilometre belt the group maintains remarkably similar lithological characters and to some extent structural ones. At the base the traditional separation from the Moine rocks reflects largely the greater diversity of Dalradian types, for instance the appearance of quartzites. In the upper divisions the very sparse fossils help to determine the age but for the most part correlation is lithological, employing especially the boulder beds (tillites) at the base of the middle division and also certain conspicuous limestones and quartzites.

The Dalradian nappes and sequence in Scotland

It is a truism that in all complex mobile belts investigations into structure and stratigraphy are bound together and this is excellently demonstrated in the Dalradian highlands. Only by determining the orientation or 'way up' of individual beds can the recumbent folds be recognized and their upper and lower limbs identified; conversely the broader picture of succession and conditions of accumulation cannot be made out until the structure is deciphered. Metamorphism has not seriously obscured the original types of the Dalradian sediments, especially in the low grade stretches such as that near the Highland Border Fault. They consist of a varied succession among which quartzites, greywackes, limestones, boulder beds and lavas still retain many of their original features; and in which slates, phyllites and schists represent finer-grained pelitic clastic rocks, or occasionally tuffs. The graded bedding and cross-bedding of the quartzites and greywackes have proved most valuable guides to orientation.

The dominant structures are large-scale overfolds, called by the Alpine term 'nappes', the most important of which is the Tay Nappe, a very large flat-lying recumbent fold found along the south-eastern margin of the Highlands. The nose of the anticline points in that direction (i.e. the fold faces south-eastwards) and near the Highland Border Fault it turns downwards (Fig. 17(c)). Like other Dalradian components the Tay Nappe has a number of lesser folds superimposed on it, but it can be recognized from Kincardineshire south-westwards to Kintyre, and since dips are low this has often been called the 'flat belt'. Over much of this stretch the upper limb has been removed by erosion and the visible rock sequences belong to the lower limb and are inverted. To the north-east in Aberdeen and Banffshire there is a similar flat belt which is probably a continuation of the Tay Nappe but here parts of both upper and lower limbs have been preserved.

In the south-western outcrops, of Islay, a somewhat similar recumbent anticline faces in the opposite direction, to the north-west, and is overthrust on to the local cover of the foreland, the Bowmore Sandstone (p. 34). Distinct from both these recumbent structures, in the area around Ballachulish and Fort William there are the two smaller Ballachulish folds. These have generally been considered as north-westerly facing structures and their limbs are partly replaced by low-angle thrusts or slides. They are separated from the Tay Nappe by some large intrusions (of Glencoe, Glen Etive, etc.) and by another major dislocation, the Iltay Boundary Slide, but also by an intermediate structural zone that is not easy to interpret.

Correspondences in Ireland

Continuations of the principal Scottish structures can be traced satisfactorily into the north of Ireland as far as central Donegal. In particular the rocks and structures of Islay can be matched in the peninsulas of Fanad and Inishowen. The Kintyre succession reappears in the small inlier of north-east Antrim where the Tay Nappe continues, also turning down slightly to the south-east. Here, and farther south-west near the Highland Border Fault in northern Tyrone, in the Sperrin Mountains, the inverted limb is exposed but over most of the main Donegal outcrops the strata are right way up.

The Creeslough succession on the north-west side of the Donegal Granite (a complex elongate Caledonian intrusion) has several analogies with the Ballachulish succession. The folds similarly face north-westwards and the strata belong to the lowest Dalradian subgroups. Moreover, although for most of the outcrop the granite forms the eastern boundary, where the Creeslough succession adjoins that of Fanad they are separated by a dislocation analogous to the Iltay Boundary Slide. Thus far the structural and stratigraphical comparison is coherent, but south-westwards from the Sperrin Mountains the Tay Nappe appears to lose its identity and other trends supervene. In north-west Mayo (Fig. 29) a medley of outcrops appears from under a Carboniferous cover; the structures are complex and not clearly related to those of Donegal. However this appears to be the one region in Ireland, on Achill Island, where there is conformity between Moine and the lowest Dalradian division. Farther south recumbent folds have been recognized in Connemara and the metamorphic grade is high; rather fragmentary accounts suggest that an east–west antiform brings up the quartzites that form the central mountains. The altered limestones are ophicalcites (Connemara Marble) and more pelitic rocks flank the central belt to the north and south.

Representative successions

From this brief and generalized description, which takes no account of secondary folds and many complex or uncertain problems, we may turn to the stratigraphical succession. This was first securely established in Perthshire but satisfactory lithological correlations have now been widely extended, for instance to Banffshire, the South-West Highlands and to Islay. A very large number of lithological units have been established, but a recent threefold grouping is adopted in Table 7. The amount of deformation the rocks have undergone does not allow thicknesses to be estimated as they may be in later systems, but in individual sections it appears that the limestones are normally thinner than the other rock types; estimated totals range up to 20 000 metres.

In the fullest sequences a gradual upward change in facies is apparent. The lowest, Appin, Group is characterized by alternations of quartzites and pelites, with thin limestones that are more widespread towards the top. The succeeding Argyll Group begins with the basal tillites, then more quartzites and pelites, and the top subgroup comprises remarkably persistent limestones with volcanic debris. The upper, Southern Highland, Group demonstrates widespread incursions of turbidite greywackes with substantial volcanic episodes; however this change was probably not synchronous throughout, for a similar greywacke facies appears in some south-westerly outcrops below the correlative limestones.

This is diachronism on a large scale but there are also many instances of lateral facies change within smaller areas – indeed possibly far more than are easily detected where almost all correlation has to be lithological. The quartzites, notably those of Islay, expand to a great thickness in some regions; proximal and distal turbidites replace one another laterally; certain 'green beds', redistributed volcanic products, appear to be only local. Stromatolites characterize some of the carbonates, such as the Islay and Lismore limestones and the dolomites of the Islay Quartzite, and in this part of the succession there are other signs of shallow water.

Outstanding among the rock types are the boulder beds or tillites, both for their correlative value and climatic implications. The tillite formation is thickest (750 m) at its type locality, Portaskaig on Islay. Here and elsewhere, although glaciated pavements as evidence of land ice are lacking, there is no doubt of the glacial origin, interpreted as the grounding of successive ice-sheets. Despite minor differences in composition the tillite is remarkably persistent and, apart from the Shetland Islands, can be found in virtually all sequences from north-east Scotland to

Table 7. Dalradian successions in Scotland and Ireland

	Central Highlands and Highland Border	Islay and the South-West Highlands	Donegal, north-west and north (Fanad & Inishowen)
UPPER or SOUTHERN HIGHLAND GROUP	Highland Border rocks: black shales and cherts*	Highland Border rocks: black shales and volcanics	Turbidites with lavas and tuffs
	Turbidites interspersed with Leny Limestone* / Aberfoyle Slates / Ben Ledi Grits / Green Beds (volcanic) / Pitlochry Schists	Turbidites with thin limestones and Green Beds (volcanic)	
MIDDLE or ARGYLL GROUP	Loch Tay Limestone / Ben Lui Schists / Volcanic rocks and pelites / Carn Mairg Quartzite / Killiekrankie Schists / Schiehallion Quartzite / Schiehallion Boulder Beds (tillite) / Limestones and pelites	Tayvallich Limestone* / Crinan Grit / Craignish and Ardrishaig phyllites / Easdale Slates*	Culdaff Limestone / Crana Quartzite / Pelites and limestones —discontinuity— / Malin Head and Slieve Tooey Limestones / Tillite / Fanad Limestone / Pelites and limestones / Ards Quartzite / Ards Black Shales and limestones
	(relationship uncertain)	Islay Quartzite* / Portaskaig tillite / Islay and Lismore Limestones / Cuil Bay Slates / Appin phyllites and Limestone	
LOWER or APPIN GROUP	Kinlochlaggan Limestone / Kinlochlaggan Quartzite / Monadhliath Schists / Eilde Quartzite / (Moine rocks)	Appin Quartzite / Ballachulish Slates / Ballachulish Limestone / Interbedded schists and quartzites / Eilde Quartzite / (Moine rocks)	Creeslough pelites and limestones / (base not seen)

Only a general equivalence is intended and some correlations are decidedly speculative. Fossil-bearing beds are marked with an asterisk and the more reliable marker horizons appear in Table 8 or are mentioned in the text. Adapted from Harris and Pitcher (1975, Fig. 12).

Mayo and Connemara. Moreover it agrees in character and situation with the late Precambrian, or Varangian, tillites known from several countries of Caledonian affinities – Greenland, Spitsbergen and Scandinavia. In northern Norway shales associated with this glaciation have been dated around 654 million years. The figure should perhaps be viewed with caution since glacial detritus may have heterogeneous sources, but in the present context it agrees with the inferred late Precambrian age of the lower part of the Dalradian sediments.

Dalradian fossils and correlation problems

Isotopic data do not help us very much in assessing the depositional age of Dalradian sediments, because most of them reflect late stages in meta-morphism or uplift and cooling. However, although fossils are exceedingly rare and some not at all precise in age-range, a very broad stratigraphical framework is now possible (Table 8). It will be seen that the Lower

Table 8. *Provisional stratigraphical guides in the Dalradian Supergroup*

8. Macduff Slates, Banffshire, ?chitinozoa	Arenig or possibly Llanvirn
7. Edzell (Glen Esk), Angus, ?chitinozoa	Tremadoc or Arenig
6. Aberfoyle, Perthshire inarticulate brachiopods	Upper Cambrian or Lower Ordovician
5. Clare Island, Clew Bay, Mayo *Protospongia wicksi*	around Middle Cambrian
4. Leny Limestone, Callander, Perthshire *Pagetides*	uppermost Lower Cambrian
3. Tayvallich Limestone, Kilchrenan, Argyll, acritarchs	probably Lower Cambrian
2. Islay Quartzite, dolomitic beds, Islay, acritarchs	Vendian (uppermost Precambrian)
1. Portaskaig Boulder Bed, etc.	*c.* 654 m.y. (Upper Riphean)

Acritarchs are unicellular plant planktonic microfossils; chitinozoa are hollow microfossil animals of unknown affinity. The divisions Vendian and Riphean are given in Table 3 and Cambrian and Ordovician stratigraphical terms in the text (pp. 60, 75) and Appendix (pp. 428–9). Data from Downie, Lister, Harris and Fettes (1971) and Downie (1975).

Dalradian is securely Precambrian, that the base of the Cambrian lies somewhere in the middle division and that the upper one presumably includes much that is Cambrian, and most probably some early Ordovician levels at the top. This scheme is not seriously at variance with the few isotopic data available, mostly from intrusions within Dalradian rocks, although there are difficulties in arriving at a more precise position, either stratigraphically or in terms of years, for the Cambrian base. The age of the Moine–Dalradian junction is not as yet satisfactorily settled and although 700 million years has been quoted it is little more than a rough estimate.

The records 7 and 8 in Table 8 come from the Highland Border rocks. These are small faulted outcrops along the line of the Highland Border Fault – some with black shales, cherts and spilitic lavas and more commonly serpentine; there are also a few Lower Palaeozoic fossils. Formerly an Arenig age was favoured, partly on lithological similarity to the Ballantrae outcrops (p. 110), and partly on supposed graptolites. The latter are now discredited and the two records quoted are the least unsatisfactory; there is little doubt, however, that these rocks are genuinely part of late Dalradian sediments. The *Protospongia* from Clare Island is in a similar facies and though divorced in outcrop, Clare Island being well out in Clew Bay (Fig. 29), it is generally considered to be genuinely Dalradian. It may well be that all the rocks once called 'Highland Border' are linked only in their facies and are of different ages in their various outcrops.

On a provisional review the evidence in Scotland strongly suggests that Lower Ordovician strata are involved in the major Dalradian folds, so that the deformation and metamorphism is unlikely to be earlier than Llanvirn and might conceivably be later. Here we meet a discrepancy, because a substantially earlier date has been strongly upheld in Connemara. The Dalradian rocks here are highly metamorphosed, much more so than the Ordovician immediately to the north in the South Mayo Trough (p. 121, Fig. 30) where mid-Arenig graptolites have been found. No actual junction is exposed and a major fault has been inferred, but metamorphic detritus attributed to Dalradian sources occurs in the Ordovician sediments. On both grounds a major pre-Arenig or very early Arenig metamorphism is deduced in Connemara. This pronounced chronological distinction supports the view that metamorphism and deformation were not uniform or synchronous along the Caledonian belt and argues strongly that not only were the processes polyphase but, in their dominant effects, diachronous.

Sedimentation and environment

In spite of the problems still rife in Dalradian geology a tentative history may be attempted. At first, in a rather limited basin, the poorly sorted feldspathic sediments of the uppermost Moines were succeeded by well sorted sands and silts of the lowest Appin subgroup; later, restricted circulation produced spreads of pyritic and carbonate-bearing muds. As the basin gradually silted up these were topped by extensive shallow water sands with calcareous and muddy intercalations. The Appin Group was then terminated by a more widespread sequence of laminated silts and muds, still in shallow waters, but in very quiet conditions.

This uneventful regime was then overtaken by the Portaskaig (or Varangian) glaciation. Multiple successive ice-sheets grounded on the shallow sea floor and left their residue of marine tills; in the earlier stages the boulders were locally derived but later more distant types became common. Then, after the disappearance of the ice-sheets, and possibly derived from their outwash products, great masses of cross-bedded sands invaded the trough; these were dominantly clastic wedges (e.g. the Islay Quartzite), some very thick, but with finer or dolomitic beds among them. Later, graded quartzites resulted from the redistribution of some of this detritus by turbidity flows. In the succeeding middle Argyll period (perhaps around the end of the Precambrian) a great variety of sediments were laid down among basins and swells: black shales, greywackes and slide breccias, and, possibly on the swells, very shallow water carbonates and sands. Turbidity flows became more common, with finer distal sediments tending to replace the coarser proximal versions, possibly as the latter belt moved farther to the south-east. Nevertheless a widespread lull in clastic accumulation marks the end of this stage, for the Tayvallich (or Loch Tay) Limestone and its equivalents form a most persistent marker unit, with some local lavas interbedded.

During the last stage, of the Southern Highland Group, turbidity flows were almost ubiquitous. Vulcanicity became important in both Ireland and Scotland, including the thick Tayvallich volcanics of Loch Awe. Here some 1000–2000 metres of basic lavas are known, many of them submarine with pillow structures. In the upper parts the overlapping wedges of greywackes passed laterally into silts and mudstones, among which were thin impure carbonates such as the Leny Limestone. This mixed clastic accumulation can be traced, at least in certain areas along the Scottish Border and Banffshire, into the early Ordovician period.

The external surroundings of the Dalradian trough present more than one problem. On the widest view the sediments and lavas accumulated on, or adjacent to, the north-west or American plate; palaeomagnetic evidence shows that the British outcrops were in low latitudes in Dalradian times despite the Varangian glaciation. In a more immediate context the interpretation of Dalradian geography is hampered by the faulted boundaries. However the foreland facies to the north-west, described in the next section, must have been in part contemporary. The Varangian glaciation left no trace here and the simplest explanation is that this period is represented by the unconformity between the Torridon Group and the base of the Cambrian. Later there is no doubt of the general equivalence of the Cambro-Ordovician quartzites and carbonates and the upper parts of the Dalradian sequence. Although similar rock types do occur in the Argyll Group any direct correlation is speculative; in most ways the two facies are very different and that of the foreland is much thinner and probably incomplete. Also the two regions were then some unknown distance farther apart, by the amount of the east–west movement on the Moine Thrust.

Then what can be deduced about the sources of the great bulk of Dalradian sediments? The foreland shelf carbonates are very widespread and it thus seems unlikely that the clastic detritus of the upper group was derived from the north or north-west. In the opposite direction there are hints of a pre-Dalradian basement or borderland – or that the trough was 'ensialic' (Fig. 14(*a*)), subsiding *on* continental crust and not built out beyond it. Fragments of high grade metamorphic rocks (? Lewisian-like) have been brought up in volcanic vents in the Midland Valley of Scotland and also in central Ireland. A southerly source is also suggested in the spatial relations of some Dalradian proximal and distal turbidites.

THE NORTH-WEST FORELAND: CAMBRIAN AND LOWER ORDOVICIAN

The quartzite and carbonate facies on this foreland ranges in age from Lower Cambrian to Lower Ordovician and is typical of a stable or cratonic environment, contrasting strongly with the much thicker Dalradian sequence just described. The principal outcrop is a long strip in north-west Scotland (Fig. 9) from Loch Durness on the north coast of Sutherland to the mainland opposite Skye, and there are further areas in the south of that island. The beds dip eastwards, overlying at different places Torridonian and Lewisian with marked overstep (Fig. 15); on the east the Palaeozoic rocks are overthrust by the Moine Schists and wedges of them are locally caught up among the thrust slices. Probably all the outcrops on Skye are of this type because they are overlain as well as underlain by thrust masses of Torridonian rocks.

There is a fairly simple distinction into an Arenaceous Group below and

carbonates (the Durness Group) above (Table 9). The latter is truncated
by the Moine Thrust at different levels along the outcrop, so that only a
composite estimate of thickness is possible, probably over 1450 metres.
On Skye representatives up to the Balnakiel Formation are known, but the
two other Ordovician components crop out only in the most northern
Durness area.

There is no doubt of the age of the *Olenellus*-bearing beds since this is a

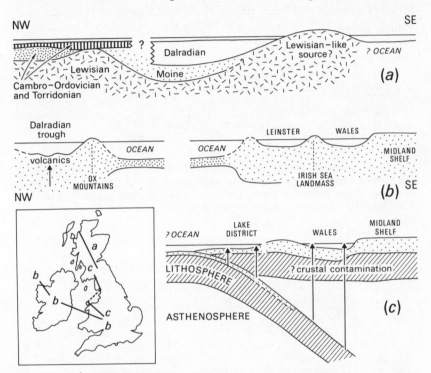

Fig. 14. Diagrammatic sections through parts of the Caledonian belt to illustrate
various views on the plate tectonic setting. (Compare also Figs. 5 and 28.) (*a*) The
north-western side in Cambrian times, to illustrate the suggested accumulation of
Moine and Dalradian sediments on continental crust, and a further south-easterly
source supplying high grade metamorphic detritus into Upper Dalradian turbidites. Any ocean crust is farther to the south-east. (Adapted from Smith, 1976,
p. 272.) (*b*) Palaeogeographical section from the north-west side (western Ireland)
to the south-east side (Wales and Midlands) in Lower to Middle Cambrian times.
The thicker continental crust is shown on either side, and thinner oceanic crust
between. (Adapted from Phillips *et al.*, 1976, p. 583.) (*c*) Section from northern
England to Wales and the Midlands to illustrate the distribution and possible
origin of middle Ordovician vulcanism. The continental crust is of normal
thickness below Wales but rather thinner below the Lake District. The magma
rising from the subduction zone is affected by some contamination from the crust
in the Welsh Province. (Adapted from Fitton and Hughes, 1970, p. 226.)

characteristic Lower Cambrian trilobite in the North American (or 'Pacific') Province; *Salterella* is a small mollusc of unknown affinity but occurs in beds of this age elsewhere. The molluscs of the Sailmohr Formation are Tremadoc in age and in the top three formations several Arenig species have been found; it is possible that the sequence just extends into the Llanvirn. From these data it seems highly probable that there is a major break or breaks in the carbonate succession, with the Middle Cambrian and lower part of the Upper Cambrian being almost or entirely absent.

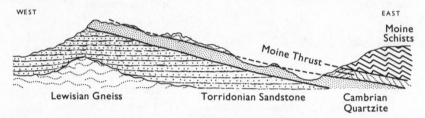

Fig. 15. Section to show the relation of Cambrian rocks (here chiefly the Arenaceous Group) to the underlying Torridonian and Lewisian, and also to the overlying Moine Schists. A subsidiary lower thrust has here carried fragments of Lewisian rocks over the Cambrian quartzites. (Redrawn after Peach and Horne, 1907, p. 541.)

Table 9. *Cambro-Ordovician of the North-West Highlands*

Formations and lithologies		Fossils	
DURNESS GROUP (1250 m)	Durine / Croisaphuill / Balknakiel	conodonts, gastropods, several cephalopods, one trilobite	TREMADOC-ARENIG
	Sangomore	no certain fossils recorded	
	Sailmohr	gastropods and cephalopods	
	Eilean Dubh	stromatolites, no fauna recorded	
	Ghrudaigh	*Salterella*	
ARENACEOUS GROUP (120 m)	Alternating dolomitic shales, sandstones and quartzites	*Olenellus* and other trilobites, *Salterella*, brachiopods	LOWER CAMBRIAN
	Massive quartzite with vertical tubes ('Pipe-Rock')	*Salterella*	
	Cross-bedded quartzites and basal conglomerate	no certain fossils recorded	

(Lewisian and Torridonian rocks)

Although the outcrop is only a narrow strip a certain amount can be deduced of the conditions of sedimentation. The Precambrian basement had been subjected to a long period of erosion and reduced to a fairly level surface before the early Cambrian seas invaded this part of the foreland. The thin beach pebble beds were succeeded by cross-bedded, well rounded and sorted quartz sands, now in the form of quartzites. The vertical tubes of the Pipe-Rock (*Skolithus*) are similar to the feeding burrows inhabited by marine 'worms' in shallow waters at the present day. The gradual decrease in the clastic components and an increase in the carbonate probably reflect a gradual extension of the seas, so that the sources of detritus became more distant. A rise in world ocean-level in early Cambrian time has been suggested, and certainly a transgressive basal facies is common in many regions.

There are many signs of shallow water in the carbonate formations, such as layers of oolites and stromatolites, the last particularly in the Eilean Dubh Formation. At present the rocks consist of both limestone and dolomite; some at least of the dolomite is secondary, as is some of the chert. Penecontemporaneous dolomitization is commonly interpreted as further evidence of very shallow and locally hypersaline conditions, and the algal growths support that deduction. Breccias occur at several levels, but as in other carbonate sequences it is not easy to determine whether any one represents a major break in sedimentation or only a minor phase of erosion and re-cementation.

The Cambro-Ordovician succession in the North-West Highlands is remarkably dissimilar in facies to those elsewhere in the British Isles or Europe, and the faunas are substantially different. The other outcrops are characterized by an Anglo-Scandinavian (or Acado-Baltic) assemblage (p. 60) but *Olenellus*, the Tremadoc and Lower Ordovician fossils of north-west Scotland, belong to a very wide American province which also includes Labrador, western Newfoundland, East Greenland and Spits-bergen. In these countries also they are found in carbonates and taken altogether form a very widespread cratonic facies typical of the large American plate. From this the Scottish fragment was split off by North Atlantic rifting in early Tertiary times. In the medley of Caledonian components it is a relatively simple sector.

THE CAMBRIAN SYSTEM ON THE SOUTH-EAST SIDE

In this section we return to Wales, where the Cambrian System was first established, and to the Welsh Borders and English Midlands, where the fossil sequence is rather more complete. Certain greywacke and slate groups in south-east Ireland include Cambrian strata. All belong to the south-eastern (European or Baltic) plate; in Britain however there is no continuing history on this side comparable to the Dalradian on the north-west and most outcrops are relatively small. An oceanic tract lay between the two plates at this time, but probably not a very wide one (Fig. 12, inset).

In the Anglo-Welsh region the three Lower Palaeozoic systems are linked together in their history, in their allied areas of outcrop and in their terminology. Although the Cambrian System was named from Wales the faunas are more prolific in the thinner, finer and less disturbed outcrops of Scandinavia, particularly near Oslo and in south Sweden. Similar faunas also occur in south-east Newfoundland, parts of south-east Canada and in New England; the whole forms the Acado-Baltic Province. The typical faunas (mostly trilobites) are as follows:

Upper Cambrian
- Tremadoc Series, dendroid graptolites, *Shumardia* and *Angelina*
- Merioneth Series, olenids and agnostids

Middle Cambrian — St David's Series, *Paradoxides* and agnostids

Lower Cambrian — Comley Series, *Holmia* and *Callavia*

Stratigraphical and zonal tables for Cambrian and later systems are given in the Appendix, page 418.

Even after 150 years the boundaries of the system are not entirely straightforward. The Tremadoc has also been taken as the basal Ordovician division (as it was in the first edition of this book and commonly is abroad), particularly on account of the incoming graptolite faunas. However, pending a complete international nomenclature, current British practice has reverted to the original definition and retained Tremadoc in the Cambrian.

Some of the problems concerning the lower boundary have appeared in the Dalradian section, but there they were associated especially with the rare microfossils and phases of deformation. There are others where modest amounts of undeformed sediments, usually quartzites, occur below the lowest trilobites. Various small shells and trace-fossils have long been known in certain such beds but they have now assumed more importance since an orderly succession of forms is recorded over the Precambrian–Cambrian boundary, especially in Russia. That boundary is under discussion, but it may well turn out that certain parts of the basal quartzites (for instance in Salop and Warwickshire) belong formally to the uppermost Precambrian. For convenience all are retained in this chapter;

in the future more integration between palaeontological and isotopic age data may prove helpful.

THE WELSH TROUGH OR BASIN

On this south-east side the most important outcrops are those of Wales, the Welsh Borders and Warwickshire. The relatively small isolated outcrops, succeeding similarly isolated Precambrian foundations, mean that the Cambrian tectonic and geographical setting is more obscure than that in Ordovician times. But casting backwards somewhat from the latter it seems that the thicker Welsh sediments accumulated in a small marginal basin or trough adjacent to the south-east plate, with those of the Welsh Borders and the Midlands farther onto that stable block, the South-East Foreland.

It is convenient here to relate the Lower Palaeozoic outcrops to the major anticlines and synclines (Figs. 16, 17(a), 21) though there is much complication on a lesser scale. On the south-east flanks of the Bangor and Padarn ridges, with their Arvonian and Cambrian outcrops, the Snowdon Syncline (or Synclinorium) brings in an Ordovician tract that extends south-westwards into the Lleyn Peninsula. South-east of Snowdonia lie the Cambrian outcrops of the Harlech Dome, whence a traverse eastwards crosses the narrow syncline of Tarrannon and enters the Berwyn Dome, a highly asymmetric anticline.

North of these structures the Silurian outcrops expand into the synclines of Llangollen and the Denbigh moors. Southwards there is a similar broadening in Central Wales, but the individual structures are complex. In South Wales the Central Wales Syncline is clearer, separating the short Teifi Anticline on the north-west from the longer Towy Anticline on the south-east. South Dyfed (south Pembrokeshire) is in the realm of Hercynian faulting and thrusting but the north of that county is largely composed of Ordovician rocks with lesser Precambrian and Cambrian outcrops.

Most of the major Welsh structures date from some part of the Caledonian movements; they also include some major faults, or fault belts, such as that along the Menai Straits or the Bala Fault. From its effects on the Ordovician outcrops the latter is deduced to have been operative during Lower Palaeozoic sedimentation, as were the faults in Salop.

Apart from the most easterly belt, for instance in the Welsh Borders and the east flank of the Towy Anticline, all the Lower Palaeozoic argillaceous rocks tend to be cleaved in some degree. In North Wales this is accentuated to the north-west to give the famous roofing slates of

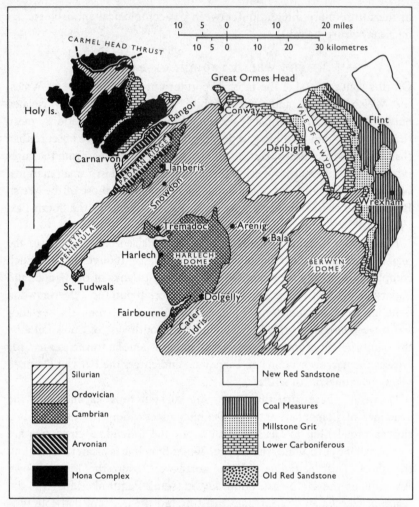

Fig. 16. Geological sketch map of North Wales; igneous rocks (largely Ordovician) not differentiated.

Llanberis; even some of the coarser pebbly rocks of the 'grit' divisions are noticeably deformed. In concomitant fashion, throughout much of Wales there has been substantial tectonic thickening of the various rock groups.

North Wales

Cambrian rocks appear in the Harlech Dome (Table 10), the Llanberis slate belt and in a small inlier along the cliffs of St Tudwal's on the Lleyn

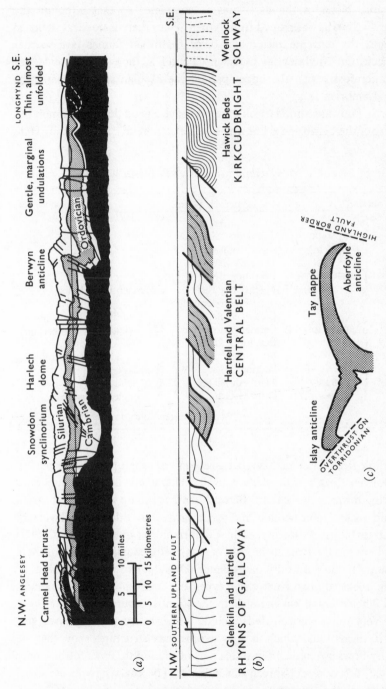

Fig. 17. Caledonian structures. (*a*) Section across North Wales from Anglesey to Salop, restored to the top of the Silurian; Precambrian basement, black. (Redrawn from Shackleton, 1954, pl. 19.) (*b*) Diagrammatic section through the western part of the Southern Uplands, showing the combination of monoclines and faults. (Redrawn from Walton, 1963, p. 91.) (*c*) Diagrammatic cross-section of Dalradian folds in the South-West Highlands to show the Tay Nappe and other structures. (Redrawn from Roberts and Treagus, 1964, p. 513.)

Peninsula. None of the exposures is complete, so that although the Cambrian System was named from North Wales there is no ideal standard here and the extreme rarity of Lower Cambrian faunas is a further drawback. On Anglesey, as noted in Chapter 2, there is some evidence from microfossils that the upper part of the Monian succession may be Lower Cambrian in age.

The St Tudwal's and Harlech outcrops have a good deal in common but the base of the Cambrian is known only from a borehole in the latter. Here

Table 10. *Cambrian succession in North Wales*

Series		Harlech Dome		Llanberis
		(Arenig)		
UPPER	TREMADOC (300 m)	Tremadoc Slates ——? non-sequence—		(Arenig)
		Dolgelly Beds		
	MERIONETH (1400 m)	Festiniog Beds Maentwrog Beds		Festiniog Beds Maentwrog Beds
		——? non-sequence—		
MIDDLE and LOWER	ST DAVID'S (1050 m)	Clogau Shales Gamlan Formation Barmouth Grits Manganese Shales Rhinog Grits	HARLECH GRITS	Bronllwyd (age Grits uncertain) —local unconformity— Llanberis
	COMLEY (1200 m)	Llanbedr Slates Dolwen Grits		Slates conglomerates, tuffs
		(Arvonian volcanic rocks)		

two lithological divisions have been recognized, the upper being part of the Dolwen Grits, with shallow water characters and conglomerates appearing towards the bottom. Beneath there is some 150 metres of varied volcanic rocks, intermediate and basic lavas and tuffs, with layers of detrital turbidite sediments. The Cambrian–Precambrian boundary is drawn between the two, at the base of a 25-metre conglomerate, but there is no angular unconformity or evidence of a major break and pebbles of volcanic rocks (though chiefly acid types) are found in the conglomerates above. The situation has parallels in the Llanberis outcrops.

The very thick Harlech Beds include numerous turbidite greywackes, the 'grit' formations, which in their sedimentary structures show that the turbidity flows swept down into this part of the Welsh Trough from more than one direction and spread out along it. The coarser rocks are often pebbly and both rock detritus and mineral assemblages can be matched in the Mona Complex of Anglesey. This affords a clue to at least one source, a landmass to the north and north-west, part of which still survives. In between the turbidity flows muds and silts accumulated. The Manganese Shales are exceptionally fine in grade, being banded grey and greenish mudstones with only thin sandy bands. The manganese ores were in part a primary precipitate, together with clay and silica. They suggest a period of relatively quiet deposition, perhaps in somewhat enclosed or lagoon-like conditions or a sheltered part of a deeper basin.

The age of the lower part of the Harlech Beds can only be deduced from the lithological correspondence with St Tudwal's, and particularly with the Hell's Mouth Grits, in which trilobites characteristic of the upper part of the Lower Cambrian have been found near the top. The junction of the Lower and Middle divisions thus presumably lies above the Rhinog Grits, but cannot be determined at all certainly.

Farther up, in generally finer beds, Middle Cambrian fossils occur in the Gamlan Shales and the black pyritous Clogau Shales; the latter contain several species of *Paradoxides*, including *davidis* and *hicksi*, as well as agnostids. Black mud sedimentation continued intermittently in Upper Cambrian times to produce much of the upper Maentwrog and Dolgelly slates, but there were also coarser incursions in the lower Maentwrog and Festiniog – grey micaceous sands and silts in which ripple-marks, cross-bedding and various trace-fossils suggest fairly shallow waters. The black shales and slates contain olenids, agnostids and brachiopods, the last including *Orusia lenticularis* and *Lingulella*.

In this Middle and Upper Cambrian sequence North Wales provides a fair example and the Tremadoc district adjoining Harlech to the north-west is the type area for that series. Here the Tremadoc overlies Dolgelly Beds with no conspicuous break but possibly a non-sequence – a layer of phosphatic nodules and the absence of a Scandinavian zone; conformity between Tremadoc and Arenig (basal Ordovician) has also been recorded from the north-easterly outcrops. The sediments are chiefly slates, with a trilobite fauna and early graptolites, of which *Dictyonema flabelliforme* is the best known. The pre-Arenig uplift acted irregularly here with a marked accentuation to the north-west, where all the Tremadoc beds are cut out a short distance from the type area. In this phase of widespread emergence the Rhobell Fawr volcano developed to the east. A cone of subaerial basaltic lavas and pyroclastic rocks was built up, overlying the Cambrian sediments with a marked local unconformity, and was later submerged by the widespread Arenig invasion.

Although the Llanberis slate belt (Table 10) and the Harlech outcrops are little more than 25 kilometres apart they show considerable differences in lithology and uncertainties in correlation. Particular problems concern the base of the system and the absence of most of the Middle Cambrian. The Arvonian rocks of the Padarn and Bangor ridges underlie certain conglomerates that were earlier held to be the unconformable basal Cambrian beds. However detailed mapping shows that there are in fact several conglomeratic layers which are interbedded with slates, like those above, and with ignimbrites, tuffs and agglomerates, like those of the Arvonian below. As at Harlech there is no evidence of a major unconformity, though otherwise there are some differences; in particular, although the volcanic formations appear to be in similar stratigraphical positions, the Harlech basic and intermediate vulcanicity contrasts with the acid types of the Arvonian.

The famous slate succession is composed of a great thickness of fine-grained rocks, rather variable in colour, with grey, purple, reddish and green components. The only fossils have come from the uppermost green beds, where the trilobite *Pseudatops viola* is a late Lower Cambrian form. These beds are thus broadly of the same age as the Protolenus Grits of St Tudwal's. The major coarse unit at Llanberis is the unfossiliferous Bronllwyd Grit, with breaks above and below. The fact that these breaks occur, and enlarge to the north-west, fits the general geographical setting for North Wales. Anglesey was part of a positive region liable to uplift and erosion; it supplied much detritus to the coarser sediments of the Welsh Trough and the uplifts are also recorded as unconformities. In the other direction the trough subsided and its formations show, intermittently, both overlap and overstep to the north-west.

South Wales

The modest Cambrian outcrops in South Wales flank the Precambrian of St David's and Hayscastle in Dyfed (Fig. 21). Much the best exposed is the coastal section south of St David's, along the north side of St Brides Bay (Table 11). Here the lower boundary is clearly unconformable with the basal conglomerate transgressing over more than one Precambrian division. The Caerfai Group is assumed to be Lower Cambrian though its very sparse fossils, crustaceans and horny brachiopods are not diagnostic; the red shales near the top contain bands of crystal tuff and may thus link up with similar pyroclastics of North Wales. The lowest zone of the Middle Cambrian, *Paradoxides oelandicus*, follows above after a local disconformity.

At certain levels Middle Cambrian trilobite faunas are fairly abundant, including zonal *Paradoxides* species. Both Middle and Upper Cambrian have some faunal resemblance to North Wales, but the whole succession is thinner and an unconformity develops eastwards within the Solva Group.

Table 11. *Cambrian succession in South Wales*

	Divisions and lithology	Thickness (metres)
UPPER CAMBRIAN	Grey mudstones and thin sandstones ('Lingula Flags')	uncertain
MIDDLE CAMBRIAN	Menevian Group: grey sandstones and mudstones	230
	Solva Group: grey, purple and green sandstones	180+
LOWER CAMBRIAN	Caerfai Group: thin red shales with crystal tuff bands, green sandstones, conglomerates	290
	(Precambrian volcanic rocks)	

Equivalents of Maentwrog and Festiniog beds have been recognized but it is doubtful whether the Dolgelly is present. The pre-Arenig erosion was vigorous throughout South Wales and only very small areas of Tremadoc beds have been recorded, near Carmarthen.

In view of the isolation of the South Wales outcrops not much can be deduced about the conditions here. A Precambrian land surface, possibly slightly irregular, had been eroded and then submerged and its debris incorporated in the initial pebbly beds of the transgressing Cambrian seas. During Middle Cambrian times argillaceous sediments were prevalent, but uplift in the easterly outcrops interrupted deposition for a time. On resumption littoral sands were laid down. The Upper Cambrian incursion included greywackes and signs of turbidity flows into rather deeper water. There is virtually no evidence, however, on the source of detritus or of any neighbouring continent. One volcanic episode probably spanned the Tremadoc–Arenig boundary; since it initiated a new major epoch in vulcanicity this group (the Trefgarn Andesites) is deferred to the next chapter.

THE EASTERN SHELF OR SOUTH-EAST FORELAND

Although small and separate at the present time there is some similarity in the principle Cambrian successions, for instance of Salop, Warwickshire and the Malvern Hills. None appears to be really thick, that is, over 2000 metres. Where the base is seen the lowest beds lie unconformably on

Precambrian rocks and are usually massive quartzites; glauconitic sands frequently occur in the succeeding Lower Cambrian and the later beds are dominantly argillaceous. Although some outcrops are strongly faulted or folded there is virtually no metamorphism. Salop and Warwickshire combined have yielded a fairly complete sequence though not all the Scandinavian zones have been found.

The Welsh Borders and the south-west

The Cambrian rocks of south Salop appear in the same general region as the Precambrian hills and ridges described earlier (Fig. 11). It is improbable that any beds older than Tremadoc are present in the Habberley valley west of the Longmynd, where exposures are poor, so the best outcrops are those on the flanks of Caer Caradoc and north-eastwards to the Wrekin. Drift is extensive here also, but the lowest beds are exposed where they dip steeply off the Wrekin and the upper ones are exposed on either side of the River Severn. In the south-west, at Comley, the details of the Lower and Middle Cambrian beds have only been recorded from excavations; nevertheless the importance of these sections (Table 12) is

Table 12. *Cambrian succession in Salop (Shropshire)*

Series	Lithological divisions	Thickness (metres)
	(Arenig–Caradoc)	
UPPER:		
TREMADOC	Shineton Shales: grey becoming sandy at top	1000 or more
MERIONETH	Shales, incomplete and poorly exposed	uncertain
MIDDLE:		
ST DAVID'S	Upper Comley Group: shales and thin sandstones	200
	———————— minor unconformity ————————	
LOWER:		
COMLEY	Lower Comley limestones	2
	Lower Comley Sandstone	150
	Wrekin Quartzite	50
	(Uriconian)	

partly due to their faunas, particularly because they include Lower Cambrian trilobites.

The basal Wrekin Quartzite was deposited on a rather uneven floor of Uriconian rocks and varies in thickness. It mostly consists of a white quartzose rock but near the base incorporates pebbles of the underlying rhyolites and tuffs. No trilobites are known from the quartzite but the Lower Comley Sandstone, glauconitic with calcareous and shaly bands, has yielded a few Lower Cambrian brachiopods, crustaceans and a trilobite (*Kjerulfia* or '*Holmia*'). The calcareous character is much more pronounced at the top producing the thin Comley limestones, which are sandy with layers of pebbles and phosphatic nodules. Several distinct faunas have been recorded, including the genera *Callavia* and *Protolenus*, and the whole is probably a condensed deposit laid down slowly and intermittently in shallow water swept by currents.

Middle Cambrian deposition followed after a phase of uplift, minor folding and erosion; fossiliferous pebbles from the limestones were then incorporated in the basal breccias above. Several zones are known, though some only from loose blocks, and isolated exposures make thickness estimates difficult; even less is seen of the Upper Cambrian shales above so that there may well be a major break over the Middle–Upper Cambrian junction. After this thin or irregular deposition the Shineton Shales are much thicker and more uniform, in company with Tremadoc sedimentation elsewhere on the south-eastern shelf.

The Malvern outcrop lies south-west of the Precambrian ridge and is overstepped on the western side by Lower Silurian sandstones. The succession is not unlike that of Salop but seems to be incomplete. The Malvern Quartzite corresponds to the Wrekin Quartzite, is arkosic at the base and lies unconformably on the Malvern gneisses. With the Hollybush Sandstone above it contains a few Lower Cambrian fossils but no trilobites. The succeeding beds, the black Whiteleaved Oak Shales and grey Bronsil Shales, are Upper Cambrian, and accordingly the whole of the Middle Cambrian is probably absent. Along the same north–south trend, the Tortworth inlier (Fig. 24) of Gloucestershire is the southernmost Cambrian outcrop in the country. It includes a substantial thickness (at least 1000 m) of Tremadoc Shales, of which the base is not seen. Thick Cambrian shales are reported 40 kilometres to the east from a deep borehole in north Wiltshire.

Warwickshire and the Midlands

Among the heterogeneous pre-Carboniferous rocks that form a basement to the Midland coalfields (Fig. 47), Cambrian rocks are known from several boreholes, a few minor outcrops and one that is more important. The quartzites and shales of Nuneaton crop out in a strip about 16

kilometres long on the north-east side of the Warwickshire Coalfield; for a
short stretch the basal beds lie on the Caldecote Volcanic Group (Charnian). There is a simple lithological distinction between the dominantly
arenaceous beds below and argillaceous above – the Hartshill Quartzite
(270 m) and the Stockingford Shales (680 m); some rather sparse Lower
Cambrian faunas have been found above and below the junction between
them. For the rest of the succession the Stockingford Shales are unusually
fossiliferous and provide the best comparison with the Scandinavian
sequence; even so there are some discrepancies. The fossil-bearing
horizons are not regularly distributed and many of them are crammed into
the thin subdivision of the Abbey Shales (Middle Cambrian) some 300
metres above the base. These are succeeded by an erosion level with
pebbles and phosphatic nodules and two Scandinavian zones seem to be
absent. The remainder is Upper Cambrian, shales with coarser bands and
an olenid fauna overlain by a small thickness of Tremadoc.

Elsewhere in the Midlands, Stockingford Shales are known from several
boreholes and there are some unfossiliferous quartzites that are presumed
to be basal Cambrian, such as those cropping out in the Lickey Hills, in
south Staffordshire. In a broad belt from Lincoln southwards to Nottinghamshire, Leicestershire, Bedfordshire and Buckinghamshire, Cambrian
beds are known or presumed at depth. Near Lincoln these include steeply
inclined to vertical quartzites, but most of the records refer to Tremadoc
shales, and again dips are steep. The question of a south-easterly extension
of the Caledonian belt is considered later (p. 99) but in terms of
Cambrian sedimentation it seems that the Warwickshire outcrops represent a fairly widespread type.

Relations between trough and foreland

These relations are best seen in comparing the facies and thicknesses in
North Wales, Salop and Warwickshire, though even here there are
extensive gaps between the outcrops. The differences are most marked in
Lower Cambrian times, when on the foreland a weathered Precambrian
surface was covered by an advancing sea and its thin sandy and pebbly
strand deposits. Wide current-swept seas persisted with environments
ranging from near-shore to offshore, producing well washed glauconitic
sands with their cross-bedding, scoured and ripple-marked surfaces and
invertebrate 'burrows'. Very different was the situation in North Wales.
Here, after the close of the basal Cambrian volcanic phase, the first major
arenaceous incursion (Dolwen Grits, etc.) also shows shallow water

characters but later on turbidites swept into the deepening trough. Gradually in Middle Cambrian times the distinction became less marked; shales of comparable thickness covered foreland and trough but there were also phases of regression and erosion at various times in various places – for instance at the base in Salop, at the top at Nuneaton and St Tudwal's and probably for a longer period at Llanberis.

In Upper Cambrian times the trough sediments of Harlech were again thicker and more varied than the shales of Nuneaton, but Tremadoc sedimentation shows an unexpected reversal in thickness relations; full sequences are few, but in the type area there is only some 260 metres, compared with over 1000 in Salop, a comparable amount in Gloucestershire and possibly farther east. By latest Cambrian times nearly all the Welsh 'trough' seems to have been uplifted.

SOUTH-EAST IRELAND: LEINSTER

In recent years there has been a recrudescence in Lower Palaeozoic research in Ireland, embodying new views and more precise data on the ages and relationships of several outcrops. Since this process is likely to continue certain Irish sections, here and in the next chapter, inevitably have a somewhat provisional character.

Apart from the Upper Dalradian formations, proven or deduced Cambrian strata are known only from the south-east province of Leinster. Two large Cambrian outcrops and two much smaller ones are known but the system retains many unanswered problems, chiefly because fossils are extremely rare. The rocks also are much deformed, their boundaries debated and the stratigraphic nomenclature is rather fluid. The main correlative data come from isolated collections of microfossils, together with the trace-fossil *Oldhamia*. These ramifying markings are known chiefly (though not exclusively) from Cambrian rocks in several countries. Lithological criteria also have had to be employed and the base of the Cambrian is nowhere seen.

The Bray Group comprises the northern large outcrop in Wicklow, south of Dublin, and two small ones north of that city (Fig. 31). The sequence in the Bray outcrop is estimated to be over 4000 metres, dominantly of turbidites; greywackes and slates form a lower division, greywackes and quartzites an upper one. In the latter a microfossil assemblage (acritarchs) indicates an age around the Lower–Middle Cambrian junction, though in such a thick sequence there may well be a considerable range above and below. Current directions from the north are also common in the upper parts and some of the coarser detritus could have been derived from Precambrian rocks like those of the Rosslare complex.

The two small outcrops probably also belong to the Bray Group – the headland of Howth north of Dublin Bay and the tiny nearby island of Ireland's Eye. Together these have supplied fine outcrops of turbidite sequences, large-scale slumping and slide breccias or olistostromes. Bentonitic beds possibly represent tuffs. Microfossils give a similar, or slightly older, age to those at Bray.

In southern Leinster the long outcrop of the comparable Cahore Group extends from the east coast of Wexford, at Cahore Point, to the south coast but between these the inland exposures are poor. In lithology the greywackes, quartzites, red sandstone and red and green shales show some resemblance to the Caerfai and Solva groups of South Wales; detailed fossil evidence is lacking but *Oldhamia* has been found at more than one level. The much thinner Askingarran Formation above, of mudstones and siltstones, has yielded a few gastropods near the top and a Tremadoc or early Ordovician age is probable.

The palaeogeographic situation of the Leinster Cambrian still remains problematical, though much work has been done on the sedimentology and structure. It seems clear that this is another 'trough sequence' with turbidity flows and slumping from several directions in the different areas. The resemblance to Wales is not very strong, and with the commonly postulated Irish Sea Landmass in between, incorporating the Precambrian of Rosslare and Anglesey, no very close association would be expected. In addition this Irish province lacks the valuable trilobite faunas (Acado-Baltic) that are diagnostic of the south-eastern, or European, plate. However Ordovician history demonstrates sufficiently clearly that Leinster belongs to this side (cf. Fig. 14(*b*)).

We have now arrived at the end of a markedly heterogeneous chapter covering some 300 million years of geological history and interpreted from a great variety of rocks, ranging from the very thick deformed and metamorphosed groups of the Highlands to the thin fossil-crammed Cambrian beds on parts of the south-eastern shelf. There has been much variation in the research and scales of approach, in the reliability of the fossil data (often very sparse in coarser clastic sediments), and the extent to which isotopic dates reflect original deposition or, more frequently, a later metamorphic event. In palaeogeographic and tectonic terms by, say, mid-Cambrian times the main outlines can be detected but not much more

(Fig. 12, inset). North-western and south-eastern plates are identifiable, and the oceanic tract between them was still a formidable barrier to benthonic faunas. A major gap, in outcrop and in data, obscures the picture across northern England, southern Scotland and most of Ireland. In the next chapter much of the gap will be filled, though not without new queries arising.

REFERENCES

Harris *et al.* 1975 (*Special Report* No. 6, Precambrian, with sections by Downie, Harris and Pitcher, Johnstone, Moorbath).
Cowie *et al.* 1972 (*Special Report* No. 2, Cambrian).
Dalradian and Moine: Anderton, 1976; Bradbury *et al.* 1976; Brook *et al.* 1976; Dewey, 1971; Dewey and Pankhurst, 1970; Downie *et al.* 1971; Harris *et al.* 1978, 1979; Kilburn *et al.* 1965; Lambert and McKerrow, 1976; Piasecki and van Breemen, 1979; Pitcher and Berger, 1972; Skevington, 1971; Spencer, 1971. (References on Caledonide tectonics are also given at the end of Chapter 4.)
Cambrian: Allen, 1968; Brück and Reeves, 1976; Brück *et al.* 1974; Cowie, 1974; Dhonau and Holland, 1974; Rushton, 1974; Wood, D. S., 1969.
British Regional Geology: Central England, The Grampian Highlands, The Northern Highlands, Northern Ireland, North Wales, South Wales, The Welsh Borderland.

4

LATER CALEDONIAN HISTORY:
ORDOVICIAN AND SILURIAN SYSTEMS

The 100 million years or so of Ordovician and Silurian history saw the closing stages of the Caledonian cycle, with the gradual convergence of the north-western and south-eastern plates and obliteration of the oceanic tract between them, the final deformation being completed probably in latest Silurian times. The two systems are thus linked together in outcrop, in many of their structures and in sedimentary characters. Vulcanicity however is a major distinction. In Silurian times it was mostly modest and restricted to a few regions, but Ordovician vulcanicity was vast and great thicknesses of lavas and pyroclastic rocks accumulated widely – the magma rising above the still-active subduction zones to form volcanoes along the continental margins or as island arcs. In a stratigraphical context it is this admixture of sedimentary and volcanic facies, intermittent earth movements, local emergence and local submergence, that makes a general statement on Ordovician history so hazardous. In its shorter duration, much reduced vulcanism and through the period increasing uniformity of facies, the Silurian is in some sense the concluding act of the Ordovician drama.

The outcrops (Figs. 18, 19) are much more extensive than those of the Cambrian and, while there is a generally agreed distinction between those typical of the north-west and south-east sides of the orogen (the south-east passing into a full complement of foreland facies), a complex, difficult belt lies across the centre – especially from southern Scotland to central Ireland – where interpretations have been more debatable. This is partly owing to incomplete or rather baffling exposures, to the search for a line of plate suture and traces of oceanic crust, but also to the inevitable complicating effects of late Silurian deformation. The distinction between the plates, their margins, rocks and faunas appears to be clearest in early to middle Ordovician times and becomes less so in the late Ordovician. Silurian history, as the surviving troughs were gradually filled, is generally more uniform. In the following systematic sections the major sequences and palaeogeographic relations are reviewed: in Wales, England, Scotland and northern and southern Ireland; the plate tectonic setting and some of the attendant problems are added at the end.

The principal stratigraphical divisions are given in Table 13 and with the zones also in the Appendix (pp. 427–8). The problems of the base and

Table 13. *Principal divisions of Ordovician and Silurian systems*

SILURIAN	Downton Series Ludlow Series Wenlock Series Llandovery Series
ORDOVICIAN	Ashgill Series Caradoc Series Llandeilo Series Llanvirn Series Arenig Series

the Tremadoc Series were dealt with in the last chapter and the Ordovician–Silurian boundary is not seriously controversial. That between the Silurian and Devonian, however, has its problems; where rocks of Downton age can be recognized in this country, in facies and distribution they form more logically the first phase in Devonian history. But in central Europe (e.g. Czechoslovakia), where graptolites continue longer than in Britain, the equivalent division is considered to be uppermost Silurian and this practice is formally adopted here.

The most important group in Ordovician and Silurian biostratigraphy is the graptolites (or strictly the Graptoloidea), which appeared in Tremadoc times. They developed rapid evolutionary changes in the Arenig and thereafter successive faunas define the Ordovician series; monograptids cover nearly all the Silurian System from just above the base and in Britain disappear within the Ludlow Series. In many shallow water sandstones, siltstones and mudstones the shelly faunas, especially trilobites and brachiopods, are abundant. In the former group trinucleids and asaphids are confined to the Ordovician; calymenids and encrinurids and appear and continue into the Silurian. Brachiopods rise to great abundance, calcareous types (orthoids, clitambonitids and strophomenoids) being characteristic.

In the limestones the shelly faunas are accompanied by corals, calcareous algae and bryozoans. Echinoderms and molluscs are present, but

Fig. 18. Ordovician outcrops. Inset above: Caledonian plates and the possible South-Eastern Caledonides. Inset below: the Solway Line, one version of the Caledonian suture.

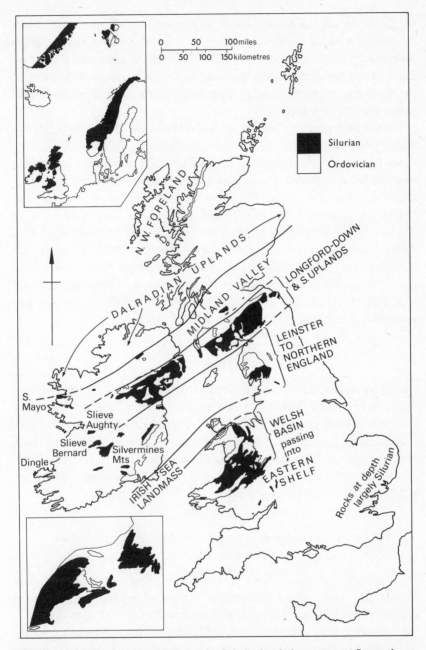

Fig. 19. Silurian outcrops and the main Caledonian belts or zones. Inset above: Caledonian countries in their present positions. Inset below: Newfoundland and part of the Northern Appalachians.

the latter not in the numbers that developed later. In Britain the first fragmental remains of vertebrates appear in Silurian rocks, as early as the Wenlock Series; in the Downton beds the primitive armoured groups are more varied and better preserved and as freshwater sediments come in they become stratigraphically important. Similarly fragments of land plants make a first appearance.

Faunal provinces are best distinguished in Lower and Middle Ordovician rocks and especially in the benthonic groups. Up till early Caradoc times they carry on the pattern of the Cambrian faunas but with many more outcrops available. Brachiopods and trilobites with American affinities, sometimes also called 'Pacific', are found south-west Scotland (e.g. Girvan, p. 113) and across into Ireland in Tyrone and Mayo, as well as in the early Ordovician of the North-West Highlands. The remaining faunas to the south and south-east have Anglo-Welsh or Baltic affinities. By Ashgill times the distinction is reduced and Silurian faunas show little provinciality, either here or in other parts of the world. Palaeolatitudes are inevitably uncertain but in late Silurian times the British area was still south of the equator, possibly around 20° to 30° S.

NORTH WALES

On the south-east side the north Welsh outcrops (Fig. 16) are outstanding, particularly in their Ordovician components. A complex array of volcanic islands and small marginal basins developed, bringing comparable facies changes, local breaks and transgressions and correlative problems. Thick clastic wedges or finer graptolitic muds pass laterally into volcanic debris or are interspersed with very thick lavas and pyroclastic rocks. In stratigraphical sequences three phases can be recognized.

Vulcanicity and sedimentation, the first phase: Arenig, Llanvirn and Llandeilo Series

In North Wales these three divisions are linked together in their sedimentary and volcanic history. After initial arenaceous incursions muddy sediments predominated during the Arenig and Llanvirn, now seen as slates, shales and mudstones. Llandeilo beds, however, are poorly represented or absent and the second phase begins with Caradoc transgressions.

The extremely varied distribution of these rocks, and particularly of the volcanic products, makes a general estimate of thickness impossible, but in the Dolgelly country the Arenig Series amounts to something like 1500 metres. In most regions an early marine transgression is marked by a shallow water or littoral facies, the Garth Grit, a conglomerate or coarse sandstone; only in a few places is the underlying junction conformable. Around much of the Harlech Dome this initial facies is under 200 metres thick, but there are marked changes to the north-west. From the Tremadoc district these basal Arenig sandstones cut down successively on to Upper, Middle and Lower Cambrian beds, until on the north-west side of the Padarn ridge they lie on Precambrian rocks. This marked uplift and erosion is emphasized on Anglesey where the facies is much more than a basal sandy phase; it was deposited on an irregular surface and varies in lithology and thickness from place to place, locally reaching 1000 metres. A rich shelly fauna has been found, chiefly of brachiopods.

Succeeding the varied littoral phase there was a period with widespread muddy marine sediments, of the upper Arenig and Llanvirn shales with several zonal graptolite species. They are varied locally by sandy bands or thin limestones, the latter with trilobites and orthids. There are also tuffaceous sediments in which sprinklings of ashes mingled with normal detritus. Some of these are fossiliferous also, like the so-called 'Calymene Ashes' of the Arenig area.

The principle early outbreak of vulcanicity was concentrated around Cader Idris (Fig. 20(*a*)) where the volcanic products and interbedded sediments amount to some 1650 metres. In age the vulcanicity is usually taken to range from upper Arenig to early Caradoc but, as is often the case, the few fossils found above and below are not wholly precise stratigraphical indicators. Welded acid tuffs or ignimbrites (commonly but not universally characteristic of subaerial ash flows) are common in the bottom extrusions of Mynydd-y-Gader and the uppermost of Craig-y-Llam. In between there are mudstones, the Cefn Hir andesitic tuffs and the Llyn-y-Gafr soda-rich tuffs, agglomerates and spilites, the last often showing pillow forms. Some of the extrusions were thus submarine. At Arenig the 'Upper Ashes' are approximately equivalent to the latest ignimbrites of Cader Idris; in both areas the vulcanicity seems to have ceased early in Caradoc times and thick mudstones continue above. In the Lleyn Peninsula there are also volcanic rocks among the Arenig and Llanvirn slates but they are not as thick as those just described. Similarly north of Snowdon and in northern Gwynnedd the rocks are chiefly slates.

To what extent the Llandeilo period is represented in the later part of this volcanic history is problematic. In much of North Wales there seems to have been emergence, or at least little sedimentation. In some areas perhaps this resulted from positive uplift, related to doming from acid magma beneath; in others the great piles of volcanic debris brought the level of accumulation, never very deeply submerged, above sea-level. The principal exception is in the Berwyn Hills, where Llandeilo beds are

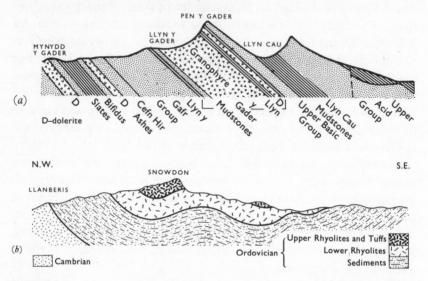

Fig. 20. (*a*) Section illustrating the Cader Idris range to true scale. (Redrawn from Cox, 1925, p. 544.) (*b*) Generalized section through the synclinorium of Snowdonia. (Adapted from several sources.)

exposed in the middle of the asymmetric dome and amount to some 600 metres. Here the dark grey graptolite-bearing shales are rare (though *Glyptograptus teretiusculus* is found) and shelly faunas are common, in siltstones, sandstones and calcareous sediments. In lithology these beds are more like the type Llandeilo of the Towy Anticline than the Ordovician slates of North Wales.

The second phase: Caradoc and Ashgill Series

Caradoc history is especially important in North Wales because it includes the great volcanic group of Snowdonia. There were also other outpourings, on only a slightly lesser scale, in various places northwards to the coastal outcrops at Conway. Volcanic rocks of Caradoc age are also known from the Lleyn Peninsula and in a small outcrop on Anglesey. To the east there are several thin bands of tuffs and lavas in the Berwyns and even as far as Salop (p. 90).

In many areas, far beyond Wales, there is a break in Ordovician history at or near the beginning of Caradoc times. The change may be marked by unconformity and erosion with subsequent transgression and submergence, or by shifts in the centres of volcanic activity. Within our present region this effect is, for instance, seen on Anglesey where shales with *Nemagraptus gracilis* overlie both the Lower Ordovician and the Mona Complex. They also contain large blocks of an earlier limestone that is not now found elsewhere, and which may have been displaced from some submarine scarp, perhaps as a tectonic slide-breccia (olistostrome) or in response to earthquake shocks. On the western mainland Caradoc rocks lie unconformably on Arenig or Llanvirn and elsewhere a lesser gap is represented by the partial or complete absence of Llandeilo beds.

The lower Caradoc sediments show rather more variation than the muds of the Arenig and Llanvirn, from the greywacke incursions on the west flanks of Snowdonia to the fine black muds (Dicranograptus Shales) of Conway. In age the vulcanicity of Snowdonia is probably wholly Caradoc and came to an end well before the end of that period. The variety of volcanic rocks, in composition, thickness and origin, the way they pass laterally into tuffaceous sediments which locally may even include fossils, and their reactions to contemporary earth movements are all characteristic of the Snowdon Volcanic Group and have built up an area of unusual complexity and varied interpretations. On Snowdon Mountain itself the group is over 1000 metres thick, composed especially of acid rhyolitic tuffs (often ignimbrites), lesser basic components, volcanic breccias and some intercalated sediments. They form the beautifully peaked summit and are exposed in the flanking corries and arêtes. To the south and north thicknesses fall off and though only a few vents are visible it is probable that the volcanic centres were largely within Snowdonia itself.

One cause of complexity was the oscillating relation of accumulation-level to sea-level. It seems unlikely that any of the sediments were laid down in deep water; rather that in different regions, and perhaps locally for considerable periods, the volcanic piles maintained themselves above sea-level. On such a view some of their ash flows accumulated on the island slopes as true subaerial products; at the shore they sank into the shallow waters, and became in part bedded tuffs; then farther out they merged with the normal detrital sediments. After vulcanicity ceased there was a return to deeper water conditions and shale deposition. Ashgill beds are not preserved in Snowdonia, but at Conway to the north there is a full succession in a dominantly graptolitic facies. The older term 'Bala' is taken from the type locality east of the Harlech Dome and refers to

Caradoc and Ashgill combined; from here southwards, where vulcanicity
was slight, the two series expand into 3000 metres of poorly fossiliferous
mudstones on the borders of North and Central Wales.

The last sedimentary phase: Llandovery, Wenlock and Ludlow Series

Long-continued erosion has removed Silurian, and particularly late
Silurian, rocks from North Wales, the most complete successions being
found from Conway south-eastwards to Denbigh, in the Clwydian range
and in the Llangollen Syncline.

Over the more westerly outcrops there is conformity between the
Ashgill and Llandovery series and at Conway the latter comprises only
about 100 metres of black shales, which nevertheless contain all the
numerous graptolite zones and must result from exceedingly slow accumulation of mud in very quiet conditions. Such conditions probably
extended westwards, for though there are no Silurian outcrops on
Anglesey comparable with the Ordovician, shales in a synclinal overfold
on Parys Mountain have yielded graptolites of several Llandovery zones.
Slightly coarser lower and middle Llandovery rocks persist southwards
through the Bala country and into the Central Wales Syncline. The upper
division swells out with arenaceous additions and passes into the
greywackes of the Aberystwyth region (p. 86). In the easterly outcrops
there are signs of the marginal uplift that is much more apparent in the
shelf facies of the Welsh Borders.

In the largest outcrop of the Denbigh Moors there is a maximum of
about 3000 metres of Wenlock and Ludlow beds, as follows:

$$\left.\begin{array}{l}\text{Elwy Group} \\ \text{Upper Nantglyn Group}\end{array}\right\}\text{middle and lower Ludlow}$$

$$\left.\begin{array}{l}\text{Lower Nantglyn Group} \\ \text{Denbigh Grit Group}\end{array}\right\}\text{Wenlock}$$

The lowest group consists of greywackes interbedded with siltstones and
mudstones – the whole formed as turbidity currents swept down into the
deeper parts of the North Wales trough. Fossils are rare but the finer beds
contain pockets of shells and occasional graptolites. The succeeding
Nantglyn groups comprise calcareous siltstones and laminated mudstones. The first have a mixed shelly and graptolitic fauna and in their
graded bedding and current-aligned structures these sediments also show
some turbidite features. The Elwy Group ('Ludlow Grits') recalls aspects
of the earlier grit group, though in general they are not so coarse. Ludlow

beds in the northern outcrops are especially noted for their extensive spreads of slumped sediments which were folded and corrugated as they slid down a submarine slope. The general direction of slumping, and therefore presumably of the slope, was to the south and south-east, and this appears to be more or less at right angles to the current direction from the uncontorted turbidites. Thus the latter presumably flowed dominantly along the axis of the Denbigh trough, whereas the unstable muds and silts slipped in contorted sheets down its sides.

In contrast to the earlier stages the latest Caledonian history is almost unknown in North Wales, though we may deduce that after vulcanicity ceased the Silurian turbidites filled in the irregular basins and spread far over the sea floors to give increasing uniformity. Since then erosion has removed any late Ludlow sediments though blocks with derived brachiopods are known in basal Carboniferous conglomerates to the east. The last stages, during or after the final plate closure, are best seen in the neighbouring Welsh Borders.

SOUTH WALES

During Ordovician and Silurian times North and South Wales were part of the same marginal basin and their histories have much in common. The main distinctions in the south are the lesser Caradoc vulcanicity, fuller Llandeilo sequence and an exceptional early Silurian volcanic phase. To the south-east the Towy Anticline (Fig. 21) is a dominant feature and complex structures and overstep appear where Lower Palaeozoic outcrops adjoin the east–west Hercynian belt. Thus westwards from Llandeilo, on the middle Towy, a major unconformity with overstep develops, partly due to late Caledonian movements and partly to Carboniferous warping, and the effects are complex. Near Carmarthen, for instance, Old Red Sandstone rests on Arenig beds but Silurian reappears farther west, and near Haverfordwest Carboniferous Limestone rests on Llandovery mudstones.

The major outcrops are considered first and the smaller ones among the Hercynian folds of the extreme south-west (south Dyfed) afterwards.

Arenig and Llanvirn: sedimentation and vulcanicity

In South Wales Ordovician vulcanicity was largely restricted to the earlier series. The late Cambrian earth movements were strong and Arenig beds rest transgressively on some part of the Upper Cambrian; as in North

Wales, there are local basal conglomerates and massive quartzites. This basal transgressive facies is succeeded by sandy mudstones with Arenig trilobites and brachiopods and thereafter black graptolite shales of late Arenig and Llanvirn age.

These sediments were interrupted by vulcanicity at various places in north Dyfed. The Trefgarn Andesite Group (?Tremadoc–Arenig) was

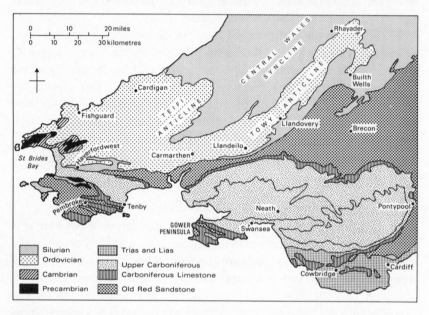

Fig. 21. Geological sketch map of South Wales. Igneous rocks not differentiated; the Upper Coal Measures of the central South Wales Coalfield are outlined.

piled up on a basement of Cambrian beds and later overlapped by the Arenig muds; at Abercastle on the north coast, keratophyres flowed out over the sea floor and incorporated the muds lying there. The Llanvirn Fishguard Group is more substantial and includes rhyolitic and basic tuffs, agglomerates and spilitic lavas; the last amount to 1000 metres, with pillow lavas and sediments rich in volcanic detritus. Ignimbrites have also been recognized near Fishguard and they may be more common than is apparent from the earlier descriptions. As in North Wales the pattern is of local volcanic islands, built up and then eroded, with a multiplicity of products both submarine and subaerial. The gentler topography of south-west Wales reduces the outcrop of these volcanic formations. Nevertheless their thickness, for instance of the Fishguard and Strumble

Head groups, make them comparable to the better known extrusives in North Wales.

A remarkably complete history of the growth and submergence of such an island has been made out from Llanvirn rocks near Builth, east of the upper Towy Anticline. At its most complete the sequence here is fourfold, with an unconformity between each unit:

Cwm Amliew Group	rhyolitic tuffs and mudstones with *Didymograptus murchisoni*
Newmead Group	beach sands and boulder beds
Builth Volcanic Group	keratophyres, breccias, tuffs, spilites and agglomerates
Llandrindod Volcanic Group	rhyolitic tuffs

(shales with *Didymograptus bifidus*)

The main component (650 m) is the Builth Volcanic Group, which was built up on the sea floor and later emerged above sea-level. Erosion set in and the lavas developed a terraced landscape and the keratophyres formed rounded knolls. The short-lived island was then gradually submerged and the sands and boulders of the Newmead Group were banked up against cliffed shores, surrounded stacks and filled in hollows and fissures. This remarkable Llanvirn topography can still be traced, laid bare by the chance effects of modern erosion. Moreover, all this local episode took place within the time of the *murchisoni* Zone, and 10 kilometres away to the north it is represented only by an unrevealing unconformity within the shales of that zone.

Farther to the south-west some Llanvirn rocks appear in the core of the Towy Anticline near Llandeilo and here also they show signs of vulcanicity. There are tuffs, volcanic clays, pebbly sands and calcareous sandstones with a shelly fauna. The volcanic components increase to the north-east and probably they are the erosion products from the Builth centre, spread farther afield over the sea floor.

Facies, troughs and the south-east margin: Llandeilo to Ludlow Series

After Llanvirn times vulcanicity was slight in South Wales; clastic sedimentation prevailed with many facies variants, of which only the more distinctive are selected here.

As elsewhere in the Welsh Basin the quietest conditions are represented by thin bottom muds, now black or grey mudstones with graptolites. They are exemplified in the Llandeilo and Caradoc of the western outcrops and continued into the Lower Silurian. This placid environment was sporadically invaded by turbidity flows, to give various silty and sandy spreads,

greywackes and occasional conglomerates; they appear in the late Ordovician of the centre and west and by inference probably had a westerly
source. The Aberystwyth Grits are a noted example; upper Llandovery in
age and up to 1500 metres thick, they are excellently exposed along
Cardigan Bay. Banded mudstones alternate with thin graded greywackes,
the latter with abundant bottom structures and trace-fossils. From both
the earlier appearance of these influxes in the south and from their
directional structures a southern or south-westerly source seems likely, the
frequent turbidity flows following the axis of the elongate trough. Another
detrital source probably lay to the east because conglomerates on the
flanks of the Towy Anticline contain Precambrian pebbles from an area
now covered by Old Red Sandstone. There is also the Caban Conglomerate of Rhayader; this filled in a deep channel gouged out of the muddy sea
floor in a form that resembles a modern submarine canyon.

A third important facies contributes to the picture of a south-easterly
shelf. Shelly mudstones and fine sandstones of several ages are characteristic of the Towy Anticline. The type of the Llandeilo Series here includes
impure limestones as well and at certain levels they are rich in
brachiopods, trilobites and bryozoans. This shelf, marginal to the south-
eastern plate, presumably extended south from Salop (p. 89) to the upper
Towy reaches, gradually curved south-westwards and then fully westwards in the fragmentary outcrops of south Dyfed.

The closing stages are imperfectly known because the distribution of
Silurian outcrops largely restricts the Wenlock and Ludlow series to
east-central Wales, adjoining the Welsh Borders. Mudstones and fine
sandstones are the dominant rock types and graptolites gradually disappear in the higher Ludlow beds. Turbidites entered from various
directions and gradually filled irregularities on the sea floor but the
detritus was less coarse than in Llandovery times. Ludlow rocks in
particular exhibit a thick basin succession in the north (Radnor and Clun
Forest) amounting to 2000 metres – a remarkable contrast to the shelf
facies at Ludlow itself (Fig. 11, section). Slumped siltstones occur at more
than one level and affect 600 metres of strata near Builth, the unconsolidated sediments sliding on the submarine slopes. In the Llandeilo district
there was another local detrital influx from the east, to form a mixed
inshore deposit with molluscs, or possibly a small delta. The final
transition from marine to marginal and coastal conditions is recorded later
in the Downton Series (p. 94).

Earth movements

In these outcrops on the south side of the Caledonian belt the orogenic phases were episodic and local; as in the Welsh Borders there was no dramatic late Silurian event and most of the effects were much earlier. The late Cambrian uplift and erosion was accompanied by tilting but, after that, folding with some faulting was largely confined to the general line of the Towy Anticline. The results are a complex combination of uncon-formity and overstep, but three major phases can be recognized, preceding Caradoc, upper Llandovery and Wenlock trangressions. Such episodic earth movements are found in other parts of the orogenic belt but are well demonstrated here where the fossiliferous shelf facies aids the identifica-tion of the small strip-like outcrops.

South Dyfed

In the extreme south-west (Fig. 22) the marked Hercynian folds affected mainly Devonian and Lower Carboniferous rocks, but among them there

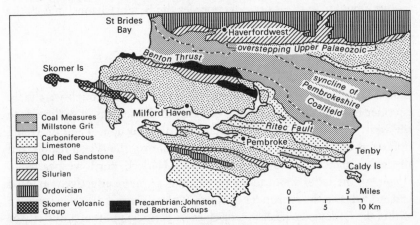

Fig. 22. Geological sketch map of south Dyfed (south Pembrokeshire), illustrating the overstepping relation of Upper and Lower Palaeozoic rocks in the north, the Benton Thrust and Lower Palaeozoic inliers farther south.

are some small Lower Palaeozoic outcrops in the cores of anticlines or as faulted strips. Isolated Llanvirn and Llandeilo sequences are known and more of Silurian age. Outstanding among the last is the Skomer Volcanic Group, an important volcanic episode that may have commenced in latest Ordovician times, but which in the upper parts includes interleaved sediments with early upper Llandovery faunas.

The principal components are thick basic lavas; a few show pillow form but most were probably subaerial. There are also local rhyolitic flows, one ignimbrite and some other tuffs. The whole amounts to 1000 metres on Skomer Island itself but thins abruptly on the mainland to the east; in petrographic type it is a late alkaline development. The remaining Silurian sediments are also unusual and largely Wenlock in age, no certain marine Ludlow being known. Calcareous mudstones include a coral-bearing limestone near the base and are overlain by grey sandstones with shallow water faunas and intertidal or deltaic characters near the top. An Old Red Sandstone facies comes in above (p. 95).

In these southernmost outcrops there is thus some evidence of a further belt of volcanic islands and irregular deposition in very shallow seas – possibly even the margins of a little-known land to the south. The facies form a continuation of the eastern shelf and Midland block, notably in the Wenlock calcareous beds, but no clear palaeogeographic picture emerges. In particular a major junction to the north is the Hercynian-trending Benton Thrust on which the country to the south, including all these Lower Palaeozoic outcrops, has been shifted northwards by some unknown amount. In the other direction lies the Bristol Channel and all the problems of South-West England (p. 99).

An Irish Sea Landmass

Before leaving the geographical setting of the Welsh Trough we should consider how far anything can be deduced of the sources of sediments to the west. Two lines of evidence point in this direction. The pre-Arenig uplift and unconformity was a major event all over the Welsh region, but the most marked overstep is that from Harlech north-westwards to Anglesey; the latter was a temporary landmass, more effectively uplifted and eroded than the remainder of North Wales. There is also some overlap within the Ordovician shales themselves. A periodic uplift of a westerly or north-westerly mass, of which Anglesey remains as a part, fits Ordovician history and, as in the Cambrian period, it supplied recognizable detritus into the neighbouring seas. The second line of evidence is more indirect. At several periods, from mid-Ordovician to early Silurian (late Silurian history is a blank on the west side of Wales) tongues of coarse detritus, especially turbidites, appear in the more westerly sediments. For their source such a landmass is also a possibility, with a south-westerly extension perhaps supplying the detritus of the Aberystwyth Grits.

The concept of an Irish Sea Landmass (cf. Fig. 14(*b*)) also receives support from the Cambrian and Ordovician outcrops in south-east Ireland, and the ridge is usually drawn to include the Precambrian rocks of Rosslare and Anglesey. In order to supply the great thicknesses of greywackes such as those of Central Wales the ridge would have had to be frequently uplifted and persistently eroded. Such a tectonic feature would not be out of place in the unstable sedimentary, volcanic island arc and back-arc basins that include Wales and Leinster.

THE EASTERN SHELF AND MIDLAND BLOCK: SOUTH SALOP

The Ordovician and Silurian outcrops of south Salop (Fig. 11), like those of earlier rocks, have a significance beyond their size. They exemplify the foreland facies of the south-eastern plate – a facies composed especially of mudstones and fine sandstones, often calcareous and rich in the typical Anglo-Welsh shelly fauna. There were also local littoral sands and pebble beds and impressive, but not very thick, limestones. That these variable and shallow marine rocks extended farther afield, at least during Silurian times, is shown by the several inliers of the western Midlands and southwards as far as Gloucestershire. Boreholes in the Midlands and south-east England carry our data a little farther.

Arenig to Caradoc Series: problems of shelf distribution

There is a remarkable difference in the Ordovician successions of the two main outcrops of the Shelve and of the Caradoc district itself (Fig. 23). At present they are separated, from west to east, by the Pontesford–Linley

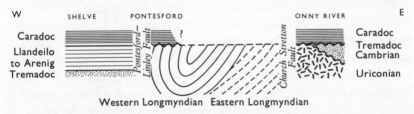

Fig. 23. Diagrammatic section to show the distribution of the various Ordovician series in south Salop (Shropshire); only the Caradoc is present at Pontesford and east of the Church Stretton Fault. The syncline and relationship of the two Longmyndian groups (Precambrian) are also indicated.

Fault, the Longmynd, the Church Stretton Fault and the earlier outcrops of the Caradoc range. Their original relationship is somewhat uncertain.

The Shelve outcrop includes a full complement of beds from Arenig to the lower part of the Caradoc Series (Table 14), the total thickness being about 4000 metres. Above the basal quartzite there is a mixed facies, ranging from the fine-grained sandstones through siltstones with shelly faunas to dark grey graptolitic shales. Only modest volcanic influence was felt in Shropshire. Thin lavas, volcanic breccias and tuffs are found in the Whittery and Hagley volcanic groups (Caradoc); the Stapeley volcanic

Table 14. *Ordovician of the Shelve district, Salop*

Series	Lithological divisions	Thickness (metres)
LOWER CARADOC (Soudleyan to Costonian)	Whittery and Hagley groups with volcanics	500
	Aldress Shales	300
	Spy Wood Grit	90
	Rorrington Beds	300
LLANDEILO	Meadowtown Beds	390
	Betton Beds	180
LLANVIRN	Weston and Stapeley beds with volcanics, and Hope Shales	} 1350
	Tankerville and Mytton Flags	900
ARENIG	Stiperstones Quartzite (Tremadoc)	120

rocks (Llanvirn) are water-laid andesitic tuffs and the Meadowtown Beds include tuffaceous additions among the flaggy sandstones. All this is a far cry from the thousand metres or more of complex volcanic products on Cader Idris or Snowdonia.

The type Caradoc or eastern succession is best exposed in the little Onny river that cuts across the southern part of the outcrop. The outstanding distinction here is the total absence of the Arenig, Llanvirn and Llandeilo beds. The basal Caradoc sands and pebbly beds are strongly transgressive and lie on various formations from Uriconian to the uppermost Cambrian. After the initial submergence there followed some 600 metres of shallow water sediments, ranging from muds to fine sands; all are calcareous in some degree and at many levels there is a profusion of brachiopods and trilobites. Graptolites are rare and do not rise above the *clingani* Zone.

A third small outcrop, the Pontesford outlier (Fig. 11), is in an apparently anomalous position. It lies on Precambrian rocks west of the Longmynd and in stratigraphy more resembles the Harnagian beds of the type outcrop than those of the nearby Shelve, from which it is separated by the Pontesford–Linley Fault. Small though the outcrop is, its position affects the interpretation of the two larger ones and of Ordovician history on the shelf. For were earlier Ordovician beds laid down east of the fault, in the Longmynd and Caradoc areas, and later removed in pre-Caradoc erosion? Or was this area and by deduction much of the eastern shelf, land from Arenig to Llandeilo times, to be submerged by the early Caradoc transgression? The second view is more usually held with the implication that through much of earlier Ordovician history a shoreline lay near the westerly Longmynd scarp. It does meet a difficulty because it demands the overlap of all the lower strata, over 2000 metres, between the Pontesford and Shelve outcrops. If, however, there had been much *lateral* movement on the intervening fault then the distance in mid-Ordovician times might have been much greater. In any event the Pontesford–Linley Fault assumes a regional significance – more at this time, probably, than that of Church Stretton itself.

The Caradoc beds of the type area are themselves not found farther east and this absence can again be interpreted either as an original absence of deposition – that is, a land area or Midland Block behind a narrow Caradoc shelf – or as the result of subsequent erosion. The distribution of all these beds in south Salop thus illustrates in a particularly cogent form the question that accompanies many unconformities in regions of repeated movement. An original absence of Ordovician from the western Midlands is usually assumed but is difficult to prove positively. Farther east (p. 90) there is sparse evidence of some shale deposition.

The upper Llandovery submergence

Whatever the answer to the questions posed in the preceding section, there was undoubtedly an important phase of earth movement in Shropshire between late Caradoc and upper Llandovery times. Faulting and folding affected the Shelve district in particular and the Ordovician rocks have a much more pronounced structure than have the Silurian, a parallel to their relations on the Towy Anticline.

When the seas returned here it was to a region of considerable topographic relief. The main upstanding masses were the Stiperstones quartzite ridge, the Longmynd and the Uriconian ridge from the Caradoc hills to the Wrekin. The littoral sands and gravels of this sea are particularly well seen around the present margins of the Shelve and Longmynd, from which they pass away into muds which were farther offshore. Modern erosion has allowed several coastal features to be

discerned, such as pebble beaches, bars, stacks and cliffed headlands at
the south end of the Longmynd. It is the same kind of glimpse of an
ancient landscape that is visible at Builth where the beach beds are seen
around the Llanvirn volcanic island. In these conditions the thickness of
the upper Llandovery beds is very variable but probably they do not
amount to more than 370 metres. Above the basal sandy facies, or away
from it farther offshore, there follow calcareous mudstones and sandstones
which include layers close-packed with *Pentamerus* valves. A local beach
bed, thick with Uriconian pebbles, is intercalated with the Pentamerus
Beds on the south side of the Wrekin.

Wenlock and Ludlow Series: shelf facies and limestones

In the western outcrops the Wenlock consists essentially of grey graptolite-bearing mudstones which merge into similar rocks in Central Wales
and the Berwyn Hills. In the east (Table 15) there is the double succession

Table 15. *Silurian succession around Wenlock Edge and Ludlow*

Series	Beds and lithologies	Thickness (metres)
	(Old Red Sandstone, Dittonian)	
DOWNTON	Red 'Downtonian', siltstones, etc. (? lower part)	
	Temeside Shales, grey-green siltstones	85
	Downton Castle Sandstone, yellow with grey-green siltstones and bone beds	30
	Ludlow Bone Bed	
LUDLOW	Whitcliffe Beds	54
	Leintwardine Beds (including Aymestry Limestone)	35
	Bringewood Beds	107
	Elton Beds	238
WENLOCK	Wenlock Limestone, or Much Wenlock Limestone Formation	29
	Coalbrookdale Formation } 'Wenlock Shales'	265
	Buildwas Formation }	40
UPPER LLANDOVERY	Purple Shales / Pentamerus Beds with local arenaceous facies	up to 364
	(Caradoc)	

The upper limit of the Downton Series is problematic; see text.

of 'Wenlock Shales' overlain by 'Wenlock Limestone' – the latter forming
the impressive scarp of Wenlock Edge. Despite the formal distinction, all

the Wenlock beds in this outcrop are somewhat calcareous and the limestones at the top are mainly an accentuation of this character. Sandy shales at the base bear occasional calcareous nodules; these increase upwards and there is a transition to the nodular type of limestone.

Two facies in particular are characteristic of the Wenlock Limestone. There are the irregular thin-bedded somewhat nodular limestones, with sporadically a rich fauna of tabulate corals (especially *Favosites*), bryozoans, brachiopods and rugose corals. Locally these are interspersed with the 'ballstones' – large masses of unstratified limestone, chiefly calcilutite but also with algae, bryozoans, stromatoporids and corals; essentially they are reef-knolls, a term more commonly used for similar Carboniferous features. As the supply of muds ceased these knolls were built up by organic frame-builders and stood up as low mounds on the shallow sea floor. At the top a metre or so of crinoidal limestones spread over both bedded and knoll facies, and the limestone phase was brought to an end by a further muddy influx, the lowest Ludlow shales. South-westwards along Wenlock Edge, towards Craven Arms and the Ludlow district, the limestone becomes thinner and the ballstones disappear (Fig. 11).

The shelf facies of the Ludlow Series also displays several shallow water and impersistent characteristics. There is a modest increase in thickness southwards from Wenlock Edge to 400 metres at Ludlow itself before the change to the basin facies farther west, which is marked by a much greater expansion. The sediments tend to be somewhat calcareous grey and yellow mudstones and siltstones. Graptolites, commonest in the finer beds, become more and more scarce upwards and fade out well below the top of the series. Even the locally rich shelly faunas become more restricted, especially the trilobites and brachiopods, though several of the divisions in Table 15 are defined by the latter.

Diachronism is also typical in these conditions. The older term 'Aymestry Limestone' referred to thick current-accumulated banks of *Conchidium* valves, with some tabulate corals and bryozoans. These shell banks reach 80 metres in thickness in the south-west but are scarcely developed at all on Wenlock Edge. West of Ludlow the middle Ludlow beds are cut by several channels, eroded by currents flowing westwards across the shelf into the deeper water of the Welsh Basin. Towards the top the calcareous siltstones become more sandy and pass upwards into the Downton Series, the remaining marine faunas being chiefly bivalves.

THE DOWNTON SERIES: A PROBLEM OF FACIES CHANGE

As noted at the beginning of this chapter, the formal Silurian–Devonian boundary in this country no longer reflects the marked change in facies from marine Ludlow to marginal or non-marine Old Red Sandstone that had been for so long the traditional basis in British stratigraphy. That change is naturally not a sharp one except where it is accompanied by unconformity. This in fact is fairly common since the late Caledonian phase (late Silurian to early Devonian) was widespread; the Welsh Borders and south-east Wales are exceptional in conformity over the boundary and the type area of the Downton Series (Appendix, p. 427), once the lowest Old Red Sandstone and now the uppermost Silurian, is six kilometres west of Ludlow. Beds of this age are also present in some of the southerly inliers.

Over much of the Welsh Borders, from Ludlow to Malvern and Usk, the Ludlow Bone Bed forms a convenient marker at the base of the Downton Series. It consists of water-worn vertebrate debris (bones, scales and spines) together with bits of brachiopod and molluscan shells, eurypterids, ostracods and phosphatic fragments. This heterogeneous assemblage marks a period of very shallow seas when the debris accumulated and was rolled and worn by currents.

Through the Downton Series there is more than one type of upward transition. At the base, above the bone bed, the sediments are dominantly grey, greenish-grey or yellow; these pass upwards into red beds, formerly called Red Marls but which are mainly siltstones. Thereafter the sandy and silty beds are reddish or brown with thin grey-green layers. There is similarly no sharp change-over in fauna: lesser bone beds are found in the fully marine series below and others in the Downton Castle Sandstone Group above. Molluscs persist into the lower Dittonian (Lower Old Red Sandstone) and with *Lingula* are often abundant in the lower Downton beds; drifted plant remains occur in red Downton and Ditton rocks. In terms of environment there is thus a gradual transition from littoral sands and silts with cross-bedding, ripple-marks and abundant shells to intertidal flats and shoals, where the shells decrease but animal tracks and burrows abound; and finally in the Ditton facies fluviatile conditions become dominant. Correlation in such facies presents problems; in the lower beds it is partly lithological and partly faunal, with more recent additions from microfossils and plant spores; vertebrates become valuable in the red upper beds and continue into the Dittonian.

Along the outcrop into South Wales there are various changes. The grey-green beds at the base of the Downton Series persist for a stretch and tend to be sandy. Farther west as the basal unconformity increases this facies disappears and a thick mass of 'Red Marls' overlies various Silurian and Ordovician formations. In south Dyfed the lower part of the red facies (the Milford Haven Group) is of this type but the situation is complicated by a debated lower junction, on the grey Silurian sandstones (p. 88). If it is conformable then red beds and fluviatile conditions set in very early, even perhaps around the Wenlock–Ludlow boundary; if it is unconformable there is less of a problem and the red beds may be equivalent to those elsewhere in South Wales – probably Downtonian and Dittonian – which they much resemble. It is a small debate in a small area, perhaps, but is recorded here to show the sort of questions that arise largely from debated field evidence in restricted or incomplete outcrops where fossils are lacking.

INLIERS AND BOREHOLES TO THE SOUTH AND EAST

The Silurian shelf extended well beyond the major outcrops and in the East Midlands and London Platform a more argillaceous Lower Palaeozoic facies makes up part of the basement. Ordovician records are sparse but some indirect evidence is derived from Triassic pebble beds in the Midlands. These include quite large boulders, up to 20 centimetres across, of an Ordovician fossiliferous quartzite. Neither rock type nor fossils resemble those of the Welsh Borders and are much more like certain quartzites now exposed in Normandy. The Midland boulders, although rounded and far-travelled, are rather large to be easily reconciled with such a distant source and others are also found in the Triassic pebble beds of Devonshire; it is possible that, somewhere now concealed or eroded since Triassic times, there were Ordovician rocks of which we have no other knowledge.

The shelf facies and limestones of the inliers

The inliers, which appear through a varied Upper Palaeozoic cover, include South Staffordshire, a group on either side of the Lower Severn valley (Fig. 24) and small areas near Cardiff and in the eastern Mendips.

The Staffordshire outcrops are only some 35 kilometres east of Wenlock Edge and a similar facies is evident. The coalfield probably lies on a Silurian foundation because several small outcrops are known and other rocks occur underground. Fossiliferous Llandovery sandstones are found, and representatives of the Woolhope Limestone (see below), Wenlock

Shales and Wenlock Limestone, the calcareous facies being conspicuous. The Wenlock Limestone is particularly well exposed in three sharp small anticlines at Dudley, north-west of Birmingham, where it has been much quarried. It is more massive than on Wenlock Edge, with both the bedded and reef-knoll facies, and is highly fossiliferous. Trilobites, which are rare

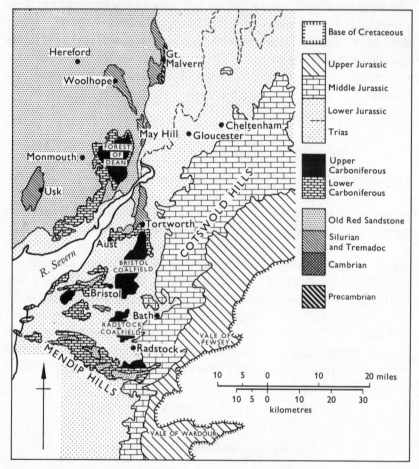

Fig. 24. Geological sketch map of the region around the Severn estuary, Gloucestershire, Avon and Somerset. Triassic and Lower Jurassic outcrops are differentiated in the north but not in the south.

in Salop, are beautifully preserved here, including the typical *Calymene blumenbachi*. The Ludlow Series also resembles that of the type area and includes nodular and impure limestones like the Aymestry Limestone facies. There is a good deal in common between the more southerly inliers (e.g. Malvern, Old Radnor, Woolhope, May Hill and Usk), though exposure varies. The upper Llandovery is seen to overlie Precambrian at Old Radnor and Malvern; it tends to be sandy and 'May Hill Sandstone' is a general term.

The Wenlock Limestone is thinner than it is farther north but some ballstone facies occurs at Woolhope and Usk. The Woolhope Limestone is a widespread calcareous development in these south-eastern inliers at the base of the Wenlock Series. Calcareous beds also appear among the Ludlow siltstones at about the Aymestry level but massive limestones and *Conchidium* banks are missing. This facies is also characterized by very thin or condensed sequences, which at their extreme result in less than three metres between May Hill and Woolhope, overlain by phosphatized pebbles.

The Ludlow Bone Bed has been found in several inliers, succeeded by sandy beds like the Downton Castle Group and above them a 'Red Marl' division. As in the main outcrop the late Caledonian movements had little effect and in many districts the principal structures are clearly Hercynian, or of that age with a north–south Malvernian trend.

If we enlarge our view for a moment to consider the Silurian over the shelf or block as a whole, certain deductions have been made from detailed analyses of the shelly communities. In upper Llandovery times there was the widespread transgression as the seas spread over the relatively stable block. Thereafter there is commonly an upward progression in the bottom faunas, especially the brachiopod assemblages, which can be related to the increasing depth of water, though it was never very deep. At this time the graptolite shales of the basin barely extended beyond the shelf edge.

Analogous patterns are discernible in the Wenlock assemblages but they are more irregular and may well be affected as much by the bottom sediments, especially the sporadic limestones and reefs, as by water depth. The conditions governing any community are likely to be multiple and the sedimentary processes on shallow shelves themselves complex.

Silurian vulcanicity

The shelf succession at Tortworth, in Gloucestershire, is similar to the other inliers in its sedimentary characters; overlying the Tremadoc shales, all the Silurian series are present and include limestones, some rich in brachiopods and corals. A much more exceptional feature is the two basaltic lava flows intercalated with the Llandovery beds, with individual thicknesses up to 60 metres. More igneous activity is known from the

Mendips, where Silurian rocks occupy the core of the most easterly anticline and include 130 metres of pyroxene andesites, lavas and tuffs. They are overlain by Wenlock mudstones and probably the vulcanicity is of that age. Silurian vulcanicity thus operated here also, although on a much smaller scale than the 1000-metre accumulations in south Dyfed and the more distant Wenlock effects in south-west Ireland. The rock types of Tortworth, the Mendips and Skomer do not suggest explosive eruptions, but ash from some source seems to have been carried as far as Staffordshire, where a succession of altered ash layers in the form of bentonites has been recorded among upper Llandovery and Wenlock sediments in a borehole section. Bentonitic clays have occasionally been noted elsewhere and Silurian vulcanicity, over a considerable period, may well have been more prevalent than commonly recognized – though not on anything comparable to the Ordovician scale.

Subsurface components in central and south-eastern England

It has already been noted that a few boreholes have penetrated Cambrian rocks, largely Tremadoc shales. Previously, Ordovician strata were thought to be lacking but there are now several records. Sediments are known from some deep boreholes: in north Salop and in Derbyshire (either Arenig or Llanvirn); near Huntingdon and offshore in the Wash (Llanvirn); in north Kent (Caradoc). Perhaps more unexpected are volcanic rocks to the south-west, especially the thick andesitic tuffs in the Vale of Evesham (Worcestershire) and probably similar ones at Banbury (north Oxfordshire). Silurian strata are also recorded in the last borehole but they are more common farther east, in Suffolk, Essex, Hertfordshire and north Kent, with one record in Surrey; the facies is largely argillaceous.

From these scattered results it seems probable that Lower Palaeozoic rocks are an important component of the London (or East Anglian) Platform, the English part of the London–Brabant Massif (p. 21). Here, south of the Precambrian ridge from Leicester to Norfolk, this eastern development may well be thick. In many boreholes steep dips have been recorded but little or no metamorphism. Geophysical surveys in East Anglia and just off the coast have indicated a 'crystalline floor' (possibly Precambrian) at considerable depth, and between this and the known Mesozoic cover an amorphous mass of strata lacking reflecting surfaces. This has been interpreted as a very thick argillaceous sequence, probably highly folded, and amounting to 7000 to 10 000 metres – a conclusion that in itself raises more palaeogeographic and tectonic questions.

First, the stable shelf or Midland Block is clearest in the western Midlands, Welsh Borders and lower Severn region, but with a possible extension in south-west Wales. The second problem is the more far-reaching one of the 'South-Eastern Caledonides'. This postulated tectonic belt (Fig. 18, inset) was originally conceived as linking the borehole data of East Anglia and Kent with similar subsurface rocks across the Channel in the Boulonnais and thence with outcrops of Belgium, in the Ardennes and along the southern edge of the Brabant Massif. The general trend was seen as north-westerly from the continent to eastern and central England.

This direction is supported by widespread structures inherited in later rocks, such as the Mesozoic in the southern North Sea (p. 270) and the Carboniferous in the East Midlands and south Yorkshire; it is also that of the Charnian rocks where these protrude through the Midland cover, but how this trend might meet, or turn into, the main Caledonian belt is obscure; any junction would probably lie under the thick Carboniferous cover of northern England, or possibly in the southern Lake District. This region, together with the Pennine inliers, is treated in the next section, but we may note here that there is a wide arc of structural trends in the southern outcrops and that those of the Yorkshire inliers are roughly north-westerly. Nevertheless the facies and unconformities of the latter do not offer much support for a linkage with any major south-easterly trough sequence.

Thus there are some scattered and somewhat inconclusive data that could indicate a further orogenic belt in a region now mostly beneath the North Sea, where, unfortunately, pre-Devonian records are very sparse indeed. To them might perhaps be added some slight palaeomagnetic evidence on the Siluro-Devonian pole positions from Britain and Baltic Russia; the positions do not appear to agree with the present geographical relations of the two countries and might suggest some movement between them. However any such belt would seriously complicate the presumed plate positions of Britain and other Caledonian countries and the concept has not been incorporated in the maps and reconstructions of this and the preceding chapter.

The Hercynian belt: Cornwall

With enigmas to the west (an Irish Sea Landmass?) and to the east (the South-Eastern Caledonides?) there yet remains another well to the south.

The Hercynian thrust zone of south Cornwall (Fig. 33) includes blocks and lenses of fossiliferous Lower Palaeozoic rocks, though their origins and some of the age ascriptions are rather obscure. The best known are large masses of an Ordovician quartzite with Llandeilo trilobites and brachiopods. The fauna has close links with similar rocks in Normandy and some affinities with Spain, Portugal and North Africa, but much less with the Anglo-Welsh Province. The Silurian identifications in limestone lenses are more dubious, but they have been compared to the uppermost faunas of Czechoslovakia and thus might be a marine equivalent of the Anglo-Welsh Downton beds. In addition the stratigraphic relations and exposures are complex and not wholly agreed. The blocks have been interpreted as part of a large breccia of nearby derivation, incorporated in Upper Devonian beds, or alternatively as unconformably overlain by low to middle Devonian slates and volcanic rocks. In either case, however, there are hints here of a marine Lower Palaeozoic province well outside the Caledonian belt and part of the early history of Hercynian Europe.

NORTHERN ENGLAND AND THE ISLE OF MAN

The several Lower Palaeozoic outcrops of this region are at present separated by tracts of later rocks, by major faults or by the sea. Nevertheless they have much in common and during Ordovician and Silurian times were part of a single area of sedimentation and vulcanism. Similarly the faunal, stratigraphic and petrographic evidence shows that they were part of the south-eastern plate, near its margin, and most probably the northernmost remnant surviving. The central and by far the largest and most important outcrop is the inlier that forms the main Lake District fells of Cumbria together with the adjoining Howgill Fells on the south-east. Small inliers occur on the western and southern margins of the northern Pennines, and the Manx Slates probably link up with the northern Lake District.

THE LAKE DISTRICT

The major groups of the lakeland hills (Table 16) have a fairly simple areal distribution, but their detailed geological structures are complex and their unravelling is hampered by a lack of recognizable horizons (often a lack of fossils altogether) through great thicknesses of rock. The oldest division,

the Skiddaw Group, occupies much of the northern section of the inlier
and the succeeding Borrowdale Volcanic Group makes up the central fells.
These are bordered on the south-east by a narrow Upper Ordovician and
Lower Silurian outcrop while the Middle and Upper Silurian compose
much of the areas around the southern lakes and Howgill Fells (Fig. 25).

The individual folds mostly have a Caledonian trend (here east-north-
east) in the north and centre, but in the south-west swing into north–
south, and in the south-east into east–west: that is, the trend becomes

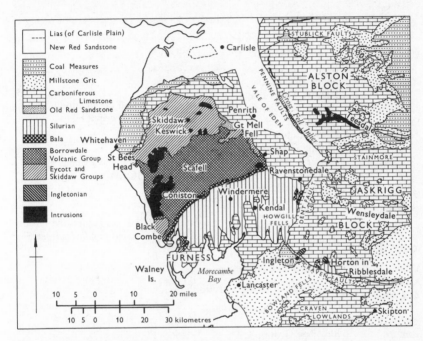

Fig. 25. Geological sketch map of the Lake District (Cumbria) and the northern
Pennines. The Whin Sill in Teesdale is a Hercynian intrusion; those of the Lake
District are Caledonian. Concealed allied granites of the Pennines are shown in
Fig. 41. Bala refers to Caradoc and Ashgill combined.

arcuate. The folding is tight and crumpled, with much faulting, in the
finer of the Skiddaw facies, whereas the more competent Borrowdale
rocks are thrown into broader folds such as the syncline with its axis
through Scafell. The more massive greywackes among the later Silurian
formations exhibit similar structures, but the relatively incompetent beds
in between, especially the Llandovery shales, have been the site of
powerful thrusting. There is evidence of several phases of earth movement, with much debate on their effectiveness.

The first phase: Skiddaw Group, Eycott Group, Borrowdale Volcanic Group

The lowest division used to be called the Skiddaw Slates, but among the varied clastic sediments well-cleaved slates are rare. Mudstones and turbidite greywackes are common, with some coarse pebbly sandstones. All fossils are sparse; they include a few trilobites and brachiopods but the

Table 16. *Ordovician and Silurian of the Lake District (Cumbria)*

Series	Lithological divisions	Thickness (metres)	
DOWNTON?	Scout Hill Flags	300	
LUDLOW	Kirkby Moor Flags	350	Varied clastics including thick greywackes
	Bannisdale Slates	1500	
	Coniston Grits	1700	
	Upper Coldwell Beds	450	
	Middle Coldwell Beds	10	
WENLOCK	Lower Coldwell Beds	180	Dominantly shales and siltstones
	Brathay Flags	400	
LLANDOVERY	Stockdale Shales	60	
ASHGILL	Ashgill Shales	113	
	Applethwaite Beds		Coniston Limestone Group (varied and incomplete)
	~~ unconformity ~~~		
CARADOC	Rhyolite	70	
	Stile End Beds		
	~~ major unconformity ~~		
CARADOC? LLANDEILO	Borrowdale Volcanic Group	(? over 6000)	
	~~ unconformity ~~~		
LLANVIRN ARENIG	Eycott Group	? 2500	
	Skiddaw Group	? 4000–	
	(base not seen)	6000*	

* All thicknesses are maximum or generalized but that of the Skiddaw Group is especially debatable owing to very varied structural interpretations.

occasional graptolites and microfossils have been the most useful. The former show that Arenig and lower Llanvirn beds are widespread but that the upper Llanvirn has been recognized only in the east. Tremadoc faunas are not known but since the base is unexposed they might be present at depth.

The complex structures and rarity of fossils have led to various stratigraphical and structural interpretations, but the sedimentary sequence set out below has been commonly accepted:

Latterbarrow Sandstone ⎞
Kirkstile Slates ⎟ total thickness
Loweswater Flags ⎟ over 4000 metres
Hope Beck Slates ⎠

Of these the two middle divisions are much the thickest and also the more widespread. In rock type grey mudstones are dominant but with marked turbidite sequences in the coarser Loweswater Flags; a southerly source is likely but little detailed evidence is available. The whole probably represents deposits of the continental margin, with much fine sediment in quiet deep water.

Another problem has concerned the age and correlation of the lavas and tuffs that form a broad strip on the north of the Skiddaw outcrop – the Eycott Group, which were usually considered to be part of the better known main Borrowdale Volcanic Group. However, the northern rocks differ in composition from the latter and at the base they are interbedded with Skiddaw-like sediments which have yielded marine microfossils of lower Llanvirn age. It is thus probable that the whole of this northern Eycott Group, which in the west is some 2400 metres thick, is a separate early volcanic development. Tuffs also occur in Llanvirn mudstones on the east side of the Lake District.

The junction between the Skiddaw and Borrowdale groups has been discussed for many years. In many outcrops it is faulted, perhaps a reflection of the differing competence of the two rock groups. Elsewhere there is growing evidence for an unconformity, locally angular. Controversy has thus been narrowed and turns principally on the scale of the break – does it reflect a full orogenic phase whereby the sediments were significantly more deformed than the volcanic rocks above, or a more gentle earth movement and prelude to vulcanicity? Much depends on the interpretation of the field evidence and the second appears to be the majority view.

In a few places there is a basal conglomerate above the unconformity and also a few mudstone lenses among the lowest volcanics, but otherwise the Borrowdale Volcanic Group consists of a great thickness, unrivalled even in Wales, of lavas, tuffs and minor agglomerates. The age can only be gauged from the beds above and below and an ascription to the Llandeilo Series is reasonable, with a possible continuation into the lowest Caradoc.

Andesites are abundant but there is, overall, a wide range from basaltic andesites to rhyolites and dacites. Flow banding and flow brecciation are frequent and thicknesses are spectacular, flow groups reaching several hundred metres. In some districts tuffs exceed the lavas in volume; the coarse ones approach agglomerates and the finest are well cleaved and form the typical green slates of the Lake District. Some of the tuffs are well bedded and show distinctive sedimentary structures, including cross-bedding and ripple-marks; some ashfalls therefore came to rest in water and in one case marine microfossils have been found in mudstones intercalated among the lower flows in the south-west. But for the most part the surroundings are obscure and there is no extensive lateral passage into marine sediments as in North Wales. Much of the vulcanicity was subaerial and fine-grained ignimbrites are important components among the tuffs.

The greatest thickness (over 5000 metres) appears to be in the south-west. Any kind of uniform succession is not to be expected in such rocks but in several regions there does seem to be a crude upward sequence of three cycles, each passing from dominantly basic extrusions to dominantly acid. Individual vents probably gave rise to local variations within this general pattern but such centres are not widely recognized. A few vents and plugs are known (e.g. the Haweswater Complex in the east) but not on a scale commensurate with the volcanic products.

In a wider context the Lake District volcano was the most spectacular part of a volcanic chain, with eruptions at different times in different sections, adjacent to the margin of the south-eastern plate. It was erupted through continental crust, although that of the present Lake District is somewhat thinner than over Britain as a whole. As the lavas and tuffs were deposited, so they must have been persistently eroded, and their debris dispersed; unfortunately we cannot tell where that debris was finally deposited since Llandeilo sediments are largely or entirely absent in this province. In Caradoc times the island was irregularly submerged beneath the sea. The type of vulcanicity in Cumbria follows a logical sequence. The Eycott Group in the north is transitional between tholeiitic and calc-alkaline, while the somewhat later main Borrowdale lavas and tuffs are typically calc-alkaline. Such a sequence would fit with magmas arising from a subduction zone dipping south from the Scottish Borders or the Solway under northern England and declining to deeper levels under Wales (Fig. 14(c)). Here the more varied extrusions have been attributed to some additional admixture from continental crust, and the most westerly and latest examples (the basal Silurian Skomer Group) are alkaline, as are some of the late Caledonian minor instrusions elsewhere. However on a more detailed examination the great range of petrographic types in Wales, the Welsh Borders and the Mendips, and also the long time-span (late Precambrian to early Wenlock in one place or another), does raise problems: the Welsh magmatism becomes less easy to interpret in simple terms. Nevertheless a south-easterly subduction zone, here and in south-east Ireland, is commonly accepted, and is one of the less controversial components in Caledonian interpretations.

The second sedimentary phase: Caradoc to Wenlock

The Coniston Limestone Group (Table 16) includes both Caradoc and Ashgill beds but the former are not widespread. Before their deposition there was a further period of earth movement; the Skiddaw and Borrowdale rocks were folded and faulted, with the greatest uplift in the south. The incoming seas thus gradually covered an irregular floor and the earliest sediments are mid-Caradoc, seen in a small northern outlier (the Drygill Shales) and in Cross Fell on the western edge of the Pennine range. The thickest and fullest succession, nearly all Ashgillian mudstones, is in the small south-easterly inliers of the Howgill Fells.

The 'main outcrop' across the fells is in fact much less complete but demonstrates the irregular invasion; the Upper Ordovician descends onto lower and lower levels in the volcanic rocks as it is traced from east to west. Finally in the extreme south-west it overlies the Skiddaw Group. In this outcrop the Coniston Limestone Group includes a miscellaneous array of shelly mudstones, limestones, ashy beds and a local rhyolite flow; all are thin and several non-sequences break the succession. The traditional 'Ashgill Shales' form only a small upper portion.

After this varied opening, Llandovery sediments succeed conformably and all through that period very slow deposition of fairly deep water muds prevailed. The lower (Skelgill) beds comprise, in some 20 to 30 metres, ten Llandovery zones of black carbonaceous mudstones with graptolites preserved in pyrite. The upper (Browgill) beds are grey rather than black, and to the muds were added silty influxes, interpreted as the distal parts of turbidite flows. Ash bands have also been recorded – a small parallel to the early Silurian vulcanicity elsewhere.

Wenlock facies consist principally of the laminated graptolitic mudstones of the Brathay Flags. However there is an upward increase in calcareous nodules and with these the entrance of a shelly fauna as the sea floor became better aerated and suited to a benthonic population. This trend culminates in the Middle Coldwell Beds at the top, impure limestones and mudstones, richly fossiliferous; they are equivalent in age to conspicuous calcareous facies elsewhere, notably in the Welsh Borders, but the Lake District was not a true shelf region either in sediments or fauna.

The Lower Coldwell Beds below are confined to the south-westerly outcrops and are the first of the coarse arenaceous flows that became a dominant feature in the last stages of the trough infilling. It is possible that the siltstones of the Brathay Flags were also distal turbidites related to the coarser versions in the Wenlock beds of Scotland. This is part of a gradual convergence between southern Scotland and northern England: the stratigraphy of the Lower and Middle Ordovician is very different, as the two sides of the Caledonian ocean belt pursued their own courses; Caradoc faunas are still largely distinct, but 'Scottish' or north-western elements begin to appear in late Ashgill times in Cumbria and when Wenlock shelly faunas become more abundant they are of widespread international genera. The general picture of a decreasing ocean and increasing cross-migration is followed here as elsewhere.

The last phase: late Silurian infilling

As shown in Table 16 the Ludlow Series is very much thicker and mostly much coarser than the two preceding ones. There are two principal facies. The first is a continuation of the banded or laminated graptolitic mudstones and siltstones, with a gradual increase in the latter component. This type is seen in the Upper Coldwell Beds and Bannisdale Slates and is again taken to represent distal turbidites. With the appearance of the Coniston Grits there is a change of regime and for the first time thick arenaceous turbidites spread widely through the trough. Sedimentary structures suggest that they came dominantly from the north-west, the major infilling in northern England perhaps derived even from the Dalradian highlands.

The Kirkby Moor Flags are also sandy, without graptolites but with a rich and varied shallow water benthonic fauna resembling the Whitcliffe Beds of Ludlow. The uppermost division (Scout Hill Flags) contain some red beds and no fossils are known except ostracods. They are probably of Downton age and may represent actual emergence, the Cumbrian or north-western trough being at last filled and obliterated. The main Caledonian orogenic phase followed and is taken to be Lower Devonian; it was succeeded soon after by the major plutonic event of northern England – the intrusion of the large granitic batholith, now seen as the

Lake District instrusions and known beneath the northern Pennines (p. 179).

The Pennine inliers

The surface rocks of the northern Pennines (Fig. 25) are largely Carboniferous but beneath this cover there is a varied basement of Lower Palaeozoic sediments and two Caledonian granites (p. 129). Exposures are few and mostly restricted to the western and southern scarp edges. In addition, where the Howgill Fells abut on the Dent Fault a number of small faulted anticlines bring up late Ordovician strata. These are nearly all Ashgill and the Cautley inlier is the type area for the series.

The Cross Fell inlier lies at the foot of the western scarp of the Alston Block farther north and is itself much faulted. Its rocks repeat in miniature those of the central and eastern Lake District from the Skiddaw Group to the Coniston Grits. In addition the Caradoc Series is better developed and in its diminutive faulted exposures somewhat resembles the contemporaneous beds of the Welsh Borders and North Wales in their calcareous mudstones with an abundant shelly fauna. The Ashgill of Cross Fell also has its peculiar facies, the Keisley Limestone, noted for its reef-like facies and trilobite faunas with affinities to those in eastern Ireland. Indeed there is a 'shelf-like' aspect to the late Ordovician facies in the northern Pennines though scarcely to the early sediments.

How far Lower Palaeozoic rocks extend eastwards is uncertain, but there is a small outcrop of Skiddaw slates in Teesdale and they were encountered in two deep boreholes on the Alston Block. Small amounts of volcanic rocks also occur in Teesdale, but it is not certain whether they represent those of the Eycott Group (which are known in Cross Fell) or the Borrowdale Group. Accordingly we do not know whether the latter episode barely reached so far east or whether its products were later removed, as they have been over parts of the Skiddaw outcrops of the Lake District.

The small exposures in the sub-Carboniferous basement of the Askrigg Block in West Yorkshire appear in the floors of valleys running southwards off the Pennine Hills. They comprise the inliers of Ingleton, Austwick and Horton-in-Ribblesdale and are cut off on the south side by the Craven fault belt; the folding in the inliers has a similar direction, around north-west. The oldest component is the Ingleton Group, isoclinally folded and deduced to be some 760 metres thick. It consists of slates, siltstones which are locally cross-bedded and a lesser proportion of greywacke and arkose. The detrital fragments of the coarser rocks suggest derivation from an igneous and metamorphic terrain. Direct evidence of age is lacking, and the older Precambrian ascription depended largely on the somewhat greater deformation and dyke injection than in the overlying late Ordovician beds. Nevertheless we have just seen that there were extensive Caradoc effects in the Lake District and some pre-Borrowdale effects as well, so that an equivalence with the lower part of the Skiddaw Group or with part of the Manx Slate Group (see below) emerges as a possibility, which is on balance supported in this account. There is also some indirect isotopic evidence that favours a Lower Palaeozoic rather than a Precambrian age.

The later beds of the inliers range from Ashgill to lower Ludlow. The former are locally unconformable on the Ingleton Group and consist of mudstones with a shelly fauna and include a minor volcanic episode and a minor unconformity. Above these succeed Llandovery mudstones and Wenlock mudstones and greywackes that correspond to the Brathay Flags and Coldwell Beds of the Lake District but are rather coarser in grade. The turbidite structures of the Wenlock greywackes suggest currents from the south-east, but by lower Ludlow times they were apparently part of the north-westerly system that characterized this northern province as a whole.

Isle of Man

The impressive central mountains of this island have long presented something of a geological enigma. They are composed of the Manx Slate Group, together with some small Caledonian granitic intrusions and dykes of several ages. A thickness of 7600 metres has been estimated for the slates, but they have been much folded. There is some resemblance to the Skiddaw Group in the greywackes, siltstones, slates and certain thick slumped beds. The very sparse fossils consist of some dendroid graptolites in a slate group, considered on structural grounds to be near the top of the sequence, and microfossils (possibly early Arenig) from a lower horizon. Although the evidence is not wholly conclusive an Arenig age is probable, with the possibility of some late Tremadoc at the base.

Probably the Manx Group passes north-eastwards into the lower part of the Skiddaw Group – a link that is supported by geophysical evidence of a basement 'ridge' in the northern Irish Sea between the two; the Caledonian granites may similarly be an offshoot from the Lake District batholith. Any correspondence in the opposite direction is more speculative, but there is some resemblance in age and rock type with outcrops on the coast south of Dublin (p. 127).

SOUTHERN SCOTLAND, NORTHERN AND WESTERN IRELAND

North of the Lake District, only some 35 kilometres away across the Solway Firth, lie the Silurian and Ordovician rocks of the Southern Uplands of Scotland, but they are clearly in a different stratigraphical and tectonic setting. They are also part of that debated belt of varied interpretations that extends across into northern Ireland. Of the outcrops there, the Longford–Down Massif has several similarities to the Southern Uplands. In Scotland the reasons for this problematic position are multiple. Not only is Cambrian history unknown, as in Cumbria, but also the Lower Ordovician rocks (pre-Caradoc) are sparse, though of much interest. On the northern side the Midland Valley – dominantly an Upper Palaeozoic tract – has only small Silurian inliers, so that the present-day relations of the Southern Uplands are obscured both on the south and north.

Graptolites are the key fossils and locally abundant; the shelly faunas occur only sporadically. Nevertheless the very important Caradoc assemblages, particularly the brachiopods, do answer one question; their dominant regional affinities are American and not Anglo-Welsh. Therefore these sediments were formed on or near to the north-western plate. At the present time there is seismic and other geophysical evidence to show that southern Scotland, like the rest of Britain, is underlain by continental crust. What the situation was in, say, early to middle Ordovician times is central to the multiple problems.

GIRVAN AND THE MAIN BELTS OF THE SOUTHERN UPLANDS

In spite of a century's mapping and study these regions still retain some stratigraphical problems, which partly arise from the complex structure. Inland the hills of the border country are often rounded and grass-covered and the streams cut through sharply folded and faulted rocks among which correlation is not easy. In the later greywackes fossils are extremely sparse. Isoclinal folding undoubtedly affects some of the less competent beds, and outcrops of the graptolite shales, for instance, may be repeated several times. The larger structure is better demonstrated by the thick greywacke sequences. These show that a dominant feature is a form of monocline – that is, a belt of steeply dipping rocks facing north-westwards is followed on the south-east by another relatively flat-lying belt and the same combination is then repeated (Fig. 17(*b*)). This arrangement should, however, result in younger beds generally appearing to the north-west, whereas in fact the opposite occurs (Fig. 26). The answer probably is that these sequences are also affected by large strike-faults or thrusts, dipping to the north-west and throwing down to the south-east, and that the latter movement more than compensates for the north-westerly down-dropping monoclines.

The southern boundary of the Southern Uplands is an irregular unconformable junction with the Upper Palaeozoic rocks from Berwickshire to the Solway Firth. On the northern side the Southern Upland Fault for most of its length throws down the Old Red Sandstone and Carboniferous of the Midland Valley. At the south-western end, however, the Girvan outcrop lies north of the fault, and where the latter traverses Ordovician rocks to the north of Loch Ryan there is a marked downthrow to the south. The following account is taken largely from outcrops in the south-west and in the Moffat district.

Arenig: the Ballantrae Volcanic Group

This important group is chiefly known from complex faulted outcrops on and near the coast near Ballantrae in Ayrshire (Figs. 26, 27), but they have much more regional significance than their 15-kilometre stretch would suggest. The Arenig components comprise spilitic lavas and tuffs, breccias, radiolarian cherts and thin seams of black shales; some of the lavas show pillow structures and graptolites (*extensus* Zone), and horny brachiopods have been found in the shales. These elements therefore include the muds and volcanic products of an early Ordovician sea floor. There are also large masses of serpentinite, associated with various intrusions and occasional metamorphic rocks, including glaucophane schists. All together they comprise the classical components of an ophiolite suite, one of the most satisfactory in the British Isles. Though now disrupted and much displaced from their place of origin they supply evidence for remnants of a Lower Ordovician oceanic crust and ocean-floor sediments not far distant in this debatable belt. The implications are returned to later.

At a few other places along the northern belt of the Southern Uplands smaller outcrops of volcanic rocks are known, some associated with fossil-bearing shales or cherts. Those at Leadhills, 100 kilometres along

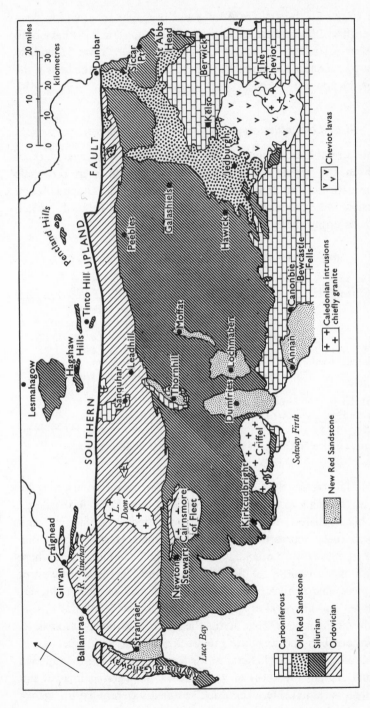

Fig. 26. Geological sketch map of the Southern Uplands of Scotland and neighbouring regions of Northumberland and north Cumbria. Only Lower Palaeozoic outcrops are shown in the Midland Valley, north of the Southern Upland Fault.

the strike to the north-east, are Arenig and probably an extension from
Ballantrae. Elsewhere spilitic, andesitic and keratophyric lavas have been
found, from Llandeilo to upper Caradoc in age. Proven Llandeilo
sediments are rare and Llanvirn doubtful; in most outcrops the base of the
Caradoc is unexposed.

Late Ordovician: sedimentation and facies variation

The use of graptolites as zone fossils dates from the pioneer work of
Lapworth in the 1870s and was based on the faunas in the shale
successions exposed near Moffat in the central belt of the Uplands. Here
three divisions were established – Glenkiln, Hartfell and Birkhill – which
with the later groups are given in Table 17.

Table 17. *Major divisions in the Southern Uplands*

Central belt, Moffat	Series	Southern belt
		? Hawick Rocks (position uncertain)
	WENLOCK	Riccarton Group
Gala Group ⎫	LLANDOVERY	Raeberry Castle Formation
Birkhill Shales ⎭		
Hartfell Shales	{ ASHGILL ⎰ UPPER CARADOC	(base not seen)
Glenkiln Shales (base not seen)	{ LOWER CARADOC ⎰ ? LLANDEILO present	

The Ballantrae Volcanic Group of the north-west belt is Arenig in age.

In the Moffat country the shales crop out in good type sections; they are
conspicuous for being extremely thin (Glenkiln 6 m, Hartfell 28 m) and
comprise black shales, dark grey mudstones (locally tuffaceous) and
cherts. Such rocks represent the deep water quiet slow sedimentation
already noted in the lowest Stockdale Shales of the Lake District and there
is little sign of life besides the floating graptolites. In the south-west of
Galloway somewhat similar rocks are found but they are rather thicker,
with siltstones and greywackes appearing in the upper Hartfell.

From Moffat the facies change across the strike is extremely impressive.
In the northern belt the turbidite greywackes appear with other coarse
facies, including local conglomerates. In the Girvan region a combined
Caradoc and Ashgill thickness of 3100 metres has been estimated, but
there is much variation. Here the thick sediments were piled up against

the steep eroded edges of the Ballantrae mass and gradually overwhelmed it. So steep are some of the junctions that they are thought to be fault margins, the faults continuing to move during Caradoc times.

From such scarps there slid, periodically, great wedges of coarse detritus (Fig. 27). The thickest is the Benan Conglomerate (200 m) but

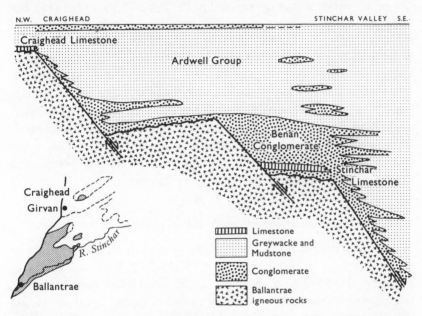

Fig. 27. Diagram to show the relation between the Ballantrae igneous rocks and the Caradoc sediments banked up against them. The coarse detritus (the Benan Conglomerate, etc.) passes laterally into greywackes and mudstones. (After Williams, 1962, p. 67.)

there are others above and below, the pebbles reflecting the erosion of the igneous mass. At some periods limestones formed in shallow waters near the shore; these include the Stinchar Limestone low in the Caradoc sequence and the Craighead Limestone, now only seen in an inlier north of Girvan, near the top. Their fossils include trilobites, brachiopods, molluscs, algae and occasional corals; it is here in particular that the American faunal affinities have been demonstrated. The coarse conglomerates appear to be rather closely restricted to the submergent Ballantrae massif but the turbidity flows extended farther, tailing off into greywackes interspersed among mudstones.

In this western sector there is thus evidence of the gradual submergence of a local landmass or island. An extension of this shore has been deduced farther to the north-east – a broad ridge called Cockburnland, just outside the main outcrop but supplying detritus into the northern belt. The directional structures of the turbidites show that flows were commonly longitudinal, parallel to such a margin.

Beyond this a major gap in the palaeogeographic reconstruction is the lack of any rocks of this age exposed in the Midland Valley (p. 116), but there are a few pointers to the interrelation between Dalradian (or Highland) uplift and sedimentation in the Southern Uplands. It was seen in the preceding chapter that Arenig rocks were involved in the Dalradian folding and that a consensus of isotopic dates suggested that the main deformation and metamorphism were relatively soon afterwards (perhaps 510–490 m.y.). Metamorphic minerals that compare well with a Dalradian assemblage characterize certain greywackes in the Rhynns of Galloway. It is thus deduced that by early Caradoc times some part of the Dalradian range was uplifted sufficiently for erosion to set in, and the isotopic and stratigraphic data combine well to support this view.

Silurian sedimentation: the turbidite infilling

In the Moffat district the slow accumulation of muds continued into lower and middle Llandovery times, as in the Birkhill Shales, which consist of 43 metres of graptolitic mudstones. Similar facies occur along the strike, but to the north-west and south-east greywackes appear. In the succeeding Gala Group these become much more widespread, with 2000 metres in the central belt. Among the conglomerate components there are probable Dalradian rocks, for instance like those of the eastern Highlands. The upper greywacke groups crop out extensively in the central and southern belts but among them many problems remain. There are differences in field interpretations and fossils are only sporadic, the most valuable being the graptolites in the finer mudstones. Much depends on 'way up' criteria in the greywackes and the importance attached to the abundant strike-faulting.

Along the south-west coast, in Kirkcudbrightshire, some continuous exposures of the Raeberry Castle Formation and the Riccarton Group have provided detailed sedimentary and structural analyses, reinforcing the monocline-plus-faulting pattern already summarized. In the type area (Roxburghshire) the Riccarton Group amounts to nearly 3000 metres of coarse and fine greywackes interspersed with variegated mudstones. Not all the standard zones have been found but the group is thought to extend over most of the Wenlock Series. As such it is approximately coeval with the Brathay Flags of the Lake District, so that the more proximal turbidites of Scotland are deduced to pass into the more distal of Cumbria, the ultimate source of both being to the north-west.

There remain the very thick Hawick Rocks – greywackes, red siltstones and mudstones, with ripple marks and cross-bedding; a calcareous matrix is prevalent. The facies appears to be unfossiliferous, its boundaries faulted and its age thus controversial or unknown. A possible Llandovery graptolite assemblage has been collected from similar rocks in the south-west, but their preservation is too poor for certainty. If the Hawick Rocks are the youngest in the Riccarton–Hawick region then they might represent the gradual shallowing of the trough, even to marginal or fluviatile conditions. However there is no positive evidence of Ludlow beds anywhere in the Southern Uplands.

The Silurian succession at Girvan is similarly curtailed and more fragmentary, though locally much more fossiliferous, being restricted to modest inliers bordered by Upper Palaeozoic rocks. Essentially the sequence comprises arenaceous turbidites and the main faunas are graptolitic. Only Llandovery and the lowest Wenlock beds are present and in the earlier part the facies contrast between the thin Moffat shales (Birkhill) and the Girvan turbidites persisted. Some shelly faunas are also found and at one level (upper Llandovery) there is an abundant shelly assemblage – a temporary shelf accumulation in shallow water. The uppermost sandstones (lower Wenlock) include red and green beds with only sparse ostracods. The shallowing waters here might be brackish or even fresh and there is a link with the inliers of the Midland Valley.

THE MIDLAND VALLEY OF SCOTLAND

At present this is a downfaulted trough, or graben, between the Scottish Highlands and the Southern Uplands. Most of it is occupied by Devonian and Carboniferous rocks but there are a number of Lower Palaeozoic inliers.

At Stonehaven (Kincardineshire), in the extreme north-east, there is an isolated outcrop of the Downton Series where some 840 metres of sandstones, siltstones, tuffs and lavas are brought up by an anticline bordered by Lower Old Red Sandstone. They are unconformable on one of the probable Arenig outcrops of the Highland Border rocks (p. 54) and have yielded eurypterids, crustaceans and a species of the pteraspid *Traquairaspis* that allows correlation with the Welsh Borders; the thick Lower Old Red Sandstone follows conformably.

The several inliers on the south-east side (Fig. 26), in Lanarkshire, Ayrshire and among the Pentland Hills near Edinburgh, have a good deal in common; the largest, with the thickest sequence (over 2000 m), is at Lesmahagow. In all except the Pentlands there is apparently a conformable junction with the Old Red Sandstone above, but that division lacks fossils, nor are Downton or Ludlow faunas found below. The presumed Silurian components are dominantly sandy and the lower parts are marine, with turbidites. The rare graptolites and rather less rare shelly faunas suggest that the lowest beds are upper Llandovery and that part of the fossiliferous sequence ranges into lower Wenlock. The silty 'fish beds' in the middle and upper levels contain bones and scales of vertebrates, but these are earlier than the armoured types of the Downton Series and more resemble the fragments of the late Silurian bone beds. The fish beds also contain arthropods, especially eurypterids.

In facies and fauna there is a gradual upward change; the marine types give place to variegated and reddish sediments with cross-bedded sandstones and conglomerates, suggesting deltaic or fluviatile conditions. Although the fish bed faunas survive for a time the uppermost sandstones and conglomerates yield no fossils. Taken together these south-westerly inliers record a rather different Silurian history from that of the main Southern Uplands outcrops, though there are some similarities with Girvan. Moreover the current directions of Lesmahagow and the Hagshaw Hills suggest a southerly source of detritus, reinforcing the idea of an intervening ridge or 'Cockburnland'. Thereafter in the north there was a gradual transition to non-marine conditions, later to pass into true Old Red Sandstone – the same kind of transition as in the Welsh Borders but earlier. At some time within the same period (late Silurian to early Devonian) the rocks of the Southern Uplands were strongly folded and faulted, then uplifted and eroded.

The interpretation of southern Scotland, its foundations and surface geology, is central to the Caledonian history of the British Isles and the plate movements that governed it. Evidence on the underlying crust is largely but not entirely geophysical. The assembled seismic data, already mentioned in northern England, confirm that continental crust extends below the Southern Uplands and Midland Valley to an average depth of 35 kilometres. In the latter, Carboniferous vents in East Lothian have brought up fragments of high grade metamorphic rocks, in a granulite facies more like a Lewisian than a Dalradian basement. Similar evidence comes from central Ireland.

Accordingly, whatever the conditions were preceding the late Silurian orogenic phase, no major tract of oceanic crust survives at present; it has been subducted and destroyed. Its previous existence, however, accords with the whole pattern of converging plate margins and the most substantial evidence and convincing reconstructions relate to Lower and Middle Ordovician times. Several schemes or models have been put forward, with a good deal in common (Figs. 14, 28) but differing particularly on the position and role of the Southern Uplands sediments.

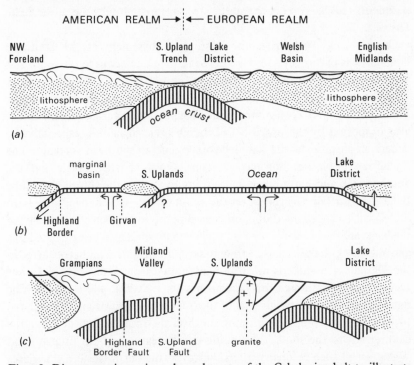

Fig. 28. Diagrammatic sections through parts of the Caledonian belt to illustrate views on the plate tectonic setting, faunas and evolution of the belt. (*a*) The faunal realms in approximately middle Ordovician times: American on the north-west and European or Baltic on the south-east. A trench is suggested in the region of the Southern Uplands with outwardly dipping subduction zones. (Adapted from Williams, 1976, p. 34.) (*b*) Another interpretation of early Ordovician times with similar subduction zones but a wider ocean between the two sides, or the convergent plates; also a small marginal basin is suggested between the Southern Uplands and the Highlands with a minor 'back-arc' spreading system at Girvan. (Adapted from Moseley, 1977, p. 767.) (*c*) The final stage across the Caledonides. An interpretation to show the thick thrust wedges of greywackes forming the Southern Uplands, which are intruded by post-tectonic granites. Coarse clastics and lavas accumulate in the Midland Valley. (Adapted from Mitchell and McKerrow, 1975, p. 314.) In all three diagrams continental crust is stippled and oceanic crust ruled vertically. Compare other versions in Fig. 14 and general plate tectonic diagrams in Fig. 5.

On one view these turbidites accumulated in exceptional thickness on the edge of the north-western plate, or partly on the neighbouring oceanic crust or as a trench filling. Too thick to be subducted *en bloc* they piled up as thrusted and folded wedges (Fig. 28(c)), ultimately to form not only the visible surface outcrops but to extend deeply enough to account for the seismically determined depth of crust. An alternative within the same general framework is that the latter resulted from major underthrusting by sediments and lavas from the south – a continuation of those in the Lake District.

On a second hypothesis the continental basement of the Southern Uplands was part of an island arc between the main ocean to the south and a small marginal (or back-arc) basin to the north (Fig. 28(b)). On either view the main closure or suture line lies approximately along the Solway Firth and the later Northumberland Trough (Fig. 18, inset). This line is also supported by the provinces of the Ordovician faunas, especially the American affinities of the Caradoc components in southern Scotland. The Ballantrae ophiolite, however, is considerably farther north and the Southern Upland Fault line has also been invoked as the suture trace. The Ballantrae (or Girvan) rocks, on the second hypothesis above, have been linked with an earlier subduction zone dipping to the north at the edge of the back-arc basin. Movement on such a zone could also be associated in a general way with the extensive though sporadic magmatism that occurred in the Dalradian highlands and Midland Valley from Cambrian to Lower Devonian times, though this igneous history is diverse and complex. The more restricted granitic plutons of the Southern Uplands are then related to a later subduction zone dipping northwards from the Solway Line, the counterpart to the southerly dipping one beneath northern England and Wales. In all such reconstructions, however, the north-western side raises more queries and complexities than that on the south-east. Further evidence comes from Ireland, though inevitably more hampered by the extensive Carboniferous cover.

IRELAND: NORTH AND WEST

In general there is an obvious continuity of Caledonian history across the Irish Sea, as is shown on several maps in this and earlier chapters. In detail there are some anomalies; some changes appear in the west and south-west, and the Carboniferous cover is inconveniently extensive in the crucial central belt. This survey begins in the north, where several analogies to Scotland are discernible, and then switches to the south-east side.

Longford–Down and Tyrone

The closest geographical and stratigraphical link between Scotland and Ireland is that between the Southern Uplands and the Longford–Down Massif; the latter however is a more subdued upland with an extensive drift cover. The Southern Upland Fault is not exposed on this side, and indeed its existence has been doubted. Most of the Lower Palaeozoic outcrops show the coarser greywacke facies but recent research has centred more on the graptolitic mudstones. Ordovician rocks occur mainly along a narrow strip in the north and on the south side of Belfast Lough, with important exposures on the coast of County Down. There is a resemblance to the graptolitic facies of the northern and central belts of the Southern Uplands and some pillow lavas, probably early Caradoc, are also known. Farther to the south-east the Silurian outcrops are more extensive with comparable parallels to central and southern belts; most horizons recognized are Llandovery but Wenlock equivalents of the Riccarton Group are found. As in Scotland the level at which the turbidite greywackes largely replace the Moffat type of graptolitic facies tends to be younger to the south-east. Within this dominantly Silurian region a few further Ordovician inliers appear (Fig. 31).

The westernmost outcrops in this province are the small Strokestown inliers in Roscommon (Fig. 29), isolated by the Carboniferous cover. However they have links with the Ordovician of County Down in their basic lavas, with pillow structures, overlying mid-Ordovician greywackes. A further extent of these volcanic rocks is suggested by a pronounced positive magnetic anomaly running parallel to the exposures on the north-west side. The outcrops of Pomeroy (Tyrone) are in a different tectonic setting, the small complex region being faulted against Dalradian rocks on the north-west side and overlapped by Old Red Sandstone on the south; within it the various outcrops are themselves dislocated. The lowest Ordovician is the Tyrone Igneous Group, extruded onto a Precambrian floor of high grade metamorphic rocks and a basic igneous complex; this basement is uncertain in age but is possibly equivalent to the pre-Caledonian (? Moine) rocks of the Ox Mountains. The Ordovician above consists of andesitic tuffs and spilitic pillow lavas with rare black shales and cherts.

This assemblage and the structural situation, against the Highland

Border Fault, previously supported an analogy with the Scottish Highland Border rocks. This seems ruled out, however, by an old record of a Caradoc graptolite (*Dicranograptus* sp.) from the shales near the top. For the same reason the Tyrone vulcanicity is not directly related to that of Ballantrae but more resembles some of the later occurrences of the Southern Uplands; it also forms a link with the Irish examples already

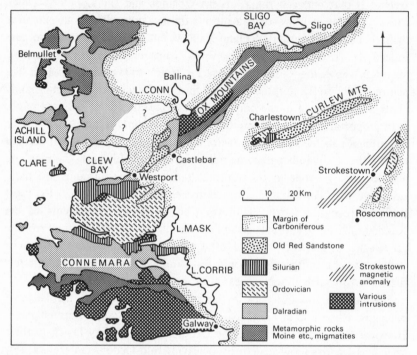

Fig. 29. Sketch map of western Ireland to show the main pre-Carboniferous outcrops.

quoted. The later sediments at Pomeroy include upper Caradoc and Ashgill shelly facies with some resemblances to the Girvan faunas and thus with some American affinities; a thin graptolitic sequence above extends into the upper Llandovery.

Some 50 kilometres to the south-west the small Silurian inlier of Lisbellaw (Fermanagh) has a slight link with Pomeroy. Among the lower Llandovery mudstones there is a substantial conglomerate with large boulders of Dalradian rocks and of the Tyrone Igneous Group; the latter is therefore deduced to be present under the intervening stretch of Old Red Sandstone. On review, while this northern and north-eastern belt has some characters that can be matched in parts of the Scottish outcrops, the correspondence is far from complete. However, both belong to the north-western side of the orogenic belt and can be seen as the remnants perhaps of a mixed volcanic and sedimentary arc on that side.

THE TROUGH OF SOUTH MAYO AND NORTH GALWAY

There is a long stretch of Carboniferous cover between the Longford–Down outcrops and those of western Mayo (Fig. 29); the small inlier south of Charlestown (eastern Mayo) has western links in its Arenig–Llanvirn

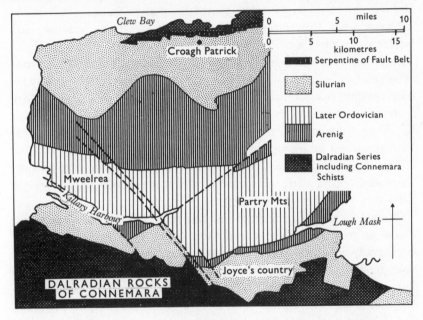

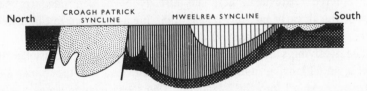

Fig. 30. Geological sketch map of the Lower Palaeozoic rocks of south Mayo and the bordering Dalradian outcrops. The section below is on the same scale and to true scale. (Simplified from Dewey, 1963, p. 315.)

shales and volcanic rocks overlain by thick arenaceous Llandovery. The main Mayo and Galway outcrops, or the South Mayo Trough, lie between the Dalradian uplands of Connemara and a strip bordering the southern shores of Clew Bay. Clare Island, at the entrance to the bay, belongs to the same province in so far as its southern part consists of Silurian strata, unconformable on Dalradian. The line of the Highland Border Fault has usually been taken to lie between the Dalradian and Silurian rocks just south of Clew Bay, accompanied by a vertical screen of serpentine as in parts of the Scottish sector. However, a more northerly course through Clare Island has also been suggested.

The Lower Palaeozoic outcrops fall structurally into three east–west belts (Fig. 30). In the northernmost of these, Silurian beds form a deep syncline, slightly overturned on its northern flank. The junction with the Ordovician outcrops to the south is partly faulted and partly unconformable; this, the widest belt, is also a complex syncline. Still farther south there are a number of faulted, and again partly synclinal, Silurian outcrops which overlap onto the Connemara Dalradian. In spite of its isolation, small area (about 35 km across) and the tantalizing way the outcrops disappear westwards under the Atlantic, the South Mayo Trough is an important part of the Caledonian complex. The succession is notable for its great thickness, few fossils and a major unconformity between the Ordovician and Silurian portions (Table 18).

The lowest beds, the Lough Nafooey Group (Arenig and possibly Tremadoc), occur along the southern edge of the central synclinal belt as black shales and cherts with occasional graptolites, and a thick sequence of spilitic pillow lavas, agglomerates and acid tuffs. The boundary against the Dalradian rocks of Connemara is not seen owing to Silurian overstep but it is generally supposed to be faulted. On one view a very large boundary fault has been deduced, the exceptionally deep trough sagging with intermittent fault movement against the neighbouring upland as deposition continued. The juxtaposition of high grade metamorphic rocks and little altered Ordovician here has already been mentioned (p. 54) as supporting an early age for the Dalradian metamorphism in Connemara – earlier than the mid-Arenig *nitidus* Zone.

The overlying sediments contain very few fossils, but occasional upper Arenig and Llanvirn faunas have been found and the uppermost formations may extend into the Llandeilo. The rocks include slates, turbidite greywackes, occasional tuff bands, and cross-bedded sandstones near the top, the whole thickening to the north-west in the Murrisk district. Some of the tuffs are graded, representing ash showers falling into the sea. The individual bands differ enough in composition and texture to be recognizable in the field – a most valuable feature in rocks so poor in fossils. The much thinner Arenig outcrops to the east along the slopes of Lough Mask include a few limestones and limestone breccias whose trilobites and brachiopods have American affinities. The Dalradian uplands supplied much recognizable detritus into the trough, particularly from the Connemara side.

At some period after the Llandeilo there was a major phase of uplift, folding and erosion and a great thickness of strata was removed before the upper Llandovery submergence. The area of sedimentation however

Table 18. *Ordovician and Silurian of south Mayo*

Northern outcrops	Series	Southern outcrops
Croagh Patrick succession (1680 m) mainly sandstones variable and cross-bedded	WENLOCK	Killary Harbour Group greywackes, slates, shales
		Upper Owenduff Group lithology as above
Lower Owenduff Group (6 m)	UPPER LLANDOVERY	Lower Owenduff Group clastics and thin limestones

~~~~~~~~~~~~~~~~ major unconformity ~~~~~~~~~~~~~~~~

Western outcrops		Eastern outcrops
Maumtrasna Group (2240 m) and Mweelrea Group (2130 m) coarse clastics, slates and ignimbrites	LLANDEILO and UPPER LLANVIRN	approximately as in western outcrops
Glenummera Group (1220 m) lithology as below	LOWER LLANVIRN	~~~~~~~~~~~
Four groups (5330 m) with conglomerates turbidites, slates and tuff bands (base not seen)	ARENIG	Glensaul Group (530 m) coarse clastics, shales, thin lenticular limestones
		~~
		Lough Nafooey Group (760 m) shales, cherts, lavas, tuffs
		(possibly some Tremadoc, base not seen)

continued to be much the same, with some extension over the northern shore to Clare Island and also to the south. The Lower Owenduff Group (upper Llandovery) is largely confined to the south-eastern outcrops, where, overlying a basal conglomerate, the beds decrease in grade through sandstones to fossiliferous limestones and shales. The Wenlock beds on the south again revert to the greywacke and slate facies with red and green shales. Graptolites and shelly faunas are known but only occur very rarely.

In the northern belt a Wenlock transgression was also effective and only six metres of Llandovery beds can be found, the Croagh Patrick succession lying either on these or on Arenig. On this side the Wenlock beds lack the turbidite greywacke facies and show some signs of belonging to a littoral facies belt, in which much of the detritus was derived from the northern Dalradian uplands. Cross-bedded sandstones are important with some finer beds, including tuff bands, and much small-scale facies change. No Ludlow strata have been proved in the main outcrops, but the uppermost faunas are middle Wenlock so some may well be present at the top. Moreover a composite succession from Clare Island and Louisburgh, on the northern strip of the mainland, makes up a thick arenaceous and conglomeratic sequence that has been assessed as Ludlow and possibly even Downton, but it also lacks fossils. The facies range from shallow marine to fluviatile at the top. The last main Caledonian effects here are thus probably Devonian; on the east the eroded folds are overlain by Lower Carboniferous.

Compared with the other Lower Palaeozoic troughs, that of south Mayo is in some ways simpler and better defined. In no other are the marginal landmasses so clearly seen. Some may be fairly safely deduced and fragments of others, such as Anglesey, still survive, but the picture presented in western Ireland of a gulf of deeply subsiding sediments, hemmed in on either side by Dalradian uplands, is particularly graphic. The general alignment and Lower Ordovician shelly faunas of the south Mayo outcrops show that they belong to the north-west side of the orogen, and despite the major break and some stratigraphical uncertainties they form an important part of the complex sedimentary and volcanic arc bordering that plate. The problems of geography, tectonics and history already apparent in Scotland are accentuated in western Ireland by the major upland of the Connemara Dalradian, which, it seems, must have lain on the oceanward side.

IRELAND: EAST AND SOUTH-EAST

In Ireland the clearest indications of the south-east side of the orogen come chiefly from the province of Leinster, from the east coast and as far south as Waterford. Here there are some significant parallels with northern England and Wales. Even so the crucial faunas are not abundant.

Inliers of eastern Ireland

These include a number on or near the coast north of Dublin (Fig. 31) where Ordovician volcanic rocks are particularly important. They represent a number of individual volcanic centres, building up small cones or islands on the sea floor. The lavas are mainly basalts and andesites, mostly calc-alkaline but with some tholeiitic types. There seems to have been little coarse detritus, the accompanying sediments being mudstones, locally graptolitic, or fringing limestones.

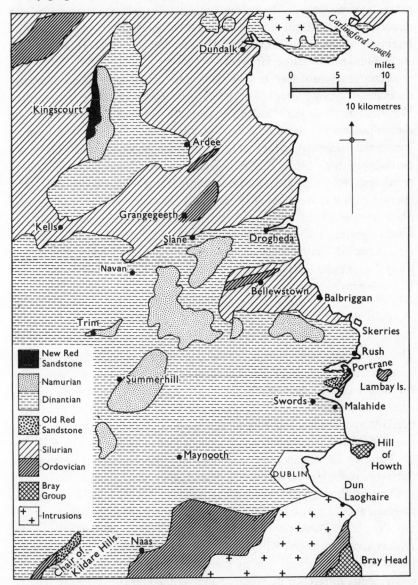

Fig. 31. Geological sketch map of eastern Ireland (part of Leinster) from Dublin northwards to the southern part of the Longford–Down Massif and the Kingscourt outlier.

The largest of the outcrops is around Balbriggan where the andesites reach 1500 metres in thickness, extruded into marine muds with layers of pillow lavas and locally breaking up into breccias. Similar rocks occur at Portrane and Lambay Island, the latter comprising a volcanic island surrounded by lavas, tuffs and agglomerate. Overall, the ages range from Llanvirn to Ashgill and at Portrane the thin limestone above the volcanics resembles the Keisley Limestone of Cross Fell (p. 107). In general the shelly faunas have Baltic affinities.

Silurian rocks are seen best at Balbriggan where there is an unusually complete range of graptolitic mudstones from Llandovery to upper Wenlock; much of the Wenlock resembles the banded mudstones of the contemporaneous Brathay Flags of the Lake District, the coarser greywacke facies appearing only at the top of that series. The most northerly of this group is the Ordovician inlier of Grangegeeth and Collon, which lies across the Meath–Louth border; here there are tuffs interspersed with a Baltic type of shelly fauna (probably basal Caradoc). Any plate margin therefore presumably lies between here and the north-western outcrops of the Longford–Down Massif – that is, within the area now occupied by the less informative Silurian greywackes. There is a general alignment with the Solway, but that line lies south of the Southern Uplands greywackes, not within them.

Kildare

The inliers here (the Chair of Kildare) are some 15 kilometres from the north-western margin of the Leinster Massif, and thus are somewhat divorced from the previous group but allied to it. The steeply dipping succession includes Llanvirn shales at the base, which are unconformably overlain by andesites and ashy mudstones of Caradoc age and these again by the Kildare Limestone (Ashgill). The vulcanicity is reminiscent of that on Lambay Island and the trilobites and brachiopods, like those of Portrane, resemble those of the Keisley Limestone.

There is thus a belt of scattered outcrops which have a number of resemblances in petrographic type and faunal affinities linking the northern Pennines, Cumbria, the Isle of Man, eastern Ireland (especially Dublin and Meath) and Kildare. It presumably formed an outer volcanic island arc on the south side of the early Ordovician oceanic tract; the conspicuous difference in volcanic and sedimentary thicknesses may be related to the size and duration of individual centres – with the Borrowdale volcanics of Cumbria outstanding – and the local inflow of clastic sediments. Near the end of Caledonian history the greater spread of Silurian turbidites produced some smoothing out and more uniformity.

The Leinster Massif

This is a major Lower Palaeozoic block in Ireland but not one of the most informative. The two unequal sedimentary outcrops are separated by the 100-kilometre stretch of the Leinster Granite and inland exposures tend to be poor. A sequence of greywackes and quartzites in the smaller north-westerly outcrop have yielded only sparse faunas but enough to suggest that both Ordovician and Silurian rocks are present. On the south-east side only Ordovician is recorded, but with excellent exposures on the south coast (Wexford and Waterford) and some at the north end (Wexford and Wicklow).

Rather sporadic fossil-bearing sediments provide a stratigraphical framework within which the Middle to Upper Ordovician vulcanicity is famous. Arenig graptolites have been found in the north, in the variegated or banded slates that comprise the upper part of the Ribband Group; the lower part is presumably Cambrian, and it is with this group and type of rock that a parallel has been drawn with the Manx Slates. The latter lie along the north-easterly strike from Wicklow and both belong to the same tectonic belt, but since both consist chiefly of a great thickness of unfossiliferous slates and greywackes no very precise implications follow.

The later Ordovician volcanic sequence is extensive but can be exemplified from the dramatic outcrops seen in the cliffs of Waterford on the south coast. Here activity began in the Llandeilo with the growth of a large submarine shield volcano, producing basalts, andesites and dacites, some calc-alkaline and some tholeiitic. Abundant andesitic sills intruded the wet sediments and tuffs. The succeeding Tramore Limestone, a well known marker at the base of the Caradoc, formed during a lull; it contains a rich fauna of trilobites, bryozoans and nodular masses of corals, the faunal affinities being mainly Baltic. With renewed subsidence turbidites entered and a second major volcanic sequence followed, to reach about 4000 metres at Tramore. In addition to the earlier rock types smaller centres developed, some built up above sea-level and producing much rhyolitic ash. Still later there was a phase of more alkaline intrusions in both southerly and easterly outcrops, some of which seem to be later than the main Caledonian deformation.

Reviewing this south-east side in Ireland, there is a resemblance in petrographic sequence (tholeiitic, calc-alkaline and alkaline magmas) to that of northern England and Wales, both being attributable to a south-easterly dipping subduction zone below the plate margin on this side. A direct continuity between south-east Leinster and Wales is less likely since the Irish Sea Landmass is reasonably postulated between the two. The parallels are thus partial, as are so many along the Caledonian belt.

THE SOUTH-CENTRAL INLIERS AND KERRY

In the sections reviewed so far the regional alignments of the various Lower Palaeozoic outcrops could be reasonably assessed, but farther west, with the greater spread of the Carboniferous and Quaternary cover, many more difficulties appear, essentially through lack of outcrops.

Several hilly regions arise from the south-central Irish plain, being chiefly anticlines of Old Red Sandstone. Within these are some Lower Palaeozoic inliers, the most important being the Silvermines Mountains, Galty Mountains, Cratloe Hills, Slieve Aughty and Slieve Bernagh in the counties of Tipperary, Limerick and Clare. The last two are unusual in that Upper Ordovician beds are known and at Slieve Bernagh they include Caradoc faunas with some modest European (Czechoslovakian) resemblances. Otherwise nearly all these inliers are Silurian, with Ludlow and Wenlock well represented in both shelly and graptolitic facies. Some Llandovery beds appear, such as the abundant graptolite mudstones at Tomgraney in Clare.

Nevertheless, in the lack or extreme sparsity of Lower to Middle Ordovician evidence, there are inevitable problems in trying to define plate margins or trace a suture across southern Ireland. On geophysical and structural grounds a possible line has been deduced (Fig. 18, inset) passing among the Lower Palaeozoic inliers and continuing along the centre of the Shannon trough – a Carboniferous feature slightly north of the present estuary.

The Dingle Peninsula

Problems also surround these outcrops but of a different type. Dingle is the most westerly point of Kerry and of the Irish mainland (Fig. 36); it contains some Silurian outcrops, flanked by Old Red Sandstone, which are far removed both in distance and sequence from the centre and east. Two major groups are present – a lower one of fossiliferous Wenlock and Ludlow and an upper Dingle Group of unfossiliferous purple and reddish sandstones, slates and conglomerates, whose age ascriptions have varied.

Both groups are strongly folded and faulted but otherwise the Silurian components are fairly straightforward. They consist of some 1600 metres of clastic and impure calcareous sediments, with a rich shelly assemblage including brachiopods, crinoids and corals. Faunal affinities are closest to those of the Welsh Borders. Intercalated in the Wenlock portion is a variable set of intermediate to acid volcanic rocks, especially rhyolites, agglomerates and tuffs. This is the most westerly Silurian vulcanicity along the line from Tortworth (p. 97) and the Mendips to Skomer and Kerry; in rock type and age, however, there is some disparity and the Welsh extrusions were considerably earlier.

Above this fossiliferous sequence there follows the Dingle Group – greyish sandstones with cross-bedding, desiccation cracks and other signs of shallow water. It is presumably equivalent to the Downton Series, but being very thick (approximately 2500 m) may range up into the Lower Devonian. There was thus an early onset of continental conditions, though apparently without the transitional fauna and flora that aid stratigraphy in the Welsh Borders.

CALEDONIAN INTRUSIONS

By far the greatest part of the intrusions of the British Isles are Caledonian and the major groups are summarized here, with some isotopic data. Where the latter are derived from many determinations the ages are fairly secure, but it must be emphasized that others, especially among the older ones, are based on only a few or debated results and may well be revised in the future. The note on recalculation (Table 2, p. 3) applies to individual dates quoted below.

(1) Newer Granites, etc., of Scotland and northern England. This is a very extensive group aged around 400 (or 415–390) million years, linked together in age and in being post-tectonic, but with some major distinctions in composition and style of emplacement. In addition to true granites, granodiorites and allied rocks are common. The most abundant are the Newer Granites of the Scottish Highlands and this type also is the only one, with small exceptions, in the paratectonic belt farther south. Here are those of the Southern Uplands and the Cheviot, of the Lake District and Leicestershire, and (subsurface) in the northern Pennines. The ring complexes of the South-West Highlands (the Glencoe, Ben Nevis and Etive granites) are similar in age but different in style and are linked

with Lower Devonian lavas. Farther north the Assynt alkali suite has no clear affinities but has stratigraphic evidence of post-Cambrian age.

(2) A few small granites, with pegmatites, in north-east Scotland that are significantly earlier with dates around 460 million years, a late tectonic or post-tectonic group.

(3) The gabbros of north-east Scotland, or the 'Younger Gabbros'; it has been suggested that these are part of a single large intrusion, with an age close to 500 million years.

(4) The Older Granites of the Scottish Highlands. This is a much more diverse group, differing in age and geological relations and possibly only linked by being affected by a major metamorphic event. Isotopic evidence is rather slim but includes the following: Ben Vuirich Granite (Perthshire), $514 \pm ^6_7$; Carn Chuinneag Granite (Ross-shire), 550 ± 10; Portsoy Granite (Banff), 655 ± 17. The last may well belong to an earlier intrusion cycle and be pre-Caledonian. There are scattered lenses of augengneiss, similarly diverse in age (e.g. 498 ± 38), of doubtful significance, and also very large areas of migmatites, sometimes grouped here, in the Moine outcrops and north-eastern Dalradian.

(5) In Ireland there are some clear parallels, especially with the post-tectonic group. Here the 'Newer Granites' include the major intrusions of Donegal, Leinster, Newry and Galway. There are also some older examples with ages close to 500 million years, notably that in the Ox Mountains, which is syntectonic. The basic and ultrabasic intrusions of Connemara so far have a debated age-range, from about 510 to 480 million years. There remains the Oughterard Granite of Galway, which with a probable age of 459 ± 7 falls in between these two groups and is also distinct in its form and occurrence.

In Scotland a few tentative deductions may be made on the relationship of intrusion to metamorphism, though more confirmatory data would usually be welcome. As mentioned in Chapter 3 the main 'peaks' shown by the Moine and Dalradian metasediments (*c.* 420 and 450 m.y.) are assumed to be uplift and cooling effects. A narrower interval, perhaps even as short as 510–490, is suggested for the main Dalradian metamorphism from the ages of the north-eastern gabbros, which are only slightly affected, and of the Ben Vuirich Granite. The latter also is intruded into Lower and Middle Dalradian rocks but is older than the Upper Dalradian MacDuff Slates, which contain Ordovician microfossils (p. 53) – a relation that suggests that in this very large and long-lasting supergroup intrusion at depth could be contemporaneous with continued sedimentation at the surface.

BRITISH CALEDONIDES IN THEIR PLATE TECTONIC SETTING

A geographical distinction was set out earlier between those north-western and south-eastern regions where remnants of the Caledonian plates could be recognized with fair success and the debatable belt in between. Here we also look at the more generally accepted conclusions and some that are more problematical – a provisional assessment of the plate tectonic hypothesis related to the Caledonides, but in simple terms and certainly not exhaustive.

On the positive side, and as a logical backward extrapolation from Mesozoic and Cainozoic orogenies, the British Caledonides can be accommodated within the plate tectonic pattern, not without many queries but without gross distortion or internal contradictions. Thus there is strong evidence from several sources for an early Caledonian ocean, and that ocean has disappeared. Many aspects of the magmatism conform to the thesis of outwardly dipping subduction zones on which that disappearance took place. Evidence for the distinction of the American and Baltic plates, much of it palaeontological, is scattered but consistent. The contrasts between shelf and trough facies (with the typical turbidites of the latter) are appropriate but scarcely diagnostic, in that such facies can be found in other tectonic situations. However, some of the marked lateral changes in sedimentary and particularly volcanic thicknesses, which do seem to be rather characteristic of the Caledonian belt, agree with mixed volcanic and sedimentary arcs, with in places one or more back-arc basins.

On the other side there are queries on certain regions or topics. Some of the vulcanicity, notably in Wales, is so long-lasting and various that on a fuller examination the simple interpretation of a single subduction zone has its problems. On the north-west side in Scotland igneous activity was also prolonged, even into Devonian times, but where are the voluminous Lower Palaeozoic andesites that in theory should be present? Some uncertainties reflect a lack of data, or simply a lack of outcrop; one could speculate on what might (or might not) be found if deep boreholes could be put down in the Solway Firth, Irish Sea or ranged across the central Irish plain.

The Dalradian outcrops of Connemara remain in frustrating isolation; the relation of the late Dalradian clastic and volcanic pile to the carbonate facies on the North-West Foreland is debatable; there is the tentative deduction that deformation at depth was, or could be, contemporaneous with late Dalradian sedimentation at the surface. The sources of many basin infillings remain obscure. The clastics of the Skiddaw and Manx groups perhaps came from some unknown Precambrian hinterland on the Baltic plate; some of the Welsh and Leinster turbidites from the rising Irish Sea Landmass; and on the north-western side a proportion of the Dalradian influxes from a barely deducible southerly source. Whereas future research may fill in many gaps, these tectonic elements, known chiefly through their erosion and destruction, are likely to remain obscure.

A major question is the line of the Caledonian suture – if it can truly be considered as a simple 'line'. The Solway Firth and its westward continuation (Fig. 18) has been supported recently from seismic evidence, reflecting differing crustal structure on either side; there are also supplementary data from the Caradoc faunas of Girvan. The line of the Southern Upland Fault is another possibility, particularly if the Ballantrae Volcanic Group is considered as a satisfactory ophiolite and as traces of an Ordovician ocean floor. Allied to this is the significance of the several other partial combinations of basic and ultrabasic rocks, pillow lavas, cherts and shales, and signs of blueschist (high pressure and low temperature) metamorphism. Petrological and geophysical research is a hopeful line in this debated belt. Already a simple but marked advance has been achieved in that seismic surveys have shown that continental crust is continuous under Britain.

There are still some curiously diverse lines of evidence on the time of closure across the British Caledonides, or between the bordering plates:

The palaeomagnetic pole positions on either side support little major closure after the Lower Ordovician, though a small amount is allowable.

Benthonic faunas are distinct till the end of Caradoc; there is some mingling in the Ashgill but little or no distinction in the Silurian.

Deformation includes a major late phase in the paratectonic belt around the end of the Silurian.

Magmatism on both sides continues well into the Devonian.

It is possible that some data are seriously incomplete and may be revised in the future. The last two items, also, may perhaps be reconciled on the supposition that the main closure was in fact late Silurian and that the later vulcanicity, including the massive early Devonian andesites of the Scottish lowlands (p. 152), arose in a different tectonic setting – within the now conjoined single plate and not on a plate margin. Some indirect evidence has been presented for major transcurrent (or strike-slip) movements at a late stage and this might be a continuing source of magma.

Even so this brief review may seem to include a formidable array of doubts, but they should not be overweighted. While the plate tectonic hypothesis is unlikely to supply all the answers, here or elsewhere, the main fabric stands and the vital areas and subjects for research are thus more clearly defined.

CALEDONIDE COMPARISONS OVERSEAS

As already noted in the text and shown in some figures (e.g. insets in Figs. 8, 12, 19) the Caledonian belt extended far beyond the British Isles, several of the sections now being ranged around the North Atlantic. Four regions are considered briefly.

In northern and central Europe several of the Hercynian massifs contain Lower Palaeozoic inliers, where the rocks are variously deformed and some are rather sparse in stratigraphical data. They extend from the Brabant Massif and southern Ardennes to the Rhenish Schiefergebirge, the Harz Mountains and Thuringia and thence to the larger outcrops bordering the Bohemian Massif. Slates and graptolitic mudstones are fairly common and the faunas Baltic in affinity. It was from such outcrops that the concept of a 'southern branch' of the Caledonides was developed, especially by continental stratigraphers – a possibility that has been revived more recently as the 'South-Eastern Caledonides' of the present chapter.

Far away in the opposite direction lie the Caledonian portions of East Greenland and Spitsbergen, particularly important in their Precambrian and Cambro-Ordovician sequences. In the latter they resemble the facies and faunas of the Scottish North-West Highlands, all being part of the American plate. Greenland is also a classic region in its earlier Precambrian assemblages and chronology (cf. p. 29).

The closest comparison between the British Isles and North America is seen in Newfoundland, whence there are more distant links with the Canadian part of the Appalachian orogenic belt. In Newfoundland there is a marked division into north-west and south-east forelands, both with platform sequences, separated by a central or axial mobile belt. The north-west platform is composed of a Precambrian basement, here of Grenville age like the nearby parts of the Canadian Shield, and a Cambro-Ordovician cover of quartzites and carbonates bearing an American trilobite fauna. To the south-east the Avalon Platform exposes a late Precambrian clastic and volcanic sequence that is not much metamorphosed; it is succeeded by Cambrian shales and a small amount of Lower Ordovician, the faunas being Acado-Baltic, or Scandinavian.

The central region is only some 240 kilometres across but in the north there is a narrow belt of highly deformed rocks, stabilized in early

Ordovician times; comparisons could be made with the Moine and Dalradian groups, though perhaps rather with those of western Ireland than with the large-scale nappes of Scotland. No Precambrian basement is known in the rest of the central region. In several outcrops the strata recognized are Lower to Middle Ordovician, but in places they extend up to Middle Silurian; clastic and volcanic rocks occur in both systems. The orogenic effects appear to be largely or entirely late Caledonian – or Acadian in American terminology. Some of the later Silurian strata are red beds, similar to an early version of Old Red Sandstone. A number of boulder beds (mélanges) are known and also ophiolite complexes; large areas are occupied by late orogenic plutons, dominantly granitic.

Much of this has a familiar ring to British geologists but there are also some substantial differences; the central region is a good deal narrower than any British analogue and ophiolites are more abundant. There is no Torridonian element in the north-west cover (a minor point) but there are two major transported masses, or klippen, probably derived from the Lower Palaeozoic of the central belt, which have no counterparts in Scotland. The most conspicuous absence is any parallel to the thick deposits of the Welsh Trough with its massive volcanic components, reinforcing the idea that this was a relatively local marginal basin.

In spite of such differences, when the various parts of the Caledonian belt bordering the North Atlantic are reassembled in their pre-Mesozoic positions they fall logically into place. Moreover at present some parts of the continental shelf are unusually wide and there are isolated remnants such as the Rockall and Porcupine banks to be fitted in. It thus seems likely that even after the Caldeonian orogeny was completed some 600 kilometres separated the present exposures of Newfoundland and Ireland.

The longest remnant of the Caledonides, the Scandinavian mountain chain, presents some of the most complex problems. In one respect the situation is complementary to that in Scotland: the eastern margin is a belt of easterly thrusting, much longer than the westerly Moine Thrust but a mirror image of it. However on the Baltic Shield, or foreland, there are several Palaeozoic outcrops that form a major part of the Acado-Baltic faunal province.

The orogenic belt is dominated by a succession of flat-lying thrust slices or nappes, far more so than in Britain, the prevalent movement being easterly. The lower ones, which are little metamorphosed, have some links with the autochthonous outcrops of the foreland and include late Precambrian to Silurian strata, though everywhere fossils are sparse. The upper western, or allochthonous, nappes are much more altered and age evidence is slim, but probably they embrace the same range. The more reliable faunas, especially Ordovician, come from the Trondheim region and from the central Caledonides (the Köli Nappe) of Sweden near the Norwegian border. The archaeocyathids of Finnmark in the extreme north belong to a group found in Lower Cambrian carbonates in many parts of the world.

In this complex situation very few direct links can be traced with the British Caledonides. The conspicuous Scottish structures, such as the Midland Valley and its bordering faults find no parallel on the other side; and although there is sporadic evidence of an early Caledonian phase, similar to the Dalradian deformation, the main climax over the whole chain appears to be late Caledonian (late Silurian to early Devonian). In that respect a comparison would be sought in the British paratectonic belt but the complex nappes and high grade metamorphism rule out any direct relationship here also. At present it seems that only the broadest comparisons are possible.

In somewhat similar fashion it is difficult to identify firmly any line of suture, or components (of crust or cover), that might belong to an American plate – suggestions of the latter being derived chiefly from American affinities reported from some of the Trondheim faunas. There are several ophiolite complexes but they occur at different levels as parts of the flat-lying nappes, with no descent to a root zone. It has also been argued that the basement gneisses in the west are Baltic in type and age, and their links are eastern rather than supporting any distant derivation from western continental crust.

Thus the Scandinavian Caledonides illustrate even more than the British sector the problems that arise in a very complex highly deformed belt where only fragmentary evidence is available and that from isolated outcrops. It also may be that the northern North Sea covers some fundamental fracture zone or line of demarcation that might account for the conspicuous differences on either side.

REFERENCES

Cocks *et al.* 1971 (*Special Report* No. 1, Silurian).
Williams *et al.* 1972 (*Special Report* No. 3, Ordovician).
General: Bassett, 1976; Rickards, 1976; Ziegler, 1970; Ziegler *et al.* 1974.
England and Wales: Allen, 1968; Bassett *et al.* 1975; Brenchley, 1969; Dean, 1964; Faller and Briden, 1978; Hurst *et al.* 1978; Ingham *et al.* 1978; Jackson, 1978; Jones and Pugh, 1949; Millward *et al.* 1978; Moseley, 1978*b*; Rast, 1969; Ridgway, 1975; Shackleton, 1954; Wadge, 1978; Wills, 1978; Wood and Smith, 1959; Ziegler *et al.* 1968.

Scotland and Ireland: Bluck, 1978a; Brück *et al.* 1974; Clarkson *et al.* 1975; Craig and Walton, 1959; Dewey, 1963; Holland, 1969; Kelling, 1961, 1962; Stillman *et al.* 1974; Walton, 1963, 1965.

Caledonides, tectonics, magmatism, dating, etc. Bamford *et al.* 1977; Bradbury *et al.* 1976; Briden *et al.* 1973; Dewey, 1969a, b, 1971, 1974; Dewey and Pankhurst, 1970; Fitton and Hughes, 1970; Harris *et al.* 1979; Lambert and McKerrow, 1976; Leake, 1978; Le Bas, 1972; McKerrow and Cocks, 1976; McKerrow and Ziegler, 1972; Mercy, 1965; Mitchell and McKerrow, 1975; Moseley, 1977; Nicholson, 1974; Pankhurst, 1970, 1974; Pankhurst and Pidgeon, 1976; Phillips *et al.* 1976; Pidgeon and Aftalion, 1978; Pitcher and Berger, 1972; Read, 1961; Smith, 1976; Strogen, 1974; Upton *et al.* 1976.

British Regional Geology: Northern England, North Wales, The South of Scotland, South Wales, The Welsh Borderland.

5

THE DEVONIAN SYSTEM

During the Devonian period the British Isles, now part of a great continental block from Russia to North America (Laurasia), was still south of the equator, moving perhaps from about 28° to 18° S. The climate was certainly warm but there are few signs of marked aridity and the vegetation, of small herbaceous plants at the outset, became much more varied and abundant later. It seems likely that there was sporadic heavy rainfall at least in the hills, for fluviatile deposits abound. The final continental collision and mountain building in this part of the Caledonian chain seems to have been largely completed; however the thick andesitic and basaltic lavas of Scotland suggest that some movement on the northern margin continued into the early Devonian. Over the country as a whole (Fig. 32) one of the classic facies contrasts is between the Old Red Sandstone – detritus accumulating in and around the Caledonian mountains – and the marine Devonian of the Hercynian belt in the extreme south-west.

THE MARINE DEVONIAN ROCKS OF DEVON AND CORNWALL

In spite of the system's name, established as far back as 1839, the international stages are mostly Belgian and German in origin; they appear in Table 19 and in the Appendix (p. 426). The most valuable zonal fossils are goniatites but these are not applicable in England until the middle of the system; they are supplemented by brachiopods, trilobites, ostracods and, in more recent years and very effectively, by conodonts, which have been extracted from some of the deformed rocks in south Cornwall.

The Devonian outcrops of the south-west are separated by the Carboniferous synclinorium of central Devon into the relatively straightforward sequence of the north and the more complex regions of South Devon and Cornwall. Here there is profound deformation and regional metamorphism, tending to increase southwards, as well as the local effects

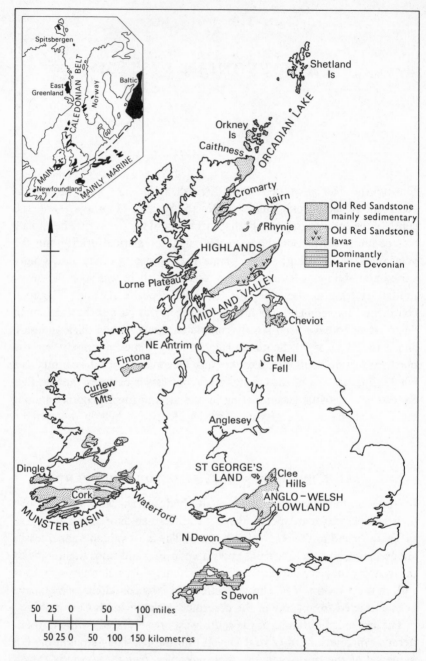

Fig. 32. Devonian outcrops. Inset: position of Old Red Sandstone outcrops (as in late Devonian times) and marine facies in northern Europe.

around the Hercynian granites. In the peneplained regions of south Cornwall extensive exposures are commonly restricted to precipitous cliffs and it is not surprising that this remains one of the most problematic geological provinces in England.

South Devon and North Cornwall

This is the largest outcrop and despite tectonic complications illustrates best the varied Hercynian marine facies and vulcanism. Almost throughout the Devonian period and well into the Carboniferous there was much

Table 19. *Devonian lithological successions in Devon*

Torquay district	Stages	North Devon
Ostracod Slates, green and purple slates; includes Saltern Cove Bed (slumped) with goniatites	} FAMENNIAN	} Lower Pilton, Baggy and Upcott Beds Pickwell Down Sandstone
Purple shales	MID-FRASNIAN	Morte Slates
Torquay Limestone and equivalents	} LOWER FRASNIAN, GIVETIAN and UPPER EIFELIAN	Ilfracombe Beds Hangman Grits
Shales with *Calceola, Anarcestes*	LOWER EIFELIAN	Lynton Beds
Meadfoot and Staddon Beds	} EMSIAN and UPPER SIEGENIAN	
Dartmouth Slates (base not seen)	? LOWER SIEGENIAN	(base not seen)

The stages can be recognized sporadically in South Devon but much less certainly in North Devon.

basic volcanic activity in one place or another, both subaerial and submarine, with spilitic types abundant.

The stratigraphical sequence around Torquay in south-east Devon (Fig. 33) is perhaps the best known, though its carbonate facies is not characteristic of the region as a whole. The lowest beds, the Dartmouth Slates (Table 19), stand out as the only non-marine formation in the south; they comprise nearly 1000 metres of grey-green and purple sandstones and siltstones, somewhat cleaved, together with some pyroclastic rocks. The Dartmouth Slates extend in a broad belt westwards to the north Cornish coast, cropping out for instance in Watergate Bay north of Newquay. The base is nowhere seen and the few fossils are poorly preserved; however molluscs, plants and vertebrates have been recognized and the last suggest

an equivalence with part of the Dittonian Stage of the Welsh Borders
(p. 147), or lower to middle Siegenian of the marine sequence. If the
volcanic horizons correspond to those of south Cornwall it is possible that
upper Gedinnian may be present towards the base. In general the
Dartmouth Slates are thought to be a southward extension from the Old
Red Sandstone detritus of the lowlands in South Wales. Thereafter the
marine facies show an irregular progress of northward invasions and
periodic deepening of the seas.

The first of these is represented by the Meadfoot Beds of south-east
Devon with a rich brachiopod fauna, the more quartzose Staddon Grits

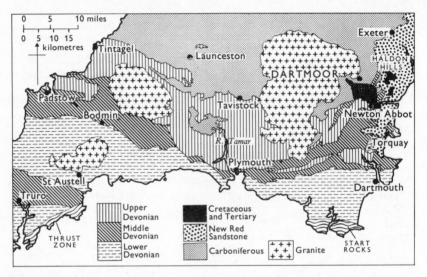

Fig. 33. Geological sketch map of South Devon and North Cornwall (outcrops in
south-western part are tentative). (Redrawn after House, 1963, p. 3.)

being mainly a lateral replacement. On the north Cornish coast this part of
the Lower Devonian is represented by slates and arenites in the cliffs
around Newquay. The basal Middle Devonian of the south-east consists of
thin shales, an intermission of deeper water and finer sediments; rare
Calceola is found among a well preserved bottom fauna and goniatites
appear. Above this the Middle Devonian and lower Frasnian is almost
wholly calcareous, forming the Torquay and Plymouth limestones in
which stromatoporoids and corals are important. Although famous and
conspicuous, this is in fact a restricted facies, perhaps built up on a
shallow clear water bank. Westwards into Cornwall there was a change to
deeper and more muddy waters, so that near Padstow there are Middle

Devonian slates with goniatites. This is not the only east–west facies change in this province, but one of the clearest. There are also abundant volcanic rocks, and near Totnes the whole of the Torquay Limestone is replaced by spilitic lavas and tuffs.

In south-east Devon the mid-Frasnian facies change demonstrates another deepening, as shales succeed the limestones. More widely the Upper Devonian rocks can be divided into two main types, interspersed with lavas and tuffs. The 'ostracod slates' accumulated relatively thickly in the basins, with a rather sparse fauna. They form all the Famennian of the Torquay area; in Cornwall the Delabole Slates are similar but also include the well known brachiopod, *Cyrtospirifer verneuili*. The cephalopod-rich impure and nodular limestones are commonly thinner and are thought to represent slower accumulation of carbonates on rises or 'swells' of the sea floor. A typical sequence occurs at Chudleigh, east of Dartmoor, where the whole of the Famennian is comprised in some 50 metres. This is one of the rare regions where there seems to be a complete sequence of quiet water sedimentation from the mid-Devonian and through the Lower Carboniferous. The Cornish coastal sections exhibit vulcanicity of more than one period, Frasnian at Pentire and from late Famennian to early Carboniferous at Tintagel; tuffs and great submarine lava flows accumulated.

South Cornwall

Many more problems arise in this, the most south-westerly, part of the peninsula, where the rocks are commonly more deformed and the faunas very rare. Much of the area south of Newquay and around the westerly granites is made up of the Gramscatho and Mylor beds, terms applied to unfossiliferous greywackes, including turbidites, and slates. They have usually been attributed to parts of the Lower and Middle Devonian.

Then there is the complex belt along the south coast headlands, formerly called the thrust or fault zone, or the Lizard–Dodman–Start line. In its most comprehensive sense this runs from the Lizard to Start Point in South Devon, but the belt includes diverse structural and stratigraphical elements, some not directly related to Devonian geology. It was noted in Chapter 2 that the Lizard, faulted against Devonian rocks on the north, is probably Precambrian; the schists of Start Point, also fault-bounded, may be, though their recorded isotopic dates are later. The Eddystone lighthouse is built on a reef of resistant metamorphic rocks, thought to have a large extent beneath the sea floor.

It is the section from the Dodman to the Roseland region east of

Falmouth Harbour that is particularly important here. The cliffs display much faulting and deformation, and some of the problems have arisen from the varied structural interpretations and, for a long time, very sparse fossils; conodonts have proved helpful recently. The Ordovician quartzite blocks (p. 100) occur near the Dodman and the phyllites of that headland itself have been considered as a further upfaulted portion of 'pre-Devonian basement' or, rather more probably, as a slightly metamorphosed Lower Devonian outcrop not fundamentally distinct from strata on the inland side.

In the present context the Roseland sequence is the most significant, since it seems to be continuous from Gedinnian to Eifelian. At the base a thin succession of slates and volcanic rocks has yielded Gedinnian conodonts; these are the oldest Devonian rocks in South-West England and overlie the Ordovician quartzites unconformably. Locally, thin limestones (previously the 'Veryan Limestones') are intercalated with the overlying Gramscatho greywackes. The sequence suggests that early in the period muddy conditions prevailed in south Cornwall, interspersed with minor vulcanicity; later the area was invaded by turbidites (Eifelian – ? Givetian), probably derived from the south. In the same way that the fine-grained facies extended irregularly northwards, so did the clastic turbidite flows. Both facies persist into the mid-Carboniferous, though with the many structural complications no one region shows the complete sequence. There are indirect suggestions of some kind of source area between Cornwall and Brittany through part of Hercynian history, and from mid-Devonian times onwards there is little equivalence in sedimentation between the two.

North Devon

The Devonian rocks here have a regional southerly dip so that there is a gradual descent in the succession from the Carboniferous junction near Barnstaple round the north-western coasts, with their fine exposures, as far as Lynmouth, where the lowest beds appear in the core of an easterly trending anticline. Farther east in Somerset and the Quantocks another anticline brings up Devonian beds in the midst of Trias. Correlative fossils are sparse, goniatites and conodonts being almost unknown; the most prolific are shelly faunas in the sandstones, siltstones and slates with some corals in the thin limestones. Trace-fossils and bioturbation are locally abundant in the shallow intertidal and deltaic facies. As a marked distinction from South Devon, vulcanicity is restricted to a few Upper Devonian tuff bands.

The lowest formation, the Lynton Beds (Table 19), are fairly typical inshore marine deposits – fine-grained sandstones and mudstones with signs of contemporary erosion. The rather sparse fauna is dominated by bivalves and as in other parts of this northern outcrop the age is not very certain (perhaps late Emsian to early Eifelian); at all events the lowest stages are not exposed. The succeeding Hangman Grits is the lower of the two Old Red Sandstone incursions (Fig. 34), neither of which reached

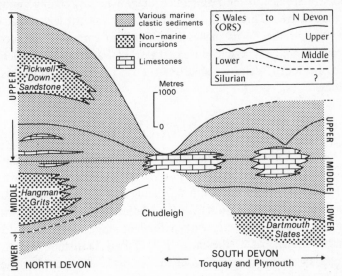

Fig. 34. Devonian facies in Devon. Inset: sketch of relations with South Wales. (Adapted from House, 1975, p. 242.)

South Devon. This one is a varied fluviatile and deltaic facies with signs of incoming marine conditions in the faunas at the top.

As in the south some limestones appear in the later Middle Devonian but not the massive carbonate banks of Torquay. The Ilfracombe Beds include slates and four thin limestones with Givetian and Frasnian corals. Regression set in with the pro-delta facies of the Morte Slates, with *Cyrtospirifer verneuili*, and this became more pronounced with the Pickwell Down Sandstone. Here variegated sandy beds include some Upper Old Red Sandstone fishes (e.g. *Bothriolepis* and *Holoptychius*) – a valuable intercalation among Famennian beds and one of the few linkages between the continental and marine facies. Thereafter followed the last and most extensive transgression to give the Upcott Beds, largely unfossiliferous, and the delta-front Baggy Beds. The latter include a range of shallow water and marginal features with cross-bedded and slumped units, intraformational conglomerates and abundant trace-fossils. *Lingula* occurs and the broken remains of marine shells, but also some fairly well preserved plants. The Pilton Beds continue this facies with inshore sediments and shelly faunas that span the Devonian – Carboniferous boundary.

As elsewhere in the south-west thicknesses are very difficult to estimate, but swelled out by the arenaceous invasions it is likely that the North Devon succession is substantially thicker than that of the south. This particularly applies to the Famennian Stage which probably amounts to some 3000 metres. Through much of Devonian time the lack of palaeontological data in the north makes it difficult to correlate the successive transgressive and regressive phases, those of Famennian age being the clearest. In South Wales we shall see the landward version of this latest oscillation.

SOUTHERN AND SOUTH-EASTERN ENGLAND

From Devon, Somerset and the lower Severn valley the Devonian rocks disappear eastwards under a Triassic cover but they are known from several boreholes from the south Midlands and the south-east. Many of them are on the London–East Anglian Platform, notably in an east–west belt through London itself, with a further range along its north-west margins. In most places the facies is Old Red Sandstone and thus far resembles an eastward extension of the Anglo-Welsh area, but there are some marine records, a few of which extend surprisingly far north. These include the boreholes in Bedfordshire, Oxfordshire, Cambridgeshire, Buckinghamshire and north-west London (Willesden). In the last over 500 metres of varied sediments, partly reddish with limestone bands, was penetrated with an Upper Devonian coral–brachiopod fauna. In general Frasnian records are the most common, perhaps in line with the Frasnian deepening or marine extension of South-West England. It is also noticeable that marine fossils, including microfaunas and microfloras, have in several cases been obtained from an Old Red Sandstone type of facies, or in beds alternating with the latter. The records are rather too scattered to encourage much palaeogeographic deduction but the margin of the Old Red Sandstone landmass is likely to have been irregular and fluctuated during the period.

Marine beds have also been met at depth in the Weald of Sussex, but here they are to be expected as they are on the Hercynian belt from

Devonshire to northern France and Belgium, lying south of the London–
Brabant Massif.

OLD RED SANDSTONE

With the closing stages of the Caledonian orogeny the uplands and
mountains arose, the seas receded and there was a shift in the main areas of
deposition. Two topographic situations can be distinguished: the internal
facies lay within the Caledonian chain, with no connection to the
Devonian seas, and the external facies on the margins where the lowlands
sloped gently down to the shores. With the sort of fluctuation noted in the
last section some marine incursions may be expected in this second
situation, which is exemplified in the Anglo-Welsh Lowlands.

Fossils are not plentiful in the Old Red Sandstone and there are great
thicknesses of strata which have so far yielded none. Plants are found
occasionally and plant spores have become increasingly useful but the
most conspicuous remains are the vertebrates. These include the ancestors
of modern fishes and the more primitive extinct early armoured groups –
agnathans and placoderms. Among the former, cephalaspids and pteraspids characterize the Lower Old Red Sandstone of the Welsh Borders and
South Wales and the small scales of thelodonts have been a valuable
adjunct.

A greater variety of groups, including placoderms and true fishes, have
been used in the middle and upper divisions and in the Caledonian
countries as a whole there is enough interdigitation between marine and
non-marine facies for a generalized correlation to be established between
them. However on a smaller or regional scale the Old Red Sandstone still
tends to be divided into lithostratigraphic units with only tentative
correlations; and of course that term reflects a very ancient traditional
lithological entity in itself, which inevitably does not coincide precisely
with the modern definition of the Devonian System.

THE ANGLO-WELSH LOWLANDS

The southernmost of the British outcrops represents a varied expanse of
lowlands, shores, shoals and coastal flats between the newly developed
uplands of Wales and the Hercynian seas to the south. The rocks are
exposed in a triangular area from the Welsh Borders (e.g. Clee Hills,
Hereford and Worcester) southwards to the Severn estuary and along the
north boundary, or 'north crop' of the South Wales Coalfield (Figs. 21,

22, 24). Minor outcrops extend across the Severn and more important ones to the west among the pronounced Hercynian folds of south Dyfed. Despite the importance of this lowland province, however, and the sedimentological studies that have been based on it, the sequence is notably imperfect. Only the Lower Old Red Sandstone is thick and extensive; the middle division is only doubtfully present (possibly not at all, no faunas being recorded) and the upper is thin and incomplete. Conformity with the Lower Carboniferous above is common but, as noted in the preceding chapter, there is a 'Downtonian problem' at the base where the international system boundary lies within the Old Red Sandstone, the lowest part of which belongs to the Silurian System as the Downton Series. A minor complication in nomenclature is that the lowest division in the Old Red Sandstone is termed the Downtonian Stage (cf. Appendix, pp. 426–7).

Lower Old Red Sandstone

As would be expected in such rocks the component divisions (Table 20) are not very well defined. Downtonian and Dittonian vertebrate faunas are fairly distinctive in the Welsh Borders but not throughout the region, while the Breconian has very little palaeontological basis.

It was seen in Chapter 4 that the Downtonian facies and faunas were decidedly varied since they reflected much of the transition from marine to continental conditions. Also that though this transition was conformable, or nearly so, in the Welsh Borders, westwards a major unconformity appeared at the base of the Old Red Sandstone – part of the greater mobility and intermittent uplift affecting south-west Wales through much of late Palaeozoic history. In this province the Dittonian is dominantly composed of 'Red Marls' (mudstones and siltstones) laid down by the streams flowing towards the southern sea off the flanks of the new Caledonian uplands of Wales, St George's Land.

A particular type of cyclic sequence has been described from Dittonian beds in several areas; the components are:

3. mudstones and siltstones, in places with carbonate nodules
2. sandstones, commonly cross-bedded
1. conglomerate, resting on an erosional or scoured surface

The cyclothem may be from three to 15 metres in thickness. The pebbles of the intraformational conglomerates are chiefly local and many consist of hardened mudstones, similar to those in the top unit; there are also ripple-marks, desiccation cracks, occasional plant remains and vertebrate debris. This cyclic facies is interpreted as the product of meandering river channels redistributing deposits from farther upstream. The whole formed on a broad alluvial plain on which muds, silts, sands and pebbles were laid down, to be intermittently shifted onwards as the streams changed their courses.

The carbonate beds (limestones or cornstones) have received a good deal of attention, and have been taken to represent the crusts or 'caliches' that

Table 20. *The Old Red Sandstone in the Anglo-Welsh Lowlands*

Stages	Welsh Borders	Brecon Beacons, etc.
FARLOVIAN	Farlow Group (150 m) grey and yellow sandstones	Grey Grits (21 m) (or Quartz Conglomerate Group) Plateau Beds (41 m)
	(Middle Old Red Sandstone probably not represented)	
BRECONIAN	Abdon and Woodbank Groups	Brownstones (490 m) and Senni Beds (380 m)
DITTONIAN	Ditton Group (450 m) red mudstones with sandstones and thin limestones	Red Marls (1300 m)
DOWTONIAN	Downton Group: Ledbury Formation (460 m) red mudstones with local thick sandstones Temeside Formation (75 m) grey-green siltstones Downton Castle Formation (30 m) yellow sandstones and greyish siltstones Ludlow Bone Bed ———local disconformity——— (Upper Ludlow beds)	locally with thin basal green beds

For relations between the Old Red Sandstone and standard Devonian stages see Appendix (p. 426) and notes.

result from soil-forming processes in arid and semi-arid climates today. They also reinforce the view that occasionally these alluvial flats emerged above water-level and dried off. When the caliche was broken up by renewed stream action it contributed to the local conglomerates and it is these that include especially the vertebrate plates and other debris. One outstanding group of carbonate beds, the 'Psammosteus Limestones', can be recognized from the Welsh Borders to south-west Wales, at the base of the Dittonian Stage.

The ill-defined Breconian Stage tends to be coarser than the red beds below and throughout much of the eastern outcrops, for instance east of the South Wales Coalfield and around the Forest of Dean, it is known as the Brownstones and consists of sandy beds with some cyclic sequences (cf. Fig. 81, p. 394), becoming coarser towards the top. For the most part the age is problematical. At the Brecon Beacons, where Old Red Sandstone forms a mountainous escarpment north of the South Wales Coalfield, the Senni Beds appear in the lower Breconian – grey, green and dull red sandstones with more plant remains than are usually found and a varied spore assemblage. Near the top *Rhinopteraspis cornubica* indicates a mid-Siegenian to early Emsian age. Normal Brownstones follow above and the Beacons are capped by the unconformable Plateau Beds – Upper Old Red Sandstone. In extreme south-west Wales (south Dyfed) other local formations appear; north of Milford Haven and of the Ritec Fault (Fig. 22) the Dittonian red marls are followed conformably by the Cosheston Group, a thick varied sequence with conglomerates and breccias, the latter including volcanic debris.

Some indication of source areas of the Lower Old Red Sandstone has been gained from the mineral and pebble contents and cross-bedding directions. In general a north-western source is dominant, as one might expect, but with changes in the rocks supplying detritus. In Downtonian and early Dittonian times there is evidence of abundant metamorphic rocks, with much mica and especially garnet. This could have been derived from a Precambrian mass like the Mona Complex, or even from farther away. Then in the later Dittonian and Breconian these components are much diminished and Lower Palaeozoic sources are conspicuous, including Ordovician volcanic rocks and Silurian sediments. A minor southerly source is suggested by the local Llanishen Conglomerate (upper Dittonian of the Cardiff district), with resemblances to the Silurian volcanic rocks of Tortworth and the Mendips.

The Ridgeway Conglomerate: ? Middle Old Red Sandstone

Almost throughout the Anglo-Welsh Lowlands there is no trace of Middle Old Red Sandstone, but south of the Ritec Fault the Ridgeway Conglomerate may be a local residue. It overlies the Dittonian marls with, it is deduced, a break and change in sedimentary regime. Some cross-bedding directions indicate a southerly source and the large pebbles and cobbles differ in content from the earlier conglomerates and the Cosheston Beds. They include an Ordovician quartzite not matched in the Welsh outcrops, and thus a ridge or source area in the Bristol Channel region or adjoining Celtic Sea becomes a possibility. No fossils are known from the Ridgeway Conglomerate.

In addition to the major stratigraphical gap the overstepping of the Upper Old Red Sandstone onto various lower divisions, and in the Mendips onto Silurian, is evidence of a late Caledonian phase of widespread uplift and gentle folding. It is possible that the instability along the Benton Thrust and Ritec Fault anticipated their more pronounced Hercynian effects; similarly there was warping on the Usk Anticline.

Upper Old Red Sandstone: Farlovian

Throughout the region the upper division is feebly developed compared with the lower, varying in thickness from only a few metres to over 400 metres south of the Severn estuary. The highest sediments record a very late Devonian to early Carboniferous transgression, so that the lower beds are commonly coarse and fluviatile and the upper ones finer; in several outcrops these become grey with a marginal marine fauna before passing up into typical marine Lower Carboniferous strata.

In this province the Upper Old Red Sandstone is not divisible on faunal grounds but the characteristic vertebrate genera, *Bothriolepis* and *Holoptychius* (chiefly the latter), have been found; there are also occasional plants and the large mussel, *Archanodon*. The transition to marine faunas is well seen in one of the thicker developments, the Skrinkle Sandstone of the extreme south-west, south of the Ritec Fault, which may have been a local boundary line. Here there are some 300 metres of interbedded sandstones, siltstones and mudstones, variegated in colour and including bands of breccia, especially near the base where *Holoptychius* has been recorded. Towards the top, but still among red and yellow beds, there is a local influx of invertebrate fossils, including spiriferids, rhynchonellids and crinoids. This is a fully marine assemblage but conditions must have been well inshore with minor oscillations because they are interbedded with plant-bearing silts and even poor coal laminae.

Along the north-east crop of the South Wales Coalfield and eastwards to the Forest of Dean the most conspicuous component is the Quartz Conglomerate – a red, yellow and grey formation with large pebbles, some of which can be matched in the Welsh uplands to the north. However compared with Dittonian or Breconian detritus the minerals and rock assemblages are more mature, derived from a more weathered land surface, and contain much vein quartz and quartzite.

Over most of the outcrop there is this fairly simple picture of a final Famennian transgression, but the full sequence may have been rather more complicated. In the centre of the north crop, on the Brecon Beacons, the equivalent of the Quartz Conglomerate is the Grey Grits; but locally below them the Plateau Beds suggest an earlier marine phase, the evidence for which elsewhere has been removed. The beds show an upward decrease in grade from conglomerates at the base to sandstones above. The latter contain not only the remains of vertebrates (*Bothriolepis*, *Holoptychius* and *Coccosteus* – presumably transported debris) but a varied marine assemblage with *Lingula*, bivalves and brachiopods, including *Cyrtospirifer verneuili*. The age is not precisely determined, but late Frasnian to early Famennian has been put forward.

A partial comparison can be made with North Devon where the Lower Pilton and Baggy Beds correspond to the late Famennian transgression (Table 19). Below these the regressive phase may perhaps be equated with the continental incursion of the Pickwell Down Sandstone. It might be logical to infer a parallel relationship between the sandy invasions of the Hangman Grits and the Middle Old Red Sandstone unconformity. However the overlying Ilfracombe Beds, with fine clastics and thin limestones, do not fit into such a framework; possibly at this time erosion was much reduced in the Welsh region, or simply the resulting detritus was carried elsewhere.

CALEDONIAN UPLANDS: WALES TO SCOTLAND

Over the long stretch of irregular uplands from South Wales to the Scottish Border little Old Red Sandstone remains. This does not mean that none was deposited, but rather that the sands, silts and gravels were rarely permanent but were shifted onwards by the streams and rivers to lower ground.

The principal uplands are the central Lake District (probably connected with the northern Pennines), the Southern Uplands of Scotland, and St George's Land. The last is a name given to the Lower Palaeozoic massif of Wales and adjacent regions. It was possibly connected in some degree with the Leinster Massif in south-east Ireland and a rather ill-defined ridge that extended eastwards into the English Midlands, the Midland Barrier. St George's Land was a persistent geographical element and an intermittent source of detritus during Upper Palaeozoic times, and its Devonian contributions to the Anglo-Welsh Lowlands have already been summarized.

The red rocks among these heterogeneous uplands are considered to be Old Red Sandstone on lithological grounds and stratigraphical position, for no fossils have been found. The most important relic is on Anglesey, where though the area is small the beds are over 500 metres thick. The basal conglomerate overlies Precambrian and Ordovician rocks and contains recognizable local pebbles; the succeeding strata include red and brown sandstones, red siltstones and concretionary limestones – a fluviatile facies with many similarities to the Dittonian and Breconian beds farther south. On this lithological basis, and because the red rocks were folded and faulted before the Carboniferous beds were laid down, they are thought to be Lower Old Red Sandstone.

On the western scarp edge of the Alston Block and around the eastern and northern edge of the Lake District, the Lower Carboniferous is locally underlain by reddish rocks, often coarse, of various ages. Most of them appear to be Carboniferous basement beds but two groups may be older: the small fans of coarse detritus along the Cross Fell escarpment and the more extensive Great Mell Fell Conglomerate of Ullswater. The latter consists of coarse red conglomerates with thin sandy layers and much cross-bedding and is estimated to be 275 metres thick. The pebbles were derived from the contemporary Lake District hills, especially the Silurian greywackes. The Lower Carboniferous Limestone follows with a probable unconformity and on this account and their thickness and coarseness the conglomerates are tentatively ascribed to the Lower Old Red Sandstone.

THE MIDLAND VALLEY AND THE SOUTH OF SCOTLAND

The Old Red Sandstone in this internal basin is exceptionally thick and variable, differing in several respects from that in the Anglo-Welsh Lowlands. Thus there was much vulcanicity early in the period; along the north-east margin in particular there are very thick coarse conglomerates and palaeontological correlation is rarely possible, the pteraspids being especially scarce. As in the Anglo-Welsh outcrops the Middle Old Red Sandstone appears to be missing.

The thick sequence of dominantly red rocks accumulated in a broad mountain-girt lowland with approximately the same trend, but not quite the same margins, as the present Midland Valley, the latter being largely due to the greater erosion of the Upper Palaeozoic rocks, less resistant than those on either side. Within the rift a central Carboniferous tract is bordered by Devonian belts (Fig. 49, p. 223) and there are many subsidiary folds and faults, especially east–west Hercynian structures. At present the Lower Old Red Sandstone is almost confined between the bordering faults of the rift; the upper division is more extensive and there is some evidence of faulting and folding between the two.

Lower Old Red Sandstone

During this period the sediments, locally with thick volcanics, accumulated in two elongate basins, adjacent to the faults on either side. Downtonian faunas have been found only in the extreme north-east of the northern belt (p. 115). The remainder of the Lower Old Red Standstone, thickest and most complete in this part of the basin, follows conformably, the total reaching some 9000 metres.

The great masses of coarse conglomerate are most abundant in the lowest (Dunnottar) group where there are 1600 metres with only minor sandy bands. The pebbles and cobbles, including much quartzite, were derived from the Dalradian uplands to the north, carried down by occasional torrential streams and dumped at the mountain foot. Higher up there are groups of lavas (chiefly basalts and andesites), sandstones, red shales and more conglomerates; the lava flows were broken up by contemporary erosion and contribute to the coarser beds. Although much of the finer sediment is water-laid there is little evidence of standing water and no pillow structures have been found in the lavas.

Rocks of the same age can be traced away from the Highland Border into Forfarshire, though the lower groups are not exposed and the sequence is somewhat thinner, with sandstones, siltstones, mudstones and local conglomerates. Some of the finer beds are grey, and from such types near the lowest exposed measures rich vertebrate faunas have been collected. They include cephalaspids and especially several acanthodian genera – small slim fishes with a delicate dentition and spine-supported fins. They and the small arthropods probably lived in shallow lakes bordered by primitive plants. Plants and spores also occur in the top group and though correlative data are sparse it appears that most of the Lower Old Red Sandstone is present.

In the Sidlaw and Ochil Hills, of the east and centre, the lowest exposed levels include very thick lavas – the Ochil Lava Formation, up to a maximum thickness of perhaps 3000 metres; this massive sequence is typically basic, a calc-alkaline assemblage dominantly of basaltic andesites. Towards the west these outcrops are truncated by the major Ochil

Fault, and the most westerly outcrops south of Loch Lomond are largely composed of the uppermost brownish sandstones and conglomerates, as yet found to be barren. Here the northern boundary is the Highland Border Fault but in places minor patches of coarse red detritus transgress onto the Highlands, as in Arran, where the fault swings round to cut across that island. The distribution of the various lithological groups thus demonstrates that the elongate basin gradually extended towards the south-west. The main axial flow of the sandy sediments was also in that direction, while the ultimate source (apart from the eroded lavas within the basin) was the contemporary highlands to the north.

In the south-eastern belt minor outcrops are found bordering the Silurian inliers of Lanarkshire and the Pentland Hills (p. 116). In the former the base is taken arbitrarily at the Greywacke Conglomerate; this overlies a disconformity and the pebbles show that the Southern Uplands were already undergoing erosion. Above there follows some 2500 metres of brown sandstone, conglomerates, lavas and tuffs – largely barren but cephalaspids and spores indicate the lower part of the Lower Old Red Sandstone. As these rocks are traced south-westwards, out of the circum-scribed area of the inliers, thick reddish conglomerates lie discordantly on a floor of folded Ordovician rocks. The distribution is analogous to the northern belt in that the rocks are thickest in the Pentlands, at the east end, and the cross-bedding of the sandstones indicate a south-westerly axial flow on this side as well.

Beyond the Southern Upland Fault only a small patch has survived, where sandstones with *Pterygotus* overlie the upturned edges of the Silurian greywackes in Berwickshire. Substantial bands of lavas, tuffs and small vents are found in these sediments and they are probably the same age as the Cheviot lavas. The Cheviot is the only Caledonian igneous mass of southern Scotland in which the extrusive rocks are preserved. The lavas overlie folded Silurian greywackes and after an early phase of agglomerates and rhyolites there followed the main flows of increasingly basic pyroxene andesites, amounting to over 1000 metres. There are virtually no associ-ated sediments and the eruption was subaerial, with only occasional ash showers. This phase was followed by the central granite intrusion and later by numerous radiating porphyrite dykes. The lavas are overlain in the west by Upper Old Red Sandstone and boulders of the granite are found in early Carboniferous rocks. Consequently, in addition to being a classic example of an ancient volcano, the stratigraphic evidence for its age is unusually precise – Lower Devonian; this agrees with the isotopic data which place the Cheviot among the Newer Granites.

Upper Old Red Sandstone

This division is thinner, normally finer in grade than the lower one and the lava flows are few and scattered. The base is everywhere unconformable, whether on the lower division or overstepping on to earlier rocks. The upper junction commonly lacks precision because it has traditionally been taken at some arbitrary lithological or colour change, diagnostic fossils on either side being rare or absent. However in southern Scotland and the Borders evidence from spores suggests that in places at least the red sandstone facies continues into the lowest Carboniferous or that there are alternations of red and grey beds.

Within the Midland Valley outcrops of Upper Old Red Sandstone are much less extensive than those of the lower division and characters of northern and southern belts may be summarized together. The principal rocks are red, pink, buff and yellow sandstones and though conglomerates occur, especially near the base, coarse conglomeratic beds are rare. After the Middle Old Red Sandstone gap, a new sedimentary basin formed in the Glasgow region and the Firth of Clyde, where four lithological groups, totalling about 1000 metres, can be recognized. Away from this centre the lowest conglomeratic division disappears, and towards the south-east in particular, into Ayrshire, the conglomerates and pebbly sandstones are also replaced by finer sediments with much caliche.

The pebbles in the lower beds indicate that the earlier sources continued, from the Dalradian highlands on the north and the Southern Uplands on the south, though the previously eroded Lower Old Red Sandstone contributed some detritus as well. With the upward increase in fineness – a cycle well displayed in the Upper Old Red Sandstone – the sandstones become more mature, with an increase in quartz, and braided streams were replaced by a final regime of meandering rivers, with caliche developing where slow or intermittent sedimentation allowed soil formation. In several places scales of *Holoptychius* are found and *Bothriolepis* is recorded in Ayrshire. On the whole, however, these rocks are not fossiliferous and the sandstones of Dura Den in Fife, thickly strewn with whole *Holoptychius*, or the articulated plates of *Bothriolepis* at Duns, in Berwickshire, are rare exceptions. By this late stage the Midland Valley seems to have been a single elongate basin in which the dominant flow had become reversed and was from west to east. Both basin and source-lands had become more stable and more uniform.

Overlap of Upper Old Red Sandstone is seen in minor areas on the west, for instance around Loch Lomond on to Dalradian rocks, and smaller outcrops in Bute, Arran and the Mull of Kintyre. In the opposite direction the overlap is much more extensive; the outcrop stretches from the East Lothian coast into Berwickshire and thence along the south-east side of the Southern Uplands (Fig. 26) to the Border country and margins of the Solway Firth. These last outcrops are small and separate but the whole is a major Upper Old Red Sandstone province – a Scottish Borders basin.

The rocks lie with gentle dips on the Cheviot lavas or fill in deeply-eroded hollows in the folded Silurian greywackes and shales. It is again a fluviatile complex and in the south-west tends to coarsen upwards; meandering and braided streams supplied pebbles and finer detritus from the Southern Uplands of Galloway to the west, recognizable from the greywackes and the Caledonian intrusive rocks. Opposing these current directions from the west and south-west, evidence from Berwickshire suggest a north-easterly source, so that the centre of this basin may have lain west of the Cheviot mass, where the contemporary sediments are rather finer in grain.

This is also a classic region in the history of stratigraphy: the striking junction of gently tilted Old Red Sandstone overlying steeply dipping greywackes (as seen at Siccar Point in Berwickshire and at Jedburgh) was used by Hutton to illustrate an unconformity, the forces that produced it, and to support his theories of geological processes.

CALEDONIAN UPLANDS: THE SCOTTISH HIGHLANDS

North of the Highland Border Fault the main sedimentary basin is the Orcadian Lake, the outlying southern portions of which are found in Cromarty, along the shores of the Moray Firth and in the northern section of the Great Glen. In addition, however, there are small outliers on the Grampian Highlands and larger outcrops to the south-west, in Argyll. The last include the major volcanic group of the Lorne Plateau Lavas, stretching from Oban north-eastwards to Loch Etive. They are principally andesites and basalts with minor acid lavas and ignimbrites; near the base sediments have yielded a number of vertebrates including acanthodians and *Cephalaspis*. It is a Lower Old Red Sandstone assemblage with resemblances to that in Forfar. Similarly the calc-alkaline lavas correspond to those within the Midland Valley rift and of the Cheviot beyond it.

That the Lorne plateau flows extended farther to the north-east is shown by remnants, with occasional sediments, in the cauldron sub-sidences of Glencoe and Ben Nevis. With the lavas, the granites of this region are part of the last major phase of Caledonian vulcanism (p. 129).

Nevertheless they are not easily related to the previous interpretations of subduction zones, or even to movements within the now conjoined plates. The Newer Granites as a whole cover a much wider belt, for instance from northern Scotland to the Southern Uplands and northern England, with further plutons of about the same age in Wales and southern Ireland.

The Scottish Highlands carry other small outliers of Old Red Sandstone in the north-east, especially at Rhynie and Tomintoul in Aberdeenshire. The former is best known for its fossil-bearing cherts and especially for the small early vascular plants, exquisitely preserved, some apparently in position of growth. Other organic remains include small terrestrial invertebrates, especially arthropods, and a microflora. Although previously thought to be Middle Old Red Sandstone, the spores show the cherts to belong to the lower division. Apart from its profound palaeobotanical interest the Rhynie chert also represents one of the few examples of pre-Carboniferous terrestrial floras in which the plants were preserved where they grew, as distinct from drifted remains in streams and lakes.

The full sequence at Rhynie, which includes clastic and volcanic rocks, has been estimated at 450 metres' thickness and lies at about 300 metres O.D. At Tomintoul to the west, coarse red conglomerates reach a height of 650 metres O.D., on an irregular Dalradian surface. What is impressive about these remnants, together with those in Argyll, is that they represent more than a thin transitory veneer or minor river gravels over the Scottish Highlands. During Lower Devonian times at least it is probable that variable sediments, often coarse, with local lavas, accumulated over much of this upland, later to be removed almost in their entirety.

THE ORCADIAN LAKE

Many of the remnants of this region are now far scattered, separated by the waters of the North Sea and North Atlantic, but once the sediments must have extended from the northern part of the Great Glen to the Shetland Islands, a distance of about 240 kilometres. All three divisions are present but the Orcadian province is particularly famed for its thick fish-bearing Middle Old Red Sandstone, much of which in fact is grey and fairly fine in grain.

The southern margins, along the Moray Firth

Here there is a great variety of reddish rocks – sandstones and shales, conglomerates and breccias, locally interbedded with impure carbonates.

Isolated exposures, much obscured by drift, extend along the south side of the Moray Firth where the Grampian hills slope down to the lowlands, from Inverness to Nairn and Elgin and as far east as Gamrie in Banffshire. West of the Great Glen Fault, and affected by parallel fractures, there are rather larger outcrops in Cromarty, also partly drift covered.

The three divisions are differently developed, separated by unconformities and also with local overstep onto the Highland basement. Lower Old Red Sandstone is somewhat restricted, consisting mostly of red conglomerates with some finer beds; the age is supplied by very occasional spores in the latter. The middle division, also varied in lithology, is much more extensive; it includes one major fish bed, in shales and calcareous nodules, whose assemblage resembles that of the Achanarras Limestone of Caithness (p. 158). At this period the lake waters were especially widespread carrying the fish faunas into the marginal regions of the south; lesser fish beds are found higher up the sequence.

Upper Old Red Sandstone is known principally from the Nairn and Elgin outcrops, and though these are also hampered by poor exposure they are unique in Britain for their fossiliferous horizons. Five successive faunas are known (though with minor problems of correlation), characterized especially by psammosteids, as well as the longer-ranging *Holoptychius*, *Bothriolepis* and *Asterolepis*. Comparable successions are known in north-west Russia and the Baltic states and through these a broad correlation is possible with the marine facies. Most of the faunas and beds (Nairn, Boghole, Alves and Scaat Craig) are approximately Frasnian, and the uppermost (Rosebrae) Famennian.

Caithness and the Orkney Islands

The Orkneys, Caithness and parts of eastern Sutherland are very largely made up of the grey flagstone facies of the Middle Old Red Sandstone, which reaches a thickness of over 3800 metres in Caithness. In general the lower strata are best seen on the mainland and the uppermost one on Orkney. Lower Old Red Sandstone is largely known in the coastal outcrops of the south, where red conglomerates with some sandstones and mudstones overlie a very irregular floor of the Moine Schists and associated intrusions. The middle division (Table 21), with local red incursions and some breccias at the base, followed after a phase of minor warping and erosion. The irregularities of the floor were gradually filled in, however (Fig. 35), and the grey flagstone facies prevailed very widely,

Table 21. *Old Red Sandstone of the Orcadian province*

Caithness	Orkney	
UPPER:		
Dunnet Sandstone Group (650 m)	Hoy Sandstone Group (over 1000 m) Hoy lavas and tuffs	
MIDDLE:		
John o'Groats Group, (627 m)	Eday Group, (over 1200 m)	
Upper Caithness Flagstone Group (over 1500 m)	Rousay Group (1500 m) Upper Stromness Group, (330 m)	EIFELIAN and GIVETIAN
Achanarras Limestone Lower Caithness Flagstone Group (over 2350 m)	Sandwick Fish Bed Lower Stromness Group, (215 m)	
LOWER:		
Sarclet Group	? minor coarse basal beds	
(Precambrian rocks)		

Thicknesses are generalized, as are correlations with marine stages.

with only minor lithological variants, from central Caithness to the Orkneys.

The principal rock types of the Caithness Flagstone Group are the calcareous 'flags' (laminated carbonate and quartz siltstones, dark with carbonaceous matter), mudstones with desiccation cracks and fine sandstones with abundant slumping. Flakes from the mudstones dried up to produce flake breccias. These rock types are repeated over and over again, at some levels in sufficiently orderly fashion to be called cyclic. Evidence of the lake shores is rarely preserved but in a few westerly outcrops marginal lacustrine carbonates and breccias are found with some minor fluviatile sands.

In the similar contemporary Stromness Flags of Orkney a large number of cycles has been recognized, governed by interactions of water-level and sediment supply, with ultimate tectonic and climatic influences. Here also there are levels of stromatolites in sheets and mounds, algal growths in the shallow lake waters and more signs of fluviatile incursions than in Caithness, cutting channels into the lake sediments.

The Achanarras Limestone, or fish bed, of Caithness and the equivalent Sandwick fish bed of Orkney represent a particularly widespread carbonate facies. It is similar to those of the cyclic sequences above and below but at this time the lake must have been unusually extensive and deep, with untroubled bottom waters, and reaching as far as the Moray Firth in one direction and probably to the Shetland Islands in the other. This horizon is the most effective marker in the Orcadian province but a number of successive fish and placoderm faunas of the Middle Old Red Sandstone have proved useful and give a broad correlation with the standard marine stages. On the mainland the lacustrine regime was brought to an end with the John o'Groats Sandstone. Channels of reddish sandstone begin to appear, cutting shallowly into the lake silts and muds and then fluviatile

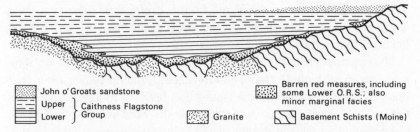

John o'Groats sandstone

Upper ⎱ Caithness Flagstone
Lower ⎰ Group

Granite

Barren red measures, including some Lower O.R.S.; also minor marginal facies

Basement Schists (Moine)

Fig. 35. Interpretative section across the Orcadian lake basin. The irregularities of the land surface (Moine Schists and granites) are filled up with breccias and sandstones of Lower Old Red Sandstone age. The Middle Old Red Sandstones consists largely of the lower and upper Caithness Flagstone groups, but the fluviatile John o'Groats Sandstone appears locally at the top. (Adapted from several sources.)

sands become dominant with much cross-bedding; on Orkney the equivalent Eday Group continues higher, with variations in grade and distribution. Together they represent a broad fluviatile regime, gradually filling in the great Orcadian Lake.

The Upper Old Red Sandstone is mainly seen in south-west Orkney as the Hoy Sandstone, on the island of that name. It was preceded by a mild phase of erosion and faulting and commenced with local eruptions of basaltic lavas. The sandstones are red, yellow and buff with abundant cross-bedding, common slumping and some lenticular conglomerates. The whole is again a fluviatile assemblage, with braided streams. No direct evidence of age is available on Orkney but the equivalent formation across the Pentland Firth in Caithness (the Dunnet Sandstone) has yielded a few, but valuable, scales of *Holoptychius*. At this point our knowledge of the history of the Orcadian province comes to an end. Similar Devonian rocks are widespread in the surrounding sea floors (p. 270).

Enigmas in the far north: the Shetland Islands

That the Old Red Sandstone of these islands was connected with the Orcadian Lake is suggested by the incursion of certain fish faunas, and more especially by the Melby fish bed. There is some lithological similarity in the sedimentary formations, but the differences from the essentially uniform and stable surroundings of the Caithness flagstone facies are also formidable and no clear palaeogeographic reconstruction can be extended to the Shetlands. The outcrops comprise three regions (west, central and south-east), distinct in their formations and separated by major north–south faults which break up the complex archipelago. Of these, the Walls Boundary Fault in the centre has sometimes been interpreted as a continuation of the Great Glen Fault (p. 20) but the linkage is not certain.

The sequences in both the western and south-eastern sectors show some similarity to the main Orcadian outcrops. The former is principally seen in isolated western headlands and peninsulas; here the Melby fish bed near the base has a rich fauna very similar to the Sandwick fish bed of the Orkneys though most of the associated sediments are more fluviatile than lacustrine. Higher up, however, the resemblance abruptly vanishes and rhyolitic lavas and tuffs are intercalated with the sediments; in the north-western headlands there are further andesitic lavas and ignimbrites. On the other side, the peninsulas of the south and east also include variable fluviatile and lacustrine facies, seen to overlie an irregular metamorphic basement. There are locally abundant plant remains and thin fish beds at more than one horizon. Most of them suggest a correlation with the Eday Beds of Orkney but the highest has no clear British equivalence and may be Upper Old Red Sandstone in age.

The central sector, west of the Walls Boundary Fault and east of the Melby Fault, is the most problematic. Faulted within themselves the outcrops comprise a very thick lower sedimentary formation (the Walls Sandstone, up to 9000 m) overlain by sediments, lavas and tuffs, including ignimbrites. The sparse fish debris among the finer sediments suggest a Middle Old Red Sandstone age but the whole may extend somewhat lower.

In so far as any deduction can be made, the Shetland Islands seem to lie near some northern limit of the Orcadian lake, or of lakes connected with it. But it was a far more unstable region, with abundant vulcanicity. The three sectors outlined above are distinct in their history and their original spatial relations are uncertain; at some later date they have been brought into their present positions by major transcurrent faulting. Among them the great thickness of the Walls Sandstone is unprecedented among British outcrops.

THE OLD RED SANDSTONE IN IRELAND

Almost all the Devonian rocks of Ireland are of the Old Red Sandstone facies. In the south there are significant resemblances to South Wales, along the same Hercynian belt. In the centre and north the outcrops are isolated and few, partly owing to the great spread of the Lower Carboniferous cover, but some links with Scotland appear.

On the east coast of Antrim the small outcrop of Cushendall resembles that of Kintyre, only 30 kilometres away across the sea. About 1000 metres of red rocks include boulders of andesite and quartzite among sandstones and mudstones. The age is probably Lower Old Red Sandstone, with perhaps a little of the upper division nearby. The largest of these northern outcrops is the Fintona block of Fermanagh and Tyrone; it is faulted against Dalradian schists on the north and detritus from these is conspicuous. Other Scottish resemblances include 500 metres of andesitic lavas. A single pteraspid plate among the red sediments confirms the Lower Old Red Sandstone age. Farther along the Caledonian strike the elongate outcrop of the Curlew Mountains is comparable (Fig. 29), the total of 1500 metres including 800 metres of lavas at the base. More unusual information comes from Mayo, where inland from Clew Bay and west of Castlebar there are several Old Red Sandstone patches. From one plant debris and spores have been collected, of most probably Middle Old Red Sandstone age – the only indication of such deposits outside the Orcadian province; the spore assemblage is similar to those found there.

South of a line between Galway and Dublin there are three main versions of Old Red Sandstone – not sharply distinct, but each with its own characters in facies or succession; all belong to the upper division and are conformable with the Lower Carboniferous above. In the south-central plain there are many inliers, from a few square kilometres to upstanding sandstone tracts such as the rims of Lower Palaeozoic outcrops, for example, Slieve Bernard, Slieve Aughty, Slieve Bloom and Silvermines. All are characterized by thin Upper Old Red Sandstone, under 300 metres thick.

South of Tipperary the same general relations hold in the Galtee and Comeragh Mountains but the thickness is much increased, to some 1500 metres in the former. This is the northern edge of the Munster Basin, which developed its own facies in late Devonian and Carboniferous times. In the east and south-east the Kiltorcan Beds appear at about the Devonian–Carboniferous junction, between typical Old Red Sandstone and the grey 'Lower Limestone Shale' above. In their type area and in the Comeraghs they consist of yellow sandstones and siltstones with a fauna of arthropods, *Archanodon*, fish and an abundant flora; the last includes lycopods, fern-like forms and the large tree-branch form *Archaeopteris hibernica*. The combination of species does not occur elsewhere in Britain and is generally taken to represent a flood-plain or lacustrine accumulation, in striking contrast to the great barren masses of rocks below.

There remain the very large outcrops of the southern belt in the counties of Cork and Kerry, south of Dingle Bay (Fig. 36). The dominant

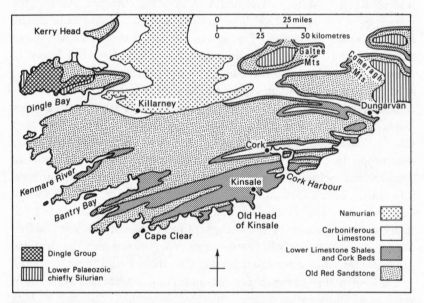

Fig. 36. Geological sketch map of southern Ireland. Lower Palaeozoic rocks include some Ordovician in the extreme south-east.

lithological types are red, purple and grey-green sandstones, often fine and interbedded with siltstones, mudstones and some conglomerates, but limestones seem to be rare; the finer rocks locally exhibit slaty cleavage. Because the sandstones, especially, are more resistant than the succeeding Carboniferous limestones and shales, the synclines of the latter have been eroded into the long inlets of west Cork and Kerry (Dingle Bay, Kenmare River and Bantry Bay), while the Old Red Sandstone forms anticlinal mountain tracts between. The highest range in Ireland, Macgillycuddy's Reeks, is among them, where over 6500 metres of strata have been estimated; if this is truly all Upper Old Red Sandstone it is a remarkable thickness. In all this southern province, south and west of the Comeraghs, the base is not exposed.

Much of the interest in this belt has centred on the outcrops of the southernmost section, of west Cork (e.g. Kenmare coast and Bantry Bay) and south Cork (from Cork Harbour to Cape Clear), largely owing to conformable sequences and facies change around the Devonian–Carboniferous junction. Along the northern margins of the Munster Basin rather incomplete evidence, principally from spores, suggests that the non-marine conditions continued slightly into the beginning of Carboniferous times, or approximately coincided with it. However, south of a line from Kenmare to Cork Harbour there was an earlier transition to marine facies, so that there are marine formations of upper Famennian age.

It is likely that the alluvial plains of southern Ireland were invaded from the south by very shallow seas or locally intertidal conditions somewhat irregularly and that this process continued, with increased submergence, into Carboniferous times. South Cork near the Old Head of Kinsale can be taken as an example of the resulting sequence:

Kinsale Formation (basal Carboniferous, 850 m)   ⎫ also called the
Old Head Sandstone Formation (? upper Famennian, 900 m) ⎬   Cork Beds
Upper Old Red Sandstone, continental facies   ⎭
             (base not seen)

The situation is similar to that along the strike in South Wales but the succession is substantially thicker, possibly because the Munster Basin was more depressed. There does not seem to be any indication of an earlier invasion comparable to that of the Plateau Beds.

The succession in the Dingle peninsula of west Kerry differs from all the other southern outcrops. Overlying the Dingle Group (p. 129) with local transgression there is a thick sequence of aeolian sands which pass laterally into alluvial fan conglomerates, the two forming the Caherbla Group. They seem to have been deposited in a down-faulted basin, bounded on the south by a ridge of metamorphic rocks from which the conglomerates were derived; in drier periods the latter were overwhelmed by desert dunes.

Aeolian sands are rare in the Old Red Sandstone, and the age of these is unknown, though it may be late Lower Devonian. The remaining red beds lie unconformably above and pass up into marine Lower Carboniferous, as in other parts of southern Ireland.

COMPARABLE FACIES OUTSIDE THE BRITISH ISLES

The two facies of this country have parallels in other parts of the Caledonian chain or on its borders (Fig. 32, inset), the various sectors now being ranged around the North Atlantic by the Tertiary rifting. The largest European portion of the Caledonides, in Scandinavia, retains very little Old Red Sandstone, possibly through its deeper erosion. Small patches are known, mainly near the western coast, bearing a sparse flora and fauna that suggest a Middle Old Red Sandstone age.

On the opposite side of the orogen and outcrops of East Greenland are much more extensive, as are those in central Spitsbergen. The red rocks are not unlike the British types but vary in their age and tectonic setting. Thus the Middle and Upper Devonian of Greenland is unconformable on the Cambro-Ordovician of the Caledonian foreland and is conformable with the Carboniferous above. In Spitsbergen only the lower divisions are in a red facies, and in Middle to Upper Devonian times all were affected by a late Caledonian orogenic phase – the Svalbardian. South-westwards along the Caledonian strike from Ireland small residual patches of all three divisions are found in Newfoundland. There are similar outcrops in the eastern Canadian provinces and a much larger one on the south side of the St Lawrence estuary (e.g. the Gaspé Peninsula). However in all this belt, the Northern Appalachians, a major feature is the pronounced middle to late Devonian Acadian orogenic phase, incorporating many large granitic intrusions that have no counterpart on the European side. Remnants of 'Old Red Sandstone' are thus spread widely along the Caledonides and each province has its variants in dominant sediments, flora and fauna (or the lack of them) and vulcanism. It is a typical late or post-orogenic accumulation, often called by the Alpine term 'molasse', and the British Isles near the centre of the belt provide a very fair example.

South-West England is part of the Hercynian belt, whose distribution has already been summarized in the Introduction (p. 21, Fig. 6). In relation to British stratigraphy (Devonian and Carboniferous) the clearest analogies are along the northern part of this belt, or along the 'sedimentary strike' from Devon and Cornwall to north-eastern France, southern Belgium (the Ardennes) and the Rhineland. As in South-West England the Devonian facies are mainly, but not entirely, marine.

The Franco-Belgian region lies south of the stable London–Brabant Massif against which there is both thinning and overlap. Along the margin of the block there are some coarse continental facies which occasionally extended farther south. A comparison can be made with South Devon, though basal Gedinnian beds are known, unconformable on slightly folded Lower Palaeozoic. Similarly the Lower Devonian includes variegated shales with pteraspids and there is a marked Middle Devonian transgression; the resulting carbonates continue up into the Frasnian, when reefs develop. After deeper water shales a Famennian regression, with fish and plant remains, interrupts the marine facies but these reappear at the extreme top. Several parallels are evident but there is little or no vulcanicity and the carbonates are much more extensive than in South-West England.

The carbonate facies is not prevalent in the Rhineland, where the Devonian is well exposed but with much complexity, subsiding on an irregular basement. Clastic rocks are dominant and, above the Lower Devonian coarse components (both continental and marine), deep water shales spread widely, especially in the south-east. Here there is an important zonal sequence of goniatites and conodonts. In the north-west a local shelf facies of Middle Devonian carbonate appears – a lateral equivalence not unlike that of the Cornish slates and the Torquay limestones. Basic vulcanicity in both lower and middle divisions is another resemblance. Although minor variegated or reddish rocks, occasionally with vertebrates or plants, do appear sporadically in the north-western outcrops there is little sign of a clear-cut boundary to the Old Red Sandstone continent, any more than there is in the boreholes of southern and eastern England.

Devonian rocks are known or deduced in the seas adjacent to many of the British land outcrops and in parts of the Hercynian belt are presumably interfolded with Carboniferous strata. A longer stretch occurs on the sea floor from Caithness to the Orkneys and Shetlands (cf. Fig. 1A). Similarly an Old Red Sandstone facies is known from several boreholes in the northern North Sea and is thought to be extensive. A more unexpected feature was found in the Argyll oilfield (Fig. 76(a)) on the western edge of the Central Graben, where there are Middle Devonian limestones, some with corals, overlain by shales and silts with anhydrite and thin beds of dolomite. The source of the marine waters is obscure though perhaps a southern connection is more likely than a northern one.

REFERENCES

House *et al.* 1977 (*Special Report* No. 8, Devonian).
General, including palaeontology: House *et al.* 1979; Westoll, 1979.
Marine facies: House, 1963, 1975; Sadler, 1973.

Continental facies; Allen, J. R. L. 1964, 1965, 1974, 1979; Allen and Tarlo, 1963; Ball *et al.* 1961; Bluck, 1978*b*; Donovan *et al.* 1974; Horne, 1971; Leeder, 1973; Naylor, 1969, 1975; Naylor *et al.* 1974; Naylor and Sevastopulo, 1979; Waterston, 1965.

British Regional Geology: The Midland Valley of Scotland, The Northern Highlands, Orkney and Shetland, South-West England, South Wales, The Welsh Borderland.

6

LOWER CARBONIFEROUS OR THE DINANTIAN SUBSYSTEM

The scope of the Carboniferous System in the British Isles is immense. In duration it is exceeded only by the Cambrian and Ordovician; in outcrop it covers a larger area than any other system; the variation of facies is unrivalled and as a result the faunal and floral range is outstanding. The Upper Carboniferous Coal Measures are the most valuable of the country's natural resources and for this reason, if for no other, the volume of research on British Carboniferous rocks is prodigious. Comparable degrees of importance and complexity are found in many other countries and a division into two subsystems is agreed internationally. The traditional lithological groups of Table 22 are derived primarily from central and

Table 22. *Divisions of the Carboniferous System and major British lithological groups*

Subsystems	Series	Lithological groups
SILESIAN	STEPHANIAN	(? absent)
	WESTPHALIAN	Coal Measures
	NAMURIAN	Millstone Grit
DINANTIAN	VISÉAN	Carboniferous Limestone
	TOURNAISIAN	

Zones are given in the Appendix (pp. 424–5).

northern England; they are only partly applicable in Scotland, and in Devon and Cornwall not at all. However, over much of the country there was a general progression from dominantly marine conditions to partly or dominantly freshwater, and the faunas and floras follow suit.

Many regions belong to sedimentary or tectonic provinces already established in Devonian times. Devon and Cornwall continue to be part of the Hercynian orogenic belt; to the north the remainder was a small part of

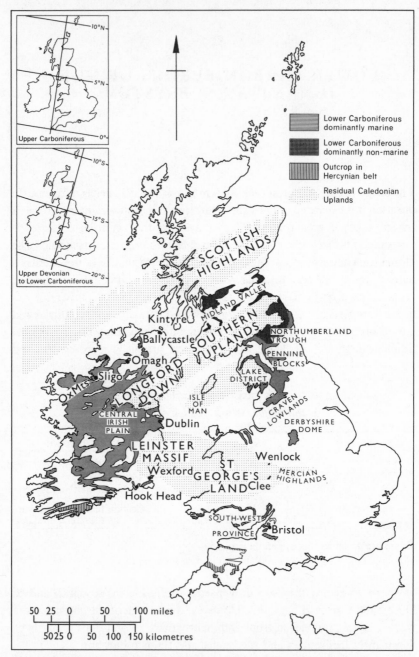

Fig. 37. Lower Carboniferous (Dinantian) outcrops and the main residual Caledonian uplands. Inset: generalized and tentative palaeolatitudes in Lower and Upper Carboniferous times. (Redrawn from Faller and Briden, 1978, p. 19.)

the great landmass of Laurasia and here accumulated various shelf facies – shallow marine to deltaic and partly emergent. This portion of Laurasia was in low latitudes, with a tropical to sub-tropical climate, not far south of the Carboniferous equator (Fig. 37, inset). In some regions sedimentation continued unbroken from the Devonian, as in parts of Ireland and especially the Anglo-Welsh Lowlands, the latter being known in Carboniferous stratigraphy as the South-West Province. On the other hand in the Midlands and north of England the invading Carboniferous seas covered several areas where Devonian erosion had been active, and the base of the Carboniferous rocks, where visible, is often unconformable on Lower Palaeozoic strata.

The remainder of this chapter is concerned with Dinantian rocks and history. The traditional stratigraphical fossils in the limestone facies have been corals and brachiopods and in the shales, goniatites; from these a fairly satisfactory zonal system has been compiled, though in the coral–brachiopod sequence, based on the South-West Province, increasing anomalies have arisen. More recently plant spores, foraminifera and particularly conodonts have been used. In the Belgian type sections, near Dinant on the river Meuse, or Maas, the Dinantian is divided into two series, Tournaisian and Viséan, and these have been widely applied elsewhere. In this country five stages for the British Isles have been defined within the Viséan Series (Table 23), based on local stratotypes;

Table 23. *Dinantian stages and certain zones employed in the British Isles*

Series	Stages	Coral–brachiopod zones, slightly generalized	Goniatite zones
VISÉAN	BRIGANTIAN		P_2
		D_2	P_1
	ASBIAN	D_1	$B_1 + B_2$
	HOLKERIAN	S_2	
	ARUNDIAN	C_2S_1 upper part	
	CHADIAN	C_2S_1 lower part	
		C_1 upper part	
TOURNAISIAN	IVORIAN	C_1 lower part	
	HASTARIAN	Z	
		K	

they exemplify chronostratigraphic divisions (according to the nomenclature summarized in Chapter 1) and are recognized by their biostratigraphical or biozonal indices. The two stages within the Tournaisian Series are based on Belgian stratotypes. The rather complex nomenclature is set out in the Appendix (p. 425). It has also been suggested that far-reaching eustatic effects influenced Dinantian sedimentation and that the beginning of each stage marks a major transgression.

The following account begins with the most southerly regions of the Dinantian shelf seas, in which the limestone facies is most prevalent, and continues northwards where more clastic influxes appear, to Scotland with its pronounced vulcanism. The Hercynian belt, Lower and Upper Carboniferous together, concludes the chapter.

THE SOUTH-WEST PROVINCE: THE CARBONATE SHELF FACIES

The most extensive Dinantian outcrops in this province are those north and south of the South Wales Coalfield, in the Gower Peninsula and westwards along the Hercynian strike into south Dyfed (Figs. 21, 22). The Forest of Dean (Gloucestershire) also has a Lower Carboniferous rim, and farther north the Clee Hills bear small outliers of this age. South-east of the Severn estuary (Fig. 24) the Mesozoic cover is more extensive but isolated limestone outcrops protrude through it, such as those associated with the Bristol and Radstock coalfields. The complex ridge of the Mendips consists of four short periclines, *en echelon*, each with a core of Old Red Sandstone. One of the outcrops west of Bristol is cut through by the River Avon on its way to the Bristol Channel. The gorge so produced was the original type section of the Carboniferous Limestone in this province and here the coral–brachiopod zones were first established.

Already in late Devonian times there had been minor marine incursions over the southern margin of the Welsh Lowlands and the base of the Dinantian coincides, approximately, with a very much more extensive transgression. Some detritus was incorporated in the lowest stage, sandy at the base and muddy higher up, the latter giving the widespread lithology of the Lower Limestone Shale (Fig. 38). But thereafter there accumulated a great variety of carbonates until almost the end of Dinantian times.

To the north lay the joint emergent regions of St George's Land and the Midland Barrier, and the thickest limestones are found in the outer or southern part of the shelf, away from that land. Thus there is a total of 1300 metres in south Dyfed, 1000 in the Gower Peninsula and 1050 in the Mendips. Here also there is the greatest proportion of the more open sea facies, bioclastic or 'standard' limestone: a grey rock with brachiopods, corals, foraminifera and crinoids. It can be exemplified at Bristol by the Black Rock Limestone near the top and by correlatives elsewhere. In the more northerly outcrops, especially in South Wales, there is a marked diminution in thickness, the appearance of erosion levels, non-sequences and a few pronounced unconformities, and a greater proportion of very shallow water deposits. Part of this change is attributable to the gentle but persistent uplift of the land to the north, complementary to the down-warping of the outer parts of the shelf. However some effects have also been attributed to eustatic transgressions and regressions, a transgression marking the base of each stage, to give a succession of major cyclic events

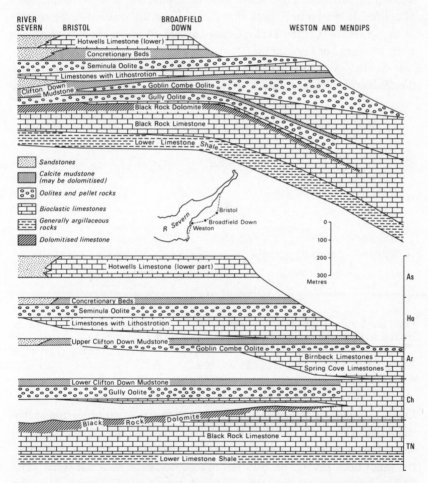

Fig. 38. Dinantian facies from the Bristol region to the more open sea of the Mendips. Above: the present distribution with approximate thicknesses, showing the repetition of facies. Below: the facies (with deduced breaks at Bristol) interpreted as a cyclic sequence. Above the Tournaisian Series (TN), each cycle corresponds to a stage indicated on the right — Chadian, Arundian, Holkerian and Asbian. (Adapted from Ramsbottom, 1977, pp. 262, 264.)

and sedimentation. On this view the regressions caused very shallow and locally hypersaline conditions, scoured or erosion surfaces and sometimes a visible unconformity but more often a non-sequence, with certain faunal horizons or indices missing. It is the last character that has led to the view that the classic Avon Gorge at Bristol is not suitable as a type section, and that the sequence in northern England is more complete (p. 175). In rock type the regressive phases are often marked by dolomitization of the underlying limestone, spreads of algal limestone and current-swept banks of oolites. It is seen in Fig. 38 that these effects are marked in the middle of the sequence (Clifton Down Group) at Bristol where there have always been problems of correlation.

In South Wales a particularly persistent transgression marks the base of the Holkerian Stage (in older terms the base of S_2 and of the Seminula

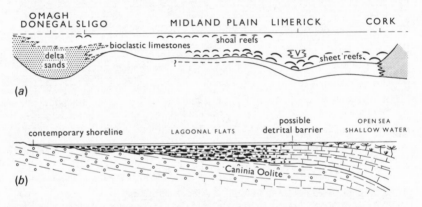

Fig. 39. Lower Carboniferous sedimentation. (*a*) Generalized section across Ireland to show the principal changes in facies. (After George, 1958, p. 280.) (*b*) The inferred conditions of sedimentation of C_2S_1 limestones (Arundian) in South Wales. (After George, 1958, p. 255.)

Oolite), though it is accompanied by northward thinning of individual units both above and below (Fig. 39(*b*)). Thus from the full succession in south Dyfed progressive overstep to the north-east results in the Holkerian limestones resting on Silurian mudstones near Haverfordwest. The underlying stages are similarly thin and irregularly developed along the northern margin of the South Wales Coalfield: at one point in the extreme north-east Namurian beds lie on some 30 metres of Lower Limestone Shale. Much farther north the last remnants of this province are represented by 50 metres of shale and limestone in the Clee Hills, though more may have been deposited and removed, since the sandstone unconformable above is Upper Carboniferous in age.

The top of the Holkerian Stage is marked by widespread algal limestones and the succeeding transgression (basal Asbian or D_1) provides a marked facies change to bioclastic fossiliferous limestones – the stage being notably more uniform than those below in many parts of Britain. At Bristol and the Mendips it forms the Hotwells Limestone; at the top the D_2 (Brigantian) limestones are only thin, and soon give place to the detrital facies that bring Dinantian sedimentation to a close. Over most of South Wales these are dark grey limestones and cherts, sometimes called Upper Limestone Shale, with brachiopods and rare goniatites; the latter belong to the P_2 zone and for the first time in this province the goniatite scale is applicable.

The Hotwells Limestone and correlatives partake of the general southward thickening, but to the north they are partly or largely replaced by a local detrital incursion – the Cromhall Sandstone east of the Severn and Drybrook Sandstone of the Forest of Dean; lower sandstones also interrupt the limestone facies in this region. This detritus did not reach the Mendips nor extend westwards over the Usk Anticline into Wales and it probably represents a small-scale deltaic deposit, some 200 metres thick, derived from the north. Its isolation however is part of the general mystery that surrounds St George's Land since otherwise it seems to have supplied almost no detritus into the Dinantian seas. Probably we should not envisage it as an upland of considerable relief, but as a relatively low swell gently uplifted from time to time, as the carbonate shelf to the south subsided, and subject to recurrent regression and transgression. The most marked uplift in mid-Dinantian times was in the west, producing the pronounced thinning and erosion of mid-Dyfed.

SOUTHERN AND CENTRAL IRELAND

The various obstacles that hinder our reconstruction of Dinantian conditions in Wales and southern England – the blank areas of North Devon and the Bristol Channel, the positive barrier of St George's Land – are only partly reflected in southern Ireland. St George's Land has a counterpart, possibly a continuation, in the Leinster Massif, but west of this there seems to have been a continuous stretch of sea, much of it shallow, covering all the southern and western parts of the country. This impressive extent of Dinantian rocks ought to provide us with a much more continuous picture of conditions and facies change from south to north, but unfortunately much of it is hidden under the drift and peat of the central Irish plain.

The tectonic setting has comparable differences. The main Hercynian folded belt of the south has a northern margin in the Hercynian front, a line of steep dips and thrusting extending from south of Dingle Bay eastwards to Dungarvan (Fig. 36). Beyond this the Upper Palaeozoic rocks were folded, though more gently, by Hercynian movements that were not deflected by earlier trends as much as they were in Wales and the Midlands. In the southern belt the Carboniferous facies most resembles that of Devon (p. 193) and Dinantian shelf limestones only appear north of a line from Kenmare to Cork Harbour – that is, parallel to the Hercynian front but well south of it. There appears to be no direct relationship between the facies change and the later structure, though both may well be related to deeper crustal lineations.

The closest comparison with the limestones of south Dyfed is in the small peninsula of Hook Head, east of Waterford Harbour, 120 kilometres away along the westerly strike. Some 350 metres are known here, probably all in the Tournaisian Series. There are fuller sequences to the west and north, through Tipperary and Limerick and across into Clare, with a maximum of nearly 1900 metres in north-west Limerick – an exceptional thickness in almost pure carbonates. A Lower Limestone Shale facies is recognizable, usually conformable on Upper Old Red Sandstone or Kiltorcan Beds.

It might be expected that in the more uniform conditions of Ireland the gentle warping movements and accentuated regressive phases seen on the margins of St George's Land were less in evidence, and this appears to be the case. However at Wexford, the most easterly outcrop, there is some overlap of the lower Dinantian limestones onto the Leinster Massif and regressive features include thin oolites succeeded by algal limestones and calcilutites with dessication cracks. The age is mid-Dinantian (or C_1–C_2) like the most marked effects in Wales.

Over most of the southern and central outcrops the outstanding feature is the great stretches of reef limestones (Fig. 39(*a*)), which conspicuously swell the thicker sequences. Typically these developed in the upper parts of the lowest stage but in places continue higher up. Near Cork 1200 metres of reef facies have been recorded, and 750 metres in parts of Limerick; the total spread was of the order of 8000 square kilometres. The most abundant reef-dwellers and reef-builders were algae and bryozoans but there are pockets rich in shells. As in similar facies elsewhere goniatites appear, whereas (for obscure reasons) they are extremely rare in bedded limestones; in the latter correlations have been effected by the macrofaunas and conodonts. The reefs appear abruptly along the line of junction between the southern trough facies (p. 196) and the northern shelf, but if they did form some kind of barrier or marginal fringe it was not accompanied by any recognizable back-reef or lagoonal facies on the northern side. The succeeding limestones are mostly of the standard bioclastic type already noted as being widespread in the late Dinantian of many outcrops. Around Limerick two substantial lava flows, totalling 350 metres, occur in the limestones above the reefs, similar to basaltic lavas of the English Midlands.

THE MIDLANDS AND THE NORTH: BASINS AND BLOCKS

The Lower Carboniferous rocks and structures of this extensive region have not the coherence of the south-west and there is no one province to which they are customarily assigned. Essentially they comprise the outcrops around the Irish Sea, north of St George's Land and south of the Southern Uplands; comparable beds also continue under the Mesozoic cover of the eastern counties. The basement rocks, unconformable below the Carboniferous, are often strongly diversified in structure and topography. This irregular foundation not only influences the post-Carboniferous movements, which include both Caledonoid and Charnoid trends, but also the history of Dinantian sedimentation and subsidence.

It is common in this Midland and northern province to find a number of partly connected troughs or basins, with somewhat differing facies and thicknesses; when contrasted with the intervening or bordering areas of lesser subsidence this pattern has been called 'block and basin sedimentation', but not all the positive elements are fault-bounded blocks. The age of the lowest beds varies from place to place, according to the relief of the residual Caledonian uplands undergoing submergence. In general the sediments become less calcareous towards the north, bioclastic limestones almost vanishing in Northumberland; here 'Carboniferous Limestone' is a misleading title for rocks of Dinantian age.

Transgressions and regressions affected this province as they did the shelf carbonates of the south-west, but in some of the deeper basins the regressive effects were modified and the sequence is fuller, without the gaps of the Bristol section. The biozones are correspondingly better developed, and so northern England has replaced the south-west as the representative Dinantian succession in this country although there is no conformity with Devonian strata at the base.

The southern margins: North Wales to the East Midlands

Lower Carboniferous rocks appear only irregularly along this belt, but in Derbyshire and the East Midlands deep boreholes have extended our knowledge farther. Both basin and block facies approach the southern margin, with overlap and thinning, and the variations in facies and thickness suggest a complex irregularly subsiding basement of Precambrian and Lower Palaeozoic rocks.

On the northern flank of St George's Land the North Wales outcrops record only a late Dinantian submergence. Limestones crop out patchily from Anglesey to the Vale of Clwyd (Fig. 16) and then form a westerly rim to the North Wales Coalfield, and share in the general north-easterly dip under the Dee estuary and Cheshire plain. A basal detrital division locally resembles Old Red Sandstone, but it contains a few late Dinantian fossils. The succeeding limestones belong very largely to the upper stages (Asbian and Brigantian, or D_1–D_2) and as such are remarkably thick, with a maximum of over 1100 metres. This is more than the total Carboniferous Limestone at Bristol and there must have been marked subsidence along this North Wales shelf; there is also a similar lack of detritus from St George's Land, apart from that reworked at base of the transgression. A late Dinantian invasion is also recorded in the small outcrops at Little Wenlock, south-east of the Wrekin, where 60 metres of strata include a basaltic lava flow.

At the southern end of the Pennines limestones of block facies are brought up by an asymmetric anticline known as the Derbyshire Dome. The base is not exposed but possible Precambrian rocks, resembling the Uriconian, were met in a borehole near Buxton and Llanvirn (Ordovician) at Eyam, near the north-east margin. Further block facies continue well to the east (p. 231). The almost complete Eyam sequence is remarkably thick, over 1800 metres, with anhydrite at the base and dolomites among the lower limestones. Signs of evaporites, commonly as solution breccias, have been found at outcrop in block facies elsewhere and interpreted as one of the regressive effects.

Around the northern margin of the Derbyshire Dome, for instance at Castleton, reef limestones with steep original dips are banked up against the bedded limestones of the massif. Farther north a more shaly basin facies is known from borehole evidence beneath a Namurian cover; the three types thus form a local apron-reef belt, back-reef and basin complex. Bedded and reef limestones are late Dinantian in age, the latter containing B and P_1 goniatites (Table 23).

To the south-west there is a somewhat similar transition to an irregular reef development and then to a basin facies in north Staffordshire. The basin facies probably extends eastwards under the Trias of the Midlands to link up with the Widmerpool Gulf south of Nottingham, and thus intervene between the Derbyshire block and the southern shore, or Midland Barrier. The complex array of gulfs and blocks below the Mesozoic cover of the East Midlands are chiefly known from their Upper Carboniferous components and therefore deferred to the next chapter (p. 230).

The final approach to the southern shore is marked by thinning or overstep at several levels among the Carboniferous strata. On the south side of the Widmerpool Gulf the Hathern borehole penetrated 260 metres of limestones low in the sequence, underlain by anhydrite and overstepped by Namurian beds. In the small inlier of Breedon Cloud (Leicestershire) four stages are crammed into less than 200 metres of red dolomitized limestones with shales, reef beds and breccias; and as a last southern outpost, in the Whittington Heath borehole (south-east Staffordshire), less than a metre of sandy and muddy limestone, high in the sequence, is all that remains.

There was a phase of igneous activity in this part of the Midlands in late Dinantian times, producing chiefly olivine basalts, with tuffs and dolerites. They are seen in the northern part of the block among the upper limestones and also in anticlinal inliers to the east, such as that at Ashover. More substantial volcanic products occurred later in the East Midlands (p. 231).

The Craven Lowlands and reef belts

The long outcrop up the centre of northern England comprises five structural and stratigraphical units; these are, from south to north, the Craven Lowlands, the Askrigg Block, the Stainmore Trough, the Alston Block and the Northumberland Trough. The northern Pennine uplands are formed of the twin blocks (Figs. 40, 41) in between which the deep east–west trough of Stainmore is covered by Namurian strata; however its western end is probably represented by the outcrops of Ravenstonedale in eastern Cumbria, which at the same time are part of the incomplete rim of Carboniferous rocks around the Lake District.

The Dinantian sequence in the Craven Lowlands is probably a continuation of the large basin that abutted on the reefs and massif of Derbyshire but there is little positive information in the region of thick

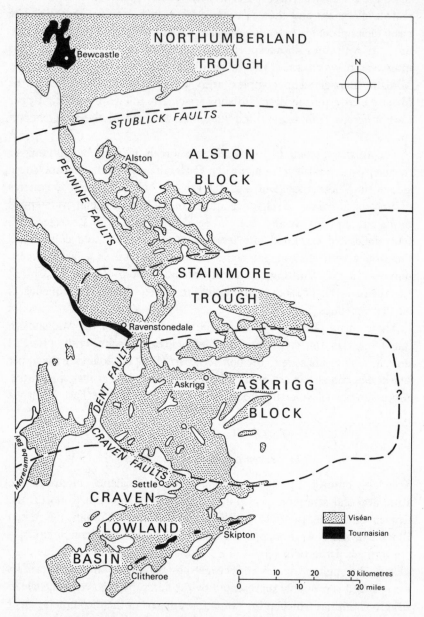

Fig. 40. Dinantian outcrops of the northern Pennines and neighbouring areas, showing blocks and basins.

Millstone Grit and Coal Measures that lies between. The dominant structures are short north-easterly trending anticlines; these are terminated on the north by the Craven Faults, above which rises the Askrigg

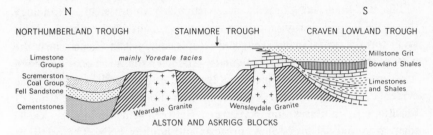

Fig. 41. Diagrammatic section from the Craven Lowlands to the Northumberland Trough, showing basins, the two Pennine blocks and their granitic intrusions. Upper surface is approximately that of E_2 (lower Namurian).

Table 24. *Summary of Dinantian succession, Clitheroe*

Stages	Divisions and lithology	Zones
	(Upper Bowland Shales, Namurian, E_1)	
BRIGANTIAN	Lower Bowland Shales (275 m) including some sandstone	P_2 P_1
ASBIAN	Worston Shale Group (920 m) shales and limestones; Pendleside Limestone near top, knoll-reefs in lower part	B, D_1 S C_2
to	Chatburn Limestone Group (830 m) dark grey well-bedded limestones:	
CHADIAN	Bold Venture Beds Bankfield East Beds	C_1
IVORIAN	Horrocksford Beds	
	(base not seen)	

Block. The lowest stages are only exposed in the cores of a few folds, of which the Clitheroe Anticline may be taken as typical. The pre-Carboniferous basement is nowhere exposed, but rocks resembling an Old Red Sandstone facies have been met in boreholes and are probably very near the base, this being a region of early Dinantian marine invasion.

In several ways the sequence (Table 24) is typical of basin sedimentation. It is exceptionally thick (nearly 1800 metres) and almost or entirely complete; much of it is argillaceous, including the Bowland Shales at the top which straddle the Dinantian–Namurian boundary; the bedded limestones tend to be dark grey and crinoidal; some show signs of a turbidite origin, as does the only major sandstone, in the Lower Bowland Shales. Stratigraphical fossils include corals, brachiopods and goniatites.

The reef-knolls at Clitheroe are particularly impressive; in lithology they have a good deal in common with the reefs of Derbyshire (where a knoll form is occasionally seen) and with a later belt at Cracoe, near the edge of the Askrigg Block. Particularly characteristic is the poorly bedded 'lime mud' or calcilutite that forms the core of the knolls; there is some evidence for steep outward depositional dips and a local sporadic fauna of brachiopods, molluscs and bryozoans. Between and around the knolls there are crinoidal limestones, breccias and boulder beds. The origin of the calcilutite is still controversial; frame-building organisms such as algae have been suggested but visible algal remains are rare. Whether the knolls formed as hillocks on a shallow sea floor or resulted from the erosion of larger lime-mud sheets is also debated, but the succeeding erosion probably accentuated the topographic form. This erosion was part of the most important regressive phase recorded in the basin. With the succeeding Worston Shale transgression a thick sequence of muds buried the Clitheroe knolls, and at the same time the seas spilled over onto the Askrigg Block. Along the edge of that block another belt of reef-knolls developed later (D_1). They show clear signs of growth in shallow water, outward depositional dips, some algal growths in the topmost shallow conditions and the accumulation of breccias, tailing away downslope and into the margins of the basin.

The lateral extent of the Craven Lowland (or Bowland) Basin is problematical but it was large, extending out into the Irish Sea region and eastwards under an Upper Carboniferous and later cover (p. 231).

The Askrigg and Alston Blocks, and Stainmore Trough

Most of the faults bounding the blocks are post-Carboniferous but some correspond broadly with earlier 'hinge-lines' where the block sequence thickens as it passes over into the neighbouring basin. The North and South Craven faults separate the Craven Lowlands from the Askrigg Block; for a short distance there is also a Middle Craven Fault, which, unlike the others, was a fault-scarp in later Dinantian times; the Bowland Shales are seen banked up against it and locally override the scarp top. The western boundary is the Dent Fault, and the more complex Pennine faults (Outer and Inner) hold the same position for the Alston Block; the northern margin of the latter is formed by the Stublick faults (Fig. 40). To the east the gentle dip carries the Lower beneath the Upper Carboniferous.

The Dinantian succession on the Askrigg Block comprises the Great Scar Limestone – only some 250 metres despite its name – overlain by Yoredale rocks (Table 25). The Great Scar is a typical block or massif

Table 25. *Block and trough sequences of the northern Pennines*

Askrigg Block	Stainmore and Ravenstonedale	Stages
(Main Limestone = Great Limestone, E_1)		
Underset Limestone Three Yard Limestone Five Yard Limestone Middle Limestone Simonstone Limestone Hardraw Limestone Gayle Limestone Hawes Limestone YOREDALE FACIES D_2 and P_2	Similar succession (with more clastic Components, continuing onto Alston Block)	BRIGANTIAN
Great Scar Limestone Group D_1, S and C zones	Knipe Scar Limestone Potts Beck Limestone Ashfell Limestone	ASBIAN and HOLKERIAN
Lower Palaeozoic rocks)	Ashfell Sandstone Limestones with detrital, shaly, dolomitic and algal beds near base	ARUNDIAN to IVORIAN
	(Lower Palaeozoic rocks)	

facies, a white to pale grey bioclastic limestone, with corals, brachiopods, foraminifera and crinoids, though these latter are often rather rare. The base of the limestone is decidedly irregular, the invading seas covering an uneven plateau, and the lowest beds are found only locally. Very shallow waters, and locally even emergence, characterize the formation; regressive phases are shown by algal layers, pebble beds and desiccation cracks, and there are even two small local coal seams. To the north the Great Scar Limestone does not maintain its thickness and purity and on the Alston Block is only partly represented by the Melmerby Scar Limestone, the detrital Yoredale facies appearing earlier here.

The succession in the Stainmore Trough (Table 25) is basin-like in its thickness and the presence of the lower strata rather than in its facies, for this is largely calcareous. This is probably because the only satisfactory outcrops (at Ravenstonedale) are at the western head of the gulf, but geophysical evidence suggests a thick sequence continuing eastwards, probably to link up with a subsurface basin in north Yorkshire.

Yoredale facies

The type area of the Yoredale succession is in the centre of the Askrigg Block, in the valley of the Ure, but the facies has a wide extent in the north of England and in Scotland. In its simplest form the Yoredale cyclothem comprises the following:

> coal, always thin and sometimes absent
> sandstone, often cross-bedded and with a rootlet bed at the top
> shales and silty shales, sparse in fossils
> shales with marine fossils
> limestone with marine fossils

This sequence is often complicated by minor intercalations and is itself on a smaller and more rapidly changing scale than the earlier Dinantian transgressions and regressions, though the cyclicity is much more evident.

On the Askrigg Block the Yoredale facies is almost confined to the uppermost Dinantian stage, but farther north on the Alston Block the detrital incursions bring in the cyclothems somewhat earlier. In Durham, Northumberland and Scotland the limestones continue up higher into the Namurian. Over much of the Pennine blocks the correlation of the Yoredale beds is frankly lithological; the cyclothems are named after the basal limestones and there is a fairly certain scheme of equivalence (Table 25). The Main or Great Limestone is the thickest and most persistent and the only one that can be traced as far as the Scottish Border. Elsewhere, and at other levels, correlation is dependent on the rare goniatites in the marine shales, supplemented by the limestone faunas.

It is fairly easy to interpret the local conditions of the Yoredale rocks but the ultimate origin of the cyclic facies as a whole is much more problematical. The limestones, which are largely bioclastic (though some algal rocks are known), represent a continuation of the shallow clear Dinantian seas. Those seas were periodically invaded by muds, then silts and sands and finally the sediment surface emerged above water-level when rootlet beds and coals were established. Then the supply of detritus ceased and with the persistent general subsidence, clear limestone-forming seas were re-established. This is a satisfactory picture for any one area, with minor alternations accounting for minor lithological variants. The source of the clastic sediments is deduced to be somewhere to the north or north-east; it fails, largely, to the south and in west Cumbria and clastic beds form an increasing proportion of the cyclothems northwards. Moreover a northern landmass is implicit through much of Carboniferous times – the ranges of the Caledonian chain from Ireland to Scotland and presumably to Norway.

It is the intermittent supply of detritus that is the puzzle and many suggestions have been made. Chief amongst these are the periodic diversion of the large river that was the transporting agent, periodic rejuvenation in the northern landmass, or a combination of the latter with eustatic rise and fall of sea-level. Eustatic causes have already been invoked for the transgression–regression pattern in the earlier Dinantian, though with a longer time oscillation. It is also noticeable that a Yoredale-scale cyclicity occurs in other countries towards the end of Dinantian times, so this idea gains some support.

The Cumbrian Block, Northumberland Trough and Isle of Man

West and north of the Pennine blocks two further units present versions of block and basin sedimentation. Indeed the Cumbrian and the Alston blocks were probably a continuous ridge or elongate upland, since the downfaulted Vale of Eden in between is a later structure. There is strong geophysical evidence that the several Caledonian intrusions of the Lake District are part of a single major batholith, and the Weardale Granite of Alston may be an outlying continuation of it.

The positive upward tendencies of both Pennine blocks and the Lake District probably relate to this basement of low density rock at depth. The Southern Uplands were similarly intruded by Caledonian granites and the Northumberland Trough developed as a marked downwarp in between – much narrower than that of the Craven Lowlands and accumulating different facies. At present the Dinantian strata form only an outer rim to the Cumbrian Block but that is the result of later uplift and erosion. The rather modest upland or plateau was probably submerged by late Asbian, or D_1, times.

After the deposition of local basement beds the plateau supplied relatively little detritus and clear water limestones are common, with corals, brachiopods and foraminifera. Thicknesses in the southern out-crops, for instance of Furness, are around 700 metres, noticeably less than in the Ravenstonedale Gulf. The situation in west Cumbria, inland from the Whitehaven Coalfield, is comparable but here the basement beds include alkali basalts – the Cockermouth lavas. These are succeeded by a disconformity and then a number of limestones separated mainly by thin shales. Little of the north-easterly detritus reached so far but, as the Dinantian outcrop is traced round to the north-east, Yoredale intercalations appear, till there is a clear resemblance to the Alston sequence, reinforcing the deduced continuity of the two blocks.

In the Northumberland Trough, which developed out of the alluvial plains of the Upper Old Red Sandstone, the deposits are much thicker, reaching up to about 2200 metres; among a medley of facies the westerly tend to be more marine, leading into the Irish Sea region which was probably a major complex basin in Lower Carboniferous and later times. Over most of the trough the base is unexposed but on the north-west, along the Scottish Borders, there is an upward passage from Old Red Sandstone; with few fossils available the base of the Dinantian is not precisely known but in places at least it seems to be within the uppermost red facies. On this northern side there are also early extrusions of alkali basalts (the Kelso and Birrenswark lavas) at or near the base and a string of volcanic plugs.

In the north-east the outcrops extend from the Tweed valley and Northumberland coast and then swing round the Cheviot lavas. The divisions (Table 26) are primarily lithological and all are cyclic in some

Table 26. *Dinantian and Lower Namurian in central Northumberland*

Upper Limestone Group ⎫	⎧ E_1 and part of E_2
Middle Limestone Group ⎬ Yoredale facies	⎨ P_2 approximately
Lower Limestone Group ⎭	⎩ D_1, upper part
Scremerston Coal Group, shales, sandstones and coals with a few marine limestones	⎫
Fell Sandstone, variegated and massive or cross-bedded	⎬ Age not known precisely
Cementstones, mudstones with thin sandstones and impure limestones	⎭
(Old Red Sandstone facies)	

degree. The argillaceous cementstones, with thin impure limestones from which the name is derived, accumulated in brackish lagoons on the coastal plain; they contain bivalves, ostracods and some plants. As the basin subsided there was a deltaic incursion from the north-east (the Fell Sandstone) and then more varied conditions developed but these were sufficiently stabilized from time to time for vegetation to flourish and form coal seams. With further subsidence the Yoredale facies spread very widely over the basin, thickening from the Alston Block over the Stublick hinge-line.

The westerly development is seen in north-east Cumbria (Bewcastle) and just over the Scottish Border in Eskdale. In the lower part a great range of marginal facies have been recognized – fluviatile, deltaic, lagoonal and offshore. The many algal limestones suggest that the very shallow waters periodically became hypersaline. Most of the sediments are fine-grained but the Whita Sandstone at the base represents a deltaic incursion from the north-west and the Fell Sandstone also reached these outcrops. Again the highest groups are in a Yoredale facies. Scattered outcrops along the Solway shores continue this westerly progression; they overlie the Silurian greywackes and Caledonian intrusions (which provided local detritus and pebbles) and include substantial fossiliferous limestones, some with corals.

Detritus entered the Northumberland Trough from the two main directions – locally at many points from the Southern Uplands and from more distant sources in the north-east. It was noted earlier (p. 132) that one suggested position of the Caledonian suture lay along the Solway Firth and the Northumberland Trough. On such a view the deep Carboniferous depression might be an inheritance and the initial basaltic activity on either side related to it.

In the Isle of Man Dinantian outcrops parallel those of northern England. On the north, below Trias and drift, the succession resembles that of west Cumbria; on the south there are reef and bedded limestones, dark grey with shales. The affinity here is with the Craven Lowlands, and the Manx outcrops are a small section in the east–west belt that extends from the Pennines to Ireland and includes the adjoining parts of the Irish Sea.

Eastern Ireland: Dublin to Kingscourt

Whether St George's Land was continuous across the Irish Sea is doubtful but the northern margins of Wales and the Leinster Massif lie on the same latitude. The Longford–Down and Southern Uplands Massif however is Caledonian in plan as well as in age, and trends south-westwards from Scotland into northern Ireland. Thus whereas the two major Upper Palaeozoic landmasses are about 200 kilometres apart on the east, with all the complex blocks and basins in between, the gap is much less on the west. A further distinction arises from the greater spread of Dinantian seas in central Ireland and often a greater uniformity of facies.

The most striking exposures near Dublin are on the coast to the north, as far as the Balbriggan Massif (Fig. 31). Most of the Dinantian rocks consist of dark grey impure limestones (Table 27). They are diversified inland by

Table 27. *Dinantian succession north of Dublin*

Lithological divisions	Stages
Loughshinny Black Shales Posidonomya Limestone	BRIGANTIAN
Cyathaxonia Beds, dark grey cherty limestones	ASBIAN
Limestone, including pebbly and oolitic rocks	? HOLKERIAN and ARUNDIAN
Rush Conglomerate, with sandstones, shales and limestones	? ARUNDIAN and CHADIAN
Rush Slates, with shales and occasional limestones	CHADIAN and IVORIAN
(base not seen)	

abundant reef-knolls; the upper beds tend to be shaly and resemble a more calcareous version of the Lower Bowland Shales of the Craven Lowlands. On the south side of the Dublin strait the Cyathaxonia Beds contain blocks of granite and metamorphic rocks that have slid seawards off the edge of the Leinster upland. The Rush Conglomerate, which crops out on the coast 25 kilometres north of Dublin, is a striking example of a submarine landslide or breccia, in which the angular boulders have ploughed into and contorted the bottom muds. The Rush Slates also include boulders from the Leinster Granite. This coarse detritus diminishes westwards, passing into a more normal calcareous sequence with some reef facies; although the base of the Carboniferous is nowhere seen, there is some faunal evidence of Tournaisian beds, at least in places.

The Kingscourt outlier (Fig. 31) belongs to this province in its Dinantian history although it overlies the low southern edge of the Longford–Down Massif. The sequence is dominantly calcareous, largely complete, and with its intercalated reef facies resembles the inland parts of the Dublin region; both seem to have been in free communication with the seas of south-central Ireland, as well as eastwards with those of the Craven country.

SCOTLAND: NON-MARINE SEDIMENTS AND VULCANICITY IN
THE MIDLAND VALLEY

Scottish Carboniferous stratigraphy is rendered particularly intricate by a heterogeneous tectonic foundation, marked lateral changes in facies and thickness, and varied and vigorous volcanic outbursts. At many levels there is a paucity of zonal fossils, and correlation with Northumberland (similarly handicapped) and farther south is not simple or wholly agreed upon; much of it is frankly lithostratigraphic. One result has been a confusing similarity of names for strata of different ages largely because they contain similar types of rocks. The main economic coal-bearing groups are the Productive Coal Measures of Westphalian age and the lower Namurian Limestone Coal Group (Table 28). The latter is flanked above

Table 28. *Divisions in the Carboniferous rocks of Scotland*

Upper Coal Measures, or 'Barren Measures'			
Middle and Lower Coal Measures, or 'Productive Measures'			
Passage Group, previously Millstone Grit			SILESIAN
Upper Limestone Group			
Limestone Coal Group			
Lower Limestone Group			
Calciferous Sandstone Measures	locally divisible in the east into	Upper Oil Shale Group / Lower Oil Shale Group / Cementstones	DINANTIAN

and below by the two Limestone Groups; although the Lower Limestone Group is uppermost Dinantian in age, it initiates a new phase in Scottish sedimentation and is more logically considered with the Namurian strata above in the next chapter. This chapter is thus restricted to the lowest division, which is also the most variable, the least well defined, and in which marine beds are very sparse or absent.

Calciferous Sandstone Measures: western facies

The base of the Dinantian, or of these measures, cannot be determined palaeontologically and the stratigraphical situation varies. In some places there seems to be a break between red beds below and grey above, or even overlap by the latter onto Lower Old Red Sandstone, but several instances of a conformable sequence are recorded; in one or two of these the evidence from spores suggests that the Devonian–Carboniferous boundary probably lies in the uppermost Old Red Sandstone facies, as has been deduced in some other provinces. In the Calciferous Sandstone Measures as a whole very little stratigraphical correlation is possible and that mainly in the east and in the upper beds. The various terms thus refer very largely to facies.

In the south-west of the Midland Valley the sediments are dominantly cementstones. Typically they are not unlike those of Northumberland – shales and impure dolomitic limestones and minor thin sandstones. The colour is normally grey but may show red and green mottling; the sparse fauna includes *Lingula*, bivalves and ostracods, together with fish scales. Films of gypsum and salt pseudomorphs show that although this is a non-marine facies it was also locally saline.

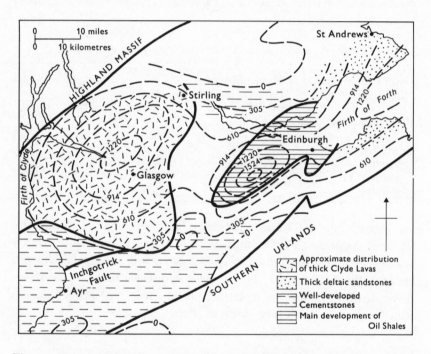

Fig. 42. Map showing the relations of the main facies in the Midland Valley of Scotland to the isopachs of the Calciferous Sandstone Measures; the thickness of the Clyde Plateau Lavas near Glasgow is uncertain and the isopachs here are conjectural. (Redrawn after George, 1958, p. 308.)

The group becomes thicker and more sandy to the north, and thinner, abruptly, southwards over the Inchgotrick Fault and on the Ayrshire shelf (Fig. 42). Here it rarely exceeds 300 metres and is often much less; along the line of the Lesmahagow inlier, for instance, it is absent altogether. The Ayrshire shelf, backed by the Southern Upland Fault, was thus a local positive 'block'. The sedimentary facies, however, is not very different from that in the Clyde basin except where coarse detritus locally accompanies erosion levels and overlap.

The west of the Midland Valley is famous for the thick widespread volcanics of this age, the Clyde Plateau Lavas. These are subaerial flows, particularly of alkali basalts. They may amount to as much as 1100 metres near Glasgow and mostly overlie cementstones, but locally overlap these onto Old Red Sandstone. They did not extend onto the Ayrshire shelf. The irregular surface of the flows was very largely submerged before the end of the period and the highly diachronous Upper Sedimentary Group deposited in the hollows. This lacks the dolomitic limestones of the south-west but includes a few coals and some marine limestones; the faunas include brachiopods and corals and are precursors of the much more extensive submergence at the base of the Lower Limestone Group above.

Calciferous Sandstone Measures: eastern facies

Farther north and east cementstones are not plentiful, though are sometimes found at the base, and two other sedimentary facies take their place – the Oil Shales of West and Midlothian and more sandy beds in East Lothian and, particularly, Fife. In general the easterly sediments are much thicker than those in the west. The typical Oil Shale Group occupies a deep basin west of Edinburgh where up to 1800 metres of fine-grained sediments accumulated. Much of them consists of shales, thin sandstones and occasional limestones; the true oil shales are thin seams of laminated bituminous muds, products of decaying vegetation in stagnant pools. Most of the calcareous beds are freshwater but there are a few marine bands, either limestones or shell beds, in the upper part. They disappear westwards, presumably against the barrier of the Clyde Plateau Lavas; such a ridge is known in the Stirling area, approximately along the division between eastern and western sections of the Midland Valley.

North-eastwards true oil shales disappear and the Calciferous Sandstone Measures become thinner on the north side of the Firth of Forth, to expand again conspicuously where the sandstones of East Fife come in. These are locally thick, coarse-grained and cross-bedded, with channelling and many other deltaic characters. Plant debris and shells occur in the finer beds. Zonal fossils are sparse, but the whole visible sequence, over

2000 metres, probably belongs to the upper two stages, or D_1 and D_2 alone. As such they represent the most rapidly accumulating sediments in the British Lower Carboniferous and are outstanding in the system as a whole.

The mineral and pebble content suggests that a source area, rich in Old Red Sandstone sediments and lavas, lay to the north or north-east. There is thus a linked series of deltaic invasions in the north of Britain (Fig. 44, p. 203) derived from the same direction – the Fell Sandstone of Northumberland (mid-Dinantian), the Fife Sandstones (late Dinantian), the Yoredale sandstones (Dinantian–Namurian) and the Millstone Grit (Namurian). The last and greatest belongs to the next chapter, but we may note here that there seems to have been a general, though not regular, advance of the deltaic facies southwards, till that of the Millstone Grit was halted by the Midland Barrier.

East of the ridge of the Pentland Hills and the deep syncline of the Midlothian Coalfield (Fig. 49) Calciferous Sandstone rocks are found again in East Lothian at North Berwick and Dunbar. Here the sediments are subordinate to the volcanic rocks; they include spreads of pyroclastic rocks and are cut by a large number of volcanic vents. The Garleton Hills are formed of another great group of lavas, basaltic and trachytic; they are probably continuous with similar flows in Fife, under the Firth of Forth, and as such a Forth Volcanic Group is comparable to the Clyde Plateau Lavas. Indeed boreholes in the central part of the Midland Valley suggest there may be lateral continuity from one to the other. All in all, vulcanicity is the foremost distinguishing character of this Scottish province, being on a scale vastly larger than elsewhere in the British Dinantian. Carboniferous vulcanicity as a whole is reviewed in the next chapter (p. 225).

The margins of the trough

In a wider context, the relation of the Lower Carboniferous depositional trough to that of the rift valley is not easy to determine. Where the cementstones approach the westerly end of the Southern Uplands they certainly become thinner but the conspicuous reduction is at the Inchgotrick Fault. In the outliers (e.g. Sanquhar, Thornhill) on the surface of the upland, the Carboniferous rocks are almost all later in age, but a few patches of Calciferous Sandstone have been found to indicate that the Southern Upland Fault was not a complete boundary to deposition. Little detritus seems to have been received from this direction, however, so that the barrier may only have been a low one.

On the northern margin the present outcrops approach the highlands also only in the west, but here it is clearer that the boundary fault had little effect. There is a substantial development of cementstones, capped with a remnant of lavas, on the highland side between Loch Lomond and the Firth of Clyde. On Arran any direct relation is even harder to find; far from the Calciferous Sandstones being thinner on the upthrow side of the boundary fault they are thicker (over 320 metres) where they lie on Dalradian rocks, and show both thinning and internal overlap on the south-east side where they lie on Old Red Sandstone. Moreover the sediments are more sandy than on the mainland, are plant-bearing and lack the marine limestones. In the Machrihanish Coalfield on the Mull of Kintyre, where Carboniferous rocks again overlie Dalradian and Old Red Sandstone, there are lavas of probable Dinantian age.

At the eastern end of the Southern Uplands a small patch of upper Calciferous Sandstones is found on the coast at Cove Harbour (Berwick-shire) and forms a link between East Lothian and the Tweed basin, farther down the coast. Presumably this also was a partial connection across a low-lying part of the upland. The marine incursions into this end of the Midland Valley probably originated from the east. The most valuable and persistent are a group called the Macgregor Marine Bands, with goniatites of the B zone (or D_1, Asbian), the lowest occurrence in Scotland. They allow correlation from West Lothian and Fife southwards to Cove Harbour, and with the lowest limestone of the Berwick succession.

IRELAND: NORTHERN SEAS AND SHORES

The general picture in north-central Ireland is of limestone-depositing seas transgressing northwards and abutting on the southern irregular edge of the Dalradian highlands of north Mayo and Donegal. The Highland Border Fault appears to have had little influence on this process, and if the Southern Upland Fault did in fact continue into Ireland it is still covered by Dinantian strata.

It seems probable that the Longford–Down Massif was more thoroughly covered by Lower Carboniferous sediments than was its Scottish partner, the Southern Uplands, which is another aspect of the 'topographic rise' to the north-east during this period. The major Kingscourt outlier has already been noted (p. 186); farther north-east limestones crop out south of Carlingford Lough and much of Strangford Lough may be floored by them, for there is a small patch of D_2 rocks on the west side.

In the north-central plain as a whole, as far as the Fintona outcrop of

Old Red Sandstone and the Precambrian of the Ox Mountains, limestones or limestones and shales are the dominant sediments, locally underlain by a basal detrital facies. The Ox Mountains seem to have persisted as a positive ridge or axis over which various formations become thinner, to expand again on the northern side. There is also an increase in terrigenous detritus to the north, coming in as irregular and diachronous wedges.

This northern facies is well exemplified by the south-to-north traverse provided by the synclines of Sligo and Donegal. The mixed facies of the former (Table 29), with about equal proportions of limestone and detrital

Table 29. *Dinantian succession in the Sligo syncline*

Stages	Lithological divisions	Thickness (metres)
LOWER BRIGANTIAN	Glendale Sandstone	60
ASBIAN	{ Limestones, massive with reefs in upper part	490
HOLKERIAN	{ Benbulben Shale, with thin limestones	105
	{ Mullaghmore Sandstone	180
ARUNDIAN and	{ Bundoran Shale	135
? UPPER CHADIAN	{ Ballyshannon Limestone and basal detrital beds	330
	(Precambrian of Lough Derg)	

formations, both shales and sandstones, passes northwards into a much more sandy thicker facies overlying the Dalradian foundations. A similar littoral or deltaic facies is found to the west across Donegal Bay in central Mayo, though it is less complete; to the east it is well developed in the Omagh syncline (Tyrone) where 1600 metres of Dinantian rocks include great wedges of sandy beds.

This belt of detritus, derived from the Dalradian highlands, shows the direct influence of the Caledonian uplands as a source of sediments more clearly than anywhere else in the British Isles. The resulting rocks include cross-bedded and slumped sandstones, boulder beds and siltstones; the organic remains include drifted wood and there are occasional calcareous beds with shells. The sandstones are poorly sorted and have mineral assemblages that can be matched in the Dalradian rocks and intrusions to the north. Usually they pass southwards, or offshore, into shales and then into limestones; sometimes the transition is remarkably rapid, from deltaic sandstones to massive bioclastic limestones in a few kilometres. There is no evidence that the Dinantian seas ever submerged these north-westerly mountains, but farther to the north-east similar detritus formed sandstones, apparently unfossiliferous, which now remain on the north-west side of the Antrim Basalts, for instance in Londonderry.

The small Ballycastle Coalfield on the north Antrim coast is only 40 kilometres from the Machrihanish Coalfield of Kintyre, and the Carboniferous rocks of the two are rather alike. The 'Calciferous Sandstone' at Ballycastle, however, is more varied; in addition to lavas and tuffs the 500 metres include a limestone, shales, sandstones, conglomerates and thin coals. The most marked lateral change, in facies and sedimentary conditions, thus lies not so much between Scotland and Ireland as in the hidden stretch under the Tertiary lavas of Antrim.

THE HERCYNIAN BELT: LOWER AND UPPER CARBONIFEROUS

Devon and North Cornwall

The outcrops of South-West England have already been mentioned in the preceding chapter. Carboniferous rocks occupy the centre of the central Devon synclinorium and, as in Devonian stratigraphy, much depends on the coastal exposures. In the centre and north small upright folds predominate, but in the south there is overturning in that direction and the outcrops are also cut into narrow strips by a series of northward-dipping thrusts. The Upper Carboniferous forms a broad belt in the centre, with the Lower Carboniferous as a narrow strip on the north and a more complex one on the south; outcrops here extend from the north Cornish coast to the western and northern margins of Dartmoor. Thickness can only be estimated locally or in very broad terms.

Despite these handicaps a general sequence has been worked out and there is probably a complete succession from the Devonian–Carboniferous junction up to some level in the Westphalian. Similarly there tends to be an upward facies progression, representing an infilling of the trough, though not an entirely regular one since the facies changes are not synchronous in the north and south. Some relationships can be deduced with South Wales and the Mendips, but only tentatively; the major geographical gap is not much filled in by the small inlier of mid-Dinantian limestones at Cannington Park in the Somerset plain; its boundaries are obscure and it may be a thrust mass.

Much of the Dinantian (Table 30) is represented by fine-grained sediments accumulating under quiet conditions and probably in fairly deep water. Thicknesses appear to be relatively small and the most valuable fossils are the goniatites, conodonts and trilobites. Faunas are best known from the bottom and top stages, especially the P_1 and P_2 goniatites. At the base there is conformity with the Devonian and in the north the Pilton Beds, grey shales with calcareous sandstones and limestones, lie athwart the junction. In the south this name is not used, but there is a gradual transition from green slates with Famennian fossils to black slates with silty bands and lenses. Above there follows a variety of

Table 30. *Carboniferous succession in West Devon and North Cornwall*

Bude Formation, sandstones, siltstones, shales with occasional marine bands (Bideford Formation partly equivalent)	Approximately WESTPHALIAN A to C
Crackington Formation, thin sandstones and shales with goniatites	Mainly NAMURIAN E_2 to G_2
Various thin shales and slates with sandy beds and impure lenticular limestones, Tintagel Volcanic Group near the base	DINANTIAN
(similar Famennian facies)	

Stages and zones of the Westphalian and Namurian are given in the Appendix (p. 424).

slates or shales interbedded with thin quartzites or sandstones, impure lenticular limestones and cherts, the last being conspicuous towards the top. On the west and north of Dartmoor over 1000 metres has been estimated, including some tuffs.

Basic vulcanicity occurs in several outcrops along the southern belt, such as the 60 metres of spilitic lavas at Brent Tor, Dartmoor, or the more varied Tintagel Volcanic Group on the Cornish coast. Here there are lavas, tuffs and agglomerates and despite marked deformation some pillow lavas are known, showing that locally in early or mid-Dinantian times volcanic products were extruded on the sea floor among the muddy sediments. No such effects occur in the northern belt, but in the most easterly outcrops (for instance at Bampton in east Devon) there is an increase in calcareous content. The impure limestones in the upper part include much turbiditic detritus, swept down by currents off the carbonate shelf that lay to the north or north-east.

The succeeding sedimentary phase is represented by the Crackington Formation, again most typically seen in the western coastal outcrops.

Primarily the formation consists of shales with many thin turbiditic sandstones; the goniatites occur in many lithological types and the whole accumulated at the bottom of a trough that was still marine. In the northern belt the quiet conditions and very thin fine sediments continued much longer, up to R_2 times, and the turbidites only reached this region later, in the upper part of the Crackington Formation. This distribution suggests that in early Namurian times at least, a northerly source of detritus is unlikely. The directional structures of the turbidites indicate a dominant axial flow along the trough, for instance from the west or north-west in the coastal outcrops. In the uppermost Crackington beds northerly flows do appear, precursors of the regime to follow.

Along the western coast of north Cornwall, and probably in much of the inland area, the highest beds present comprise the Bude Formation. Its base is taken somewhat arbitarily where thick structureless sandstones appear, accompanied by siltstones and shales with occasional plants; some beds are still turbiditic. In this formation the marine fossils are restricted to shale bands and although most of them belong to Westphalian A, the uppermost has yielded shells that are probably low in Westphalian C. North-westerly and north-easterly currents brought in the detritus, to be deposited in a deltaic regime, with the finer muds carried farther out into the basin. At times the salinity was probably reduced by the temporary influx of freshwater floods.

In much smaller outcrops in the north, beds of the same age are rather different in facies and are called the Bideford Formation. Here there is also a range from black mudstones through various silty and sandy beds to thick sandstones, locally in channels and cross-bedded, but unlike the Bude type this is a cyclic facies. Fine disseminated plant detritus is common and there are a few shells (for instance of the *lenisulcata* Zone), and an early Westphalian flora is known. This paralic sequence has been compared to that of the normal Coal Measures to the north (p. 210) but there are certain differences: rootlet beds are rare, there are no seatearths or true coals and no marine bands. The thin seams of soft friable coal-like detritus go by the name of 'culm' – a local term that once was applied to the Carboniferous of the south-west as a whole – the Culm Province.

The Bideford facies thus does not seem to be a simple southward extension from the Coal Measures of South Wales, and there may have been some low ridge between them (p. 220). Nevertheless, as in the much better known Westphalian history to the north, the silting-up process from time to time resulted in emergent conditions. Thereafter follows a stratigraphical gap and the Hercynian deformation.

Southern Ireland

Carboniferous rocks occupy the synclines in the Hercynian belt of Ireland (Fig. 36) and, being generally less resistant than the intervening Old Red Sandstone, form long inlets such as Bantry Bay and Kenmare River in the west and reach as far east as Cork Harbour. This detrital facies used to be known as 'Carboniferous Slate' but more recently as the Cork Beds (p. 163), which overlap the Devonian–Carboniferous junction; in palaeogeographic terms the region is the Munster Basin. In sediments and succession there is some resemblance to North Devon but the Irish sequence barely extends into the lowest Namurian and below that it is thicker. Stratigraphical fossils are extremely rare and consist chiefly of conodonts and spores near the base and goniatites near the top; as in Devon the middle stages are problematic or unrecognized, though the sequence appears to be complete. The current classification is very largely lithostratigraphic and the boundaries between the groups are almost certainly diachronous, in some places demonstrably so.

Including the Famennian component at the base, the thickness in south Cork reaches 2400 metres but is somewhat less in the west, at Bantry Bay. In general there is an upward progression in facies as the seas of the Munster Basin transgressed over the edge of the Old Red Sandstone coastal lowlands – a transgression which was earlier than that of the Lower Limestone Shales to the north or at Hook Head in the north-east (p. 174), where the Old Red Sandstone facies persisted into early Dinantian times. The alluvial and deltaic sandstones of the Old Head Formation were succeeded by pro-delta silts, and then by shallow water impure carbonates; the latter represents a relatively sudden change to a current-swept shelf and the base is marked by thick beds with crinoid debris. In late Dinantian times (P_1 and P_2) circulation in the deepening basin became more restricted, detritus largely failed and black goniatite-bearing mudstones appear. In a very few outcrops these extend up into the E zones (lower Namurian), but thereafter we know nothing of this trough.

There is some lateral variation in this facies progression; in particular the transgression in the west at Bantry seems to have been somewhat in advance of that in south Cork, and there may have been a low divide between where the carbonates are missing and more pro-delta silts present. However in the last stage the black mudstones supervened in both regions and the semi-stagnant basin was presumably extensive. During the penultimate stages the reef facies farther north accumulated to great thicknesses, and although detailed correlation is lacking it must have been contemporaneous at this time with a thinner sequence in the trough.

THE LOWER CARBONIFEROUS SCENE

The palaeogeographic picture reflected in the clastic belt of southern Ireland is largely a confession of ignorance and that of South-West England is not much better. The extensive shallow water carbonates to the north in both regions rule out any detrital influx from that direction in Dinantian times. In the preceding chapter a possible southerly source was mentioned, between Cornwall and Brittany, and the Carboniferous history of the latter is very different from that of Britain, with two marked orogenic phases. The first is very early in the period, so that the lower Dinantian rocks are transgressive coarse clastics, and the second resulted in an absence of most Namurian and all Westphalian strata. It is possible that this phase of uplift and erosion within the Hercynian belt also contributed to the Namurian and later turbidity flows of north Cornwall and Devon.

Over the remainder of the British Isles Dinantian history is dominated by the newly developed shallow seas and their carbonate sediments, into which the reduced Caledonian uplands supplied varying amounts of detritus, notably in the north. Because sediment-level and sea-level were not far removed from each other the regime reflected alterations in one or both with unusual delicacy. Cyclic sedimentation is thus another prevalent factor, conspicuous in the later and more clastic facies, and possibly also in the larger eustatic cycles that have been postulated throughout the subsystem.

After the long Devonian continental intermission new groups of fossils colonized the shallow seas; the tropical plant cover of the remaining lands occasionally produced thin coals or drifted plants, and even in marine rocks the plant spores become useful in correlation. With the several new invertebrate groups, stratigraphical palaeontology takes on new complexities and specializations – also new nomenclature.

As in Devonian times the main parallels are along the Hercynian belt into central Europe. The most conspicuous is the continuation of the shelf limestones from South Wales and Bristol, south of the London–Brabant Massif (subsurface Dinantian in Kent and small outcrops in the Boulonnais) to the major province of southern Belgium, or the Ardennes. All along this belt, trending east-south-east, there are strong resemblances in lithology, faunas and correlation; the main distinction is the greater structural complexity in Belgium where the Carboniferous rocks are affected by overfolding and a major northerly thrust.

Across the Rhine the strike swings round to north-east in the Rhine Massif (Rhenisch Schiefergebirge) and Harz Mountains. The calcareous shelf facies here is found only in some restricted northerly outcrops and boreholes, most of the Dinantian being in a detrital facies – the German Kulm. The rocks range from shales, with important zonal goniatites, to greywackes and there are several parallels with Devon and Cornwall. These include a very thin muddy sequence near the base, later various greywacke incursions, mainly from the south and south-east, with the principal infilling appearing in late Dinantian times and continuing into the Namurian; the basic volcanic rocks are also alike.

The main absence on the continent of Europe is in any continuation of the Scottish facies, since no Dinantian survives on the flank of the Scandinavian Caledonides; only a very little is known beneath the big Upper Carboniferous basin of the southern North Sea or on its margins. In the opposite direction, however, there are some partial similarities, especially in the Canadian Appalachians of Nova Scotia and New Brunswick, with much smaller outcrops in Newfoundland. In all this region the Acadian orogeny (Middle to Upper Devonian) was vigorous and the succeeding Lower Carboniferous is strongly transgressive and locally very thick. It seems to have accumulated in small troughs, some faulted, trending north-eastwards. There is much variation from one to another but coarse clastics are common, with local oil shales and red beds but also some marine shelly limestones and evaporites; in places the sediments are interspersed with basaltic and andesitic lavas and tuffs.

Unlike Scotland, however, there is no transition southwards to shelf carbonates. These do exist in North America, very extensively, but in a different tectonic setting. The type area of the Mississippian, almost but not quite equivalent to the Dinantian, is along the upper Mississippi and Iowa rivers. Here shallow epicontinental seas were very widespread and the relatively thin limestones overlie essentially stable continental foundations. In Britain the thicker more irregular facies reflect inherited Caledonian structures, modified by new influences.

REFERENCES

George *et al.* 1976 (*Special Report* No. 7, Dinantian).
General: Fitch *et al.* 1970; Francis, 1978; George, 1958, 1969; Mitchell, 1972; Ramsbottom, 1973; Ramsbottom and Mitchell, 1980.

S. England, Wales; De Raaf *et al.* 1965; George, 1974; Matthews, 1977; Prentice, 1962; Ramsbottom, 1970; Renouf, 1974.

N. England, Scotland, Ireland: Francis, 1965*a, b*; Leeder, 1974; Mitchell *et al.* 1978; Naylor, 1969, 1975; Naylor *et al.* 1974; Naylor and Sevastopulo, 1979; Ramsbottom, 1973; Ramsbottom *et al.* 1974.

British Regional Geology: Bristol and Gloucester District, The Midland Valley of Scotland, Northern England, North Wales, Pennines and Adjacent Areas, South Wales, South-West England.

7

UPPER CARBONIFEROUS: THE SILESIAN SUBSYSTEM AND HERCYNIAN OROGENY

In those parts of Britain where the traditional divisions Carboniferous Limestone, Millstone Grit and Coal Measures are applicable the second is approximately equivalent to the Namurian and the third to the Westphalian; it is questionable whether Stephanian beds are represented. Early Silesian history carried on from the latest Dinantian where already some clastic influences had appeared. The British area also passed slowly across the equator and the lower lands were clothed with abundant tropical forests. At their coasts a range of deltas developed, which together with the fringing lowlands were affected by a series of transgressions and regressions. In Namurian times marine muds, with deltaic influxes, were dominant; in the Westphalian the silting up proceeded farther to give very wide coastal swamps. Hercynian folding and uplift completed this major span of geological history.

NAMURIAN IN THE NORTH: MILLSTONE GRIT

South and Central Pennines

This section of the Pennine range extends from the Derbyshire Block northwards to the Craven Faults; it is also the type area of the Millstone Grit (Table 31), where millstones were once fashioned out of the coarser sandstones. The rocks range from shales to coarse pebbly sandstones or gravel conglomerates. The sandstones, standing out as thick 'gritstone edges' or scarps, are much the most conspicuous, but being products of relatively rapid deposition they probably represent less of Namurian time than their thickness at first sight suggests. Fossils are virtually absent in the sandy and silty beds; the black shales contain faunal bands locally rich in marine forms (chiefly goniatites and bivalves) and there are occasional impure silty limestones with an abundant shelly fauna.

The Namurian reaches a maximum thickness of nearly 2000 metres in

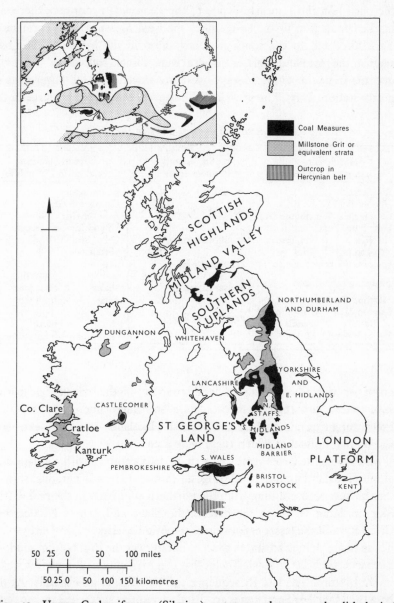

Fig. 43. Upper Carboniferous (Silesian) outcrops, shown as the lithological divisions Millstone Grit and Coal Measures – the latter showing the main coalfields. Inset: exposed and concealed coalfields of north-west Europe, omitting some of the deeper measures and those under the North Sea. (Redrawn from Wills, 1951, pl. ix.)

the north, on the Lancashire side of the Pennines. From here there is a marked reduction to the east, under the Coal Measures and Permian of Yorkshire, and to the south in Derbyshire till the Millstone Grit fails against the Midland Barrier. There were minor sources of quartzose detritus from the south or south-west, for instance in the lower grits of north Staffordshire. The great bulk, however, came from a source in the

Table 31. *Selected lithological units in the Millstone Grit*

Mid-Pennines e.g. west Yorkshire	Stages and major zones	South Pennines e.g. north Derbyshire
Rough Rock	YEADONIAN, G_1	Rough Rock
Huddersfield White Rock Guiseley and Woodhouse Grits }	MARSDENIAN, R_2	{ Chatsworth Grit Ashover Grit
Four grits } in West } Yorkshire } { U. and L. Kinderscout Grits }	KINDERSCOUTIAN, R_1	{ U. and L. Kinderscout Grits, Shale Grit, Mam Tor Sandstone
Shales and ? non-sequence	ALPORTIAN, H_2	
Shales and minor grits	{ CHOKERIAN, H_1 ARNSBERGIAN, E_2	mainly (quartzitic shales sandstones
Marchup Grit		fairly from the
Skipton Moor } { Wilpshire and Grits } { Pendle Grit Upper Bowland Shales }	PENDLEIAN, E_1	thin south in north Staffordshire)

north or north-east. The mineral content of the coarse feldspathic sandstones, with their fresh orthoclase and microcline, accords with a Precambrian terrain such as the Scottish Highlands or a more extensive Caledonian upland of which these were a part.

As a whole the facies represent a thick delta complex with a great range of individual environments. As Fig. 44 shows, there was a gradual shift in the thickest basin infilling from the northern edge early in the period (the Skipton Moor Grits) to the centre (the Shale and Lower Kinderscout Grits, R_1) and the lesser expansion of the Middle Grits (R_2) in Derbyshire. These major deltaic advances were built out over the thin bottom muds of the basin, such as the Edale Shales that comprise much of the lower stages in Derbyshire. The late R_1 sequence can be taken as an example. At the base the Mam Tor Sandstones are distal turbidites, succeeded by proximal turbidites in the Shale Grits; above, the Grindslow Shales and lower part of the Kinderscout Grits include delta-slope deposits, cut by some deep channels with turbidite infilling; later there are fluviatile channels with interdistributary delta-top sediments. A temporary stable phase of fluviatile facies extends widely as the Upper Kinderscout Grit until the whole complex was submerged by a new transgression, the basal R_2 marine band.

In the R_2 beds a thinner but unusually regular cyclic regime set in, with marine shales passing up through coarsening sediments to sandstones, locally topped with a seatearth or a poor coal. The highest sandstone, the

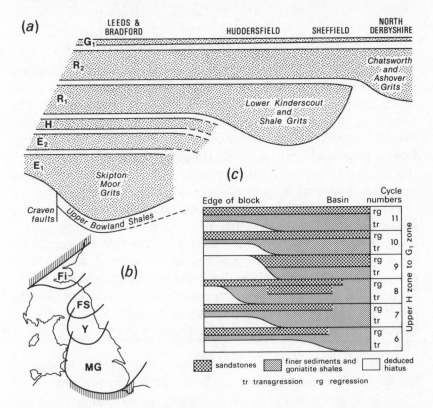

Fig. 44. The Millstone Grit in northern England. (*a*) Diagrammatic cross-section showing how three sandstone groups in particular contribute to the basin infilling in a southward progression (After Ramsbottom, 1965, pl. 4). (*b*) Late Dinantian and Namurian deltaic incursions; north to south: Fife sandstones, Fell Sandstone, Yoredale sandstones, Millstone Grit. (*c*) Suggested Namurian cycles from the Craven Lowlands to the Askrigg Block; only the upper part of each transgression reaches onto the block and is succeeded by a regressive sandy facies. (Redrawn from Ramsbottom, 1977, p. 279.)

Rough Rock, is relatively thin but the most persistent of all. In this latest levelling-up, fluviatile deposits covered much of the delta and the Lower Coal Measures follow without any great change.

It is probable that several transgressions and regressions affected

Namurian sedimentation, but with greater clastic influx the results differed from those in the Dinantian below. In the centre of the basin the regressions are chiefly seen in faunal changes, when the more fully marine goniatite faunas temporarily give place to very shallow water types indicative of reduced salinity (e.g. the bivalves *Naiadites* and *Curvirimula*).

The northern Pennines, Northumberland and West Cumbria

There is no abrupt change at the Craven Faults, but over a 30-kilometre belt the Namurian is reduced to less than a third. It consists of two facies, with a rather ill-defined and diachronous boundary between them. The lower is an upward continuation of the Yoredale type but with generally thinner limestones; on the Alston Block and into Northumberland this facies is called the Upper Limestone Group (cf. Table 28). Above it there is a rather thin 'Millstone Grit' appearing as low as E_2 near the Craven Fault belt but not till R_1 in parts of Durham. This northern version is a variable sequence of shales, siltstones, sandstones and poor coals, which chiefly lacks the limestones of the Yoredales below and the productive seams of the Coal Measures above. In this facies and on the more stable block foundations the regressive phases take the form of inconspicuous non-sequences or lacunae. The earliest faunas of each cycle, as recorded in the basin, are lacking and the marine muds extend onto the block only in the middle or upper zones, where much of the sediment comprises the uppermost sandy unit (Fig. 44(*c*)), often topped with a seatearth, ganister or a thin coal.

In west Cumbria, at Whitehaven, Namurian beds are also unimportant but for rather different reasons. Little of the clastic components of the Yoredale facies reached so far west and above a dominantly calcareous sequence there is little that can be called Millstone Grit. Above the First Limestone (equivalent to the Great Limestone of the Alston Block, E_1) there is a thin clastic sequence with only a 25-metre sandstone. The goniatites found, of the E_2 and G_1 zones, are so close together that a major non-sequence is inferred.

NAMURIAN IN SCOTLAND

The stratigraphical divisions fit the facies titles Carboniferous Limestone, Millstone Grit and Coal Measures even less well in Scotland than they do in northern England. The following groups are established for beds of

Namurian age and slightly above and below but at several levels correlative fossils are few:

Passage Group	E_2 to lowest Westphalian
Upper Limestone Group	E_1 and E_2
Limestone Coal Group	E_1
Lower Limestone Group	P_2

Lower and Upper Limestone Groups and Limestone Coal Group

These resemble a Yoredale cyclic sequence but even more than in the typical Yoredale facies there is much variation, local additions and local omissions. At the base the Hurlet Limestone marks a widespread marine invasion, covering the beds below – oil shales, sandstones and cement-stones alike – and only interrupted by occasional ridges of lavas. The Lower and Upper Limestone Groups resemble each other in that marine limestones are fairly abundant; or to put it in another way, the marine transgressions flowed freely into much of the Midland Valley and periods of emergence leading to peat growth and coal development were rare. Conversely in the Limestone Coal Group the marine rocks, which are almost all shales, are not common and the upper parts of the cyclothems (the shales, sandstones and coals) are well developed with many workable seams.

These generalizations apply where the groups are thick and typical, but thickness itself is another significant factor in variation. It was shown that the Calciferous Sandstone Measures were thin, or locally absent, in certain areas – particularly on the south Ayrshire block or shelf where the Inchgotrick Fault coincides with the thickness change. This sort of variation between areas of differing subsidence, whose margins approximate to major faults, is much easier to demonstrate in the three 'Limestone Groups' because the coals and limestones as marker beds allow better estimates of thickness. The results are well shown in Ayrshire with abrupt changes over three major east–west faults, forming a local version of block and basin effects.

Towards the east and north-east there is a general but irregular thickening into basins, such as those of Midlothian and Fife, or the deepest in Kincardine where the Namurian reaches nearly 1500 metres. An analysis of the Lower Limestone Group demonstrates a parallel in lithology, for the thinner sequences are more shaly and sandstones swell the thickest. The belts separating the basin areas are probably tectonic in origin, but ridges of lavas also formed positive dividing elements; thus the main mass of the Clyde Plateau Lavas separates Ayrshire from the Central Coalfield (Fig. 49) and lavas on the Fife coast lie between the latter and the Fife–Midlothian basin.

Passage Group

In recent years this term has replaced the rather unsatisfactory 'Millstone Grit' of Scotland. The group is relatively thin but when fullest is probably a conformable succession. The zonal goniatites only penetrated the Midland Valley occasionally but spores have proved helpful; bivalves of the *lenisulcata* Zone (lowest Westphalian) have been found 100 metres below the top in Ayrshire. In the centre and north the Passage Group amounts to over 300 metres but locally may be absent or, as in parts of Ayrshire, represented by volcanic rocks.

The sediments comprise a variable range of sandstones (for instance the Roslyn Sandstone of Edinburgh) which are often coarse, cross-bedded or pebbly, with subordinate shales, fireclays and occasional coals; the rather sparse marine incursions resulted in shales and impure limestones. Irregularity of sedimentation and subsidence, so characteristic in Scotland, is strikingly demonstrated in the tiny Westfield basin, east of the Central Coalfield; here the seams vary in an astonishing manner over a number of north-east axes and at their thickest there is 30 metres of coal in a total of 160 metres of measures, an accumulation of peat that may well be unique in Britain.

NAMURIAN IN THE SOUTH AND WEST

Even outside the most massive deltaic sequences, the seas of Dinantian times became gradually silted up, ultimately to be transformed into the lowland swamps in which the Coal Measures were formed; here also there is commonly an intervening detrital phase of Namurian age, sometimes called Millstone Grit. The beds are never so thick as in the delta complex of the Pennines, but in the changing geographical conditions they played something of the same part and many of the same local facies can be recognized. Some of the most outlying and least detrital sequences provide us with the best goniatite successions, such as the black shales of central Ireland.

Wales and the South-West Province

The North Wales outcrops represent a small part of the south-westerly margin of the large Pennine basin but the Namurian is thin and the massive feldspathic detritus scarcely reached so far. The base is conformable on the uppermost Dinantian in which local detritus had already invaded the limestone seas. Apart from some irregular sandstones at the top the sequence is largely argillaceous, the 200 metres of the Holywell Shales including representatives of all the major zones. Farther south less complete developments include calcareous sandstones with an unusual fauna of benthonic brachiopods as well as the more usual goniatites and bivalves – a population more characteristic of shallow or coastal waters. The North Wales outcrops thus represent a partly marginal marine belt which received only modest amounts of detritus, and like other residues from St George's Land these were chiefly quartzitic sands.

In South Wales the Namurian strata accumulated in a fairly simple basin and now appear as a rim to the coalfield and among the more complex structures of south Dyfed. Subsidence and deposition were less than in the Pennines and under 800 metres are known in the Gower Peninsula, the thickest sequence and nearest to the centre of the basin. Here there is probable conformity with the Dinantian below but to the east and north-east there is marked thinning and overlap, so that higher horizons in the Namurian come to lie on lower Dinantian beds; for a short stretch in the east G_I sandstones rest on a thin remnant of Lower Limestone Shale. The axis of the Usk Anticline is only a short distance away and uplift on it may have contributed to this marginal relationship. A comparable overlap takes place to the north-west, and near the centre of St Brides Bay upper Namurian beds lie on Ordovician.

With this marked overlap the basal detrital facies near the margins are highly diachronous but also fairly constant in their main constituents; in some places they are almost pure orthoquartzites, in others pebbly with much vein quartz. The main source was from the north, with some minor increments from the south-east. In this basin too there can be traced various transgressions and regressions corresponding to those of northern England; as would be expected the former are clearest in Gower and the lesser ones failed to reach the more northerly fringes.

During all this time St George's Land continued to be a governing influence, while on its southern flank the South Wales basin subsided and filled. The relation between grade-size and stratal thickness, however, is not that of the northern Namurian. There the successive depressions of the Millstone Grit delta swelled out with a greater proportion of coarse detritus, and here the coarser beds are inshore and the thicker sequence farther out is finer. This may reflect the more restricted source in South

Wales, compared with the voluminous infilling of the Pennine basin, carried persistently southwards from the Caledonian uplands.

On the eastern side of the Usk uplift there are no Namurian (or lower Westphalian) beds in the Forest of Dean and it is uncertain whether any were deposited. The Bristol Namurian basin seems to have been small and isolated, with limited marine access and few goniatites. Some 200 metres of sandstones intervene between the uppermost Dinantian and the Coal Measures; a number of plants are known but there is not much evidence to show whether the sequence is complete. To the south there is a marked thinning in the Mendips.

The major Irish basin

All Upper Carboniferous outcrops in Ireland are small and fragmentary compared with the vast stretches of Lower Carboniferous and (presumably) with their former extent. The Namurian beds can be divided into small residual patches overlying the limestones, the marginal rims of certain coalfields and the only one of any size, an outcrop in the west on either side of the Shannon estuary. The very small amount of lower Namurian shales in the southern Hercynian belt has already been noted. Over the rest of the country there was a restricted shelf in the north-east and a very much larger basin area comprising the remaining outcrops. Thinner and overlapping sequences suggest some margins to the basin, possibly in Connemara and Mayo, the Longford–Down Massif, and more clearly the Leinster Massif. Even the latter, however, seems to have supplied little recognizable detritus.

In the centre and north there is very little indeed by way of outcrops or evidence. Near Dublin and in the Kingscourt outlier (Fig. 31) there seems to be conformity with the Dinantian beneath, and the same applies to a narrow strip along the centre of the Shannon trough – a depression trending east-north-east. Elsewhere, and notably on either side of that trough, there was some basal unconformity, overlap and generally thinner sequences. Thus Namurian strata make up much of the west and south of Clare; the southern outcrops continue into Kerry and Limerick. One of the thinnest and least complete sequences is that surrounding the Leinster Coalfield, so that near this massif there was relatively little sediment or subsidence. A narrow extension westwards was perhaps associated with the ridge of the Limerick volcanic rocks.

In many places there is a rather irregular upward progression in facies: in the lowest zones basinal or pro-delta goniatite shales are prevalent (e.g.

the Clare Shales), succeeded by turbidite sandstones of the delta front, and these by cyclic or delta-top sediments. Occasionally fluviatile or subaerial channels cut through the latter and are seen to pass into turbidites at the delta edge. There was thus a gradual infilling of the basin, presumably from the north, before the paralic swamp conditions were established in the Westphalian.

The 'shelf' area of the north-east is known from two minor outcrops. The Dungannon Coalfield lies west of the Antrim Basalts (Fig. 56) in Tyrone. It is small but contains the fullest Silesian sequence in the north. Even so the Namurian is probably incomplete and goniatites seem not to have penetrated onto the shelf. Within the 400 metres of 'Millstone Grit' the facies is more varied than in the large southern basin and includes mudstones and sandstones, marine shales and coals. The clearest resemblance is with Cumbria and, as in that county, there is probably a major non-sequence in the middle. The coal-bearing strata of the Ballycastle Coalfield on the north Antrim coast correspond most clearly to the Limestone Coal Group of Scotland. They do not extend above E_2 and contain a marine band and a few seams, some up to a metre in thickness.

The several resemblances across the Irish Sea that were noted in Dinantian sedimentation continued into Namurian times. The infilling of the Irish deltaic basin, however, included more fine sediment and the sequences are much thinner than the maximum of the Pennine Millstone Grit. In general this accords with the greater uniformity and stability in Ireland, though the evidence from Namurian stratigraphy is seriously incomplete.

THE COAL MEASURES: WESTPHALIAN A, B, C AND D

The Westphalian strata in Britain contain the vast majority of the productive coal seams and in their bituminous coals the country's greatest single natural resource. The base of the Coal Measures and of the Westphalian is drawn at the Gastrioceras subcrenatum Marine Band and this is commonly about the level of, or slightly below, the lowest workable seams – apart, of course, from the Namurian coals of Scotland already mentioned. The upper limit is much less uniform, chiefly on account of the erosion following the Hercynian movements; it is questionable whether Stephanian beds are present in Britain except possibly in the south-west or as minor barren formations in the Midlands and Ayrshire. Moreover in the uppermost surviving beds, often called 'Barren Measures', the seams tend to be reduced or absent but the change is not everywhere at the same level. Some of these measures are grey with normal rock types and locally a few thin seams; others may have been red in the first place, but much of the 'red barren measures' were probably reddened by water percolating downwards (possibly from an early Permian land surface) and introducing iron oxides or oxidizing the iron compounds already present.

The normal productive measures include a relatively small number of rock types which are repeated many times over in the full succession of any major coalfield – the non-marine shales and mudstones, siltstones, sandstones, seatearths and the coal itself, and the marine bands. All these can be found in the Namurian rocks below, and to some extent the change from Millstone Grit to Coal Measures is a change in the proportion of the different sediments and the conditions they represent. They are also often associated in a regular order to form cyclic units or cyclothems; an example is given on page 215.

The marine bands are usually dark grey to black, carbonaceous shales or more rarely a mixed carbonate rock; fossiliferous limestones of Yoredale type are rare but are known in Scotland. Goniatites and bivalves are typical but there are also bands with *Lingula* and '*Estheria*', the latter being a small crustacean; lesser components include ostracods, foraminifera, gastropods, conodonts and calcareous brachiopods. Some parts (e.g. upper Westphalian B and lower Westphalian C) are rich in marine incursions and a thick 'band' may reach 10 metres. In such conditions the low-lying swamplands must have been submerged for long periods beneath a shallow muddy sea. Among the faunas, *Lingula* probably represents inshore conditions, and the more prolific assemblages, including some benthonic species, somewhat deeper and more open waters. The faunal distribution within a given band thus allows some palaeogeographic deductions on inshore and deeper waters and these agree with those already outlined. For instance in the northern coal basin 'inshore' (or perhaps brackish) faunas tend to occur round the margins while the more 'offshore' species are found in the centre (cf. Fig. 82, p. 396). The south-western province of South Wales, Somerset and Bristol developed its own regional distribution pattern, while the Scottish marine bands, fewer and less prolific, reinforce the distinction of this province. The source of the lower Westphalian marine incursions of north-western Europe remains obscure. The greater variety of many of the faunas in South Wales suggests that a westerly source is possible; moreover the incursions continued later in Britain than in the more easterly continental coalfields. But beyond that there is little firm evidence.

The non-marine grey mudstones make up the largest proportion of the productive measures; they contain plants and bivalves (the mussels) and their main variant is clay ironstone. The siltstones and fine to medium sandstones commonly lack fossils. The former are often laminated or with small-scale cross-bedding; the latter may be flaggy or cross-bedded and are usually lenticular or impersistent in distribution. Most of the productive seams are bituminous, though anthracites are known from South Wales and cannels from many sources. Coal is ultimately derived from decayed plant and woody detritus. It has usually been supposed to have originated in place ('autochthonous') as a type of lowland peat – a view supported by the many rootlet beds beneath the seams. Nevertheless an alternative 'allochthonous' origin involving a degree of transport and redeposition has some support, for any rooted or cross-cutting structures in the seams themselves are very rare. Several seams, especially in the Midlands, are associated with tonsteins – thin but extensive beds of kaolinitic mudstone, a few actually within the coal itself. Tonsteins may result from soil-forming processes or in some cases be derived from volcanic ash showers, but their wide extent suggests a more uniform sedimentary environment than the afforested swamps where the plants grew.

Conspicuous variations in the unity or thickness of a seam can often be traced to regional differences in subsidence or contemporary erosion. A thick coal sometimes splits into several thinner ones (Fig. 48), the latter resulting from greater subsidence and the periodic influx of detrital sediment. Washouts are local interruptions in a coal and its replacement by a sandstone: the layer of plant detritus has been eroded by a stream and its channel filled with fluviatile sands. The sinuous course of such a stream wandering across the swamp flats can sometimes be traced by mapping the washout in a given seam. The more massive lenticular sandstones were built up similarly by larger distributaries which shifted their position from time to time.

In spite of all these local variations a country-wide view of the Coal Measures emphasizes its uniformity. Many seams can be identified over hundreds of square kilometres and individual marine bands, with their characteristic species, over even wider areas. Those that define the stratigraphical divisions in Fig. 45 can be found, under different names, in nearly all the British coalfields and most of those of northern France, Belgium, the Netherlands and Germany. This great areal spread of what are, relatively, very thin beds is the clearest testimony to the impressive uniformity of the late Carboniferous swamplands; moreover this was only part of the great Laurasian continent. On the American side the Pennsylvanian Coal Measures of the same age show many similar characters. During lower Westphalian times only a small rise in sea-level (probably eustatic, though this is difficult to prove) drowned these lowlands over enormous areas. In upper Westphalian times, in north-western Europe as a whole, these incursions ceased.

During the Hercynian orogeny the Coal Measures were folded, uplifted and eroded. The existing coalfields are the remnants that have escaped that erosion and most of them have a synclinal or part-synclinal form.

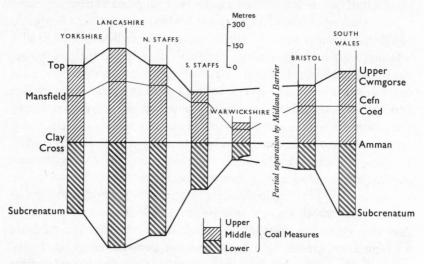

Fig. 45. Diagram showing the thickness of Lower and Middle Coal Measures in major coalfields and their correlation by marine bands. Of these Subcrenatum is the current international name; those above are Vanderbeckei, Aegiranum and Cambriense. (Adapted from several sources.)

Many of these structures follow inherited trends, especially Caledonoid and Charnoid, and in detail their complexities reflect the incompetent nature of much of the rocks. Originally they were covered by Permian and later strata which have since been gradually removed. Thus some coal-fields are open, with no cover, some partly concealed, and a few still wholly concealed.

THE GREAT NORTHERN COAL BASIN: THE PENNINE COALFIELDS

The several coalfields of the Midlands and north of England differ considerably in size and structure, but are linked together by being remnants of a single area of swamp flats stretching from St George's Land and the Midland Barrier to the Southern Uplands of Scotland. The continuity of certain major marine bands shows, however, that these residual uplands must have been low and incomplete. The three Pennine coalfields – North Staffordshire, Lancashire and the East Pennine (Yorkshire plus Nottinghamshire and Derbyshire) – lie near the centre of the basin; the East Pennine is the largest and has the greatest output of any field in the country.

Structure

There is a general tendency for the beds to dip away from the Pennine range, with fairly simple structures on the east but more complex ones on the west. From Yorkshire to Nottinghamshire the exposed measures form one side of a gentle rather irregular basin, the concealed portion of which extends far to the east (p. 230). Subsidiary folding and faulting follow both north-easterly and north-westerly trends, the latter being Charnoid and including concealed anticlines known from oil exploration in the East Midlands. Fault displacements rarely exceed 150 metres. The eastern part of the Lancashire Coalfield is governed by steep westerly dips and a series of large compensating faults; these trend north to north-east, with easterly downthrows of 300 metres or more. In the northern part the broad faulted Rossendale Anticline brings up Millstone Grit between Burnley and Manchester. The generally steeper dips west of the Pennines mean that the Coal Measures are carried deep below the Cheshire plain in a relatively short distance and less of the concealed coalfield is workable than in Yorkshire. Where the productive measures appear from beneath the Trias in North Staffordshire the dominant structure is a syncline trending north-north-east and plunging southwards.

Productive Measures: Westphalian A, B and lower C

The Lower and Middle Coal Measures in the three Pennine fields are all similar and in many ways the East Pennine Coalfield serves as a standard succession for the country as a whole. The base is conformable on the Namurian Millstone Grit and though the *lenisulcata* Zone is poorer in economic seams than those above the rock types are normal and there are several marine bands. For the remainder of Westphalian A and much of Westphalian B marine bands are sparse, the Vanderbeckei (Clay Cross or Sutton Manor) at the base of the Middle Coal Measures being an exception. There is a return of marine invasions higher up, including the widespread Aegiranum (Mansfield, Dukinfield, Gin Mine), and the Middle Coal Measures terminate with the Cambriense (Top, Lady) Marine Band in the middle of Westphalian C. The thickest seams such as

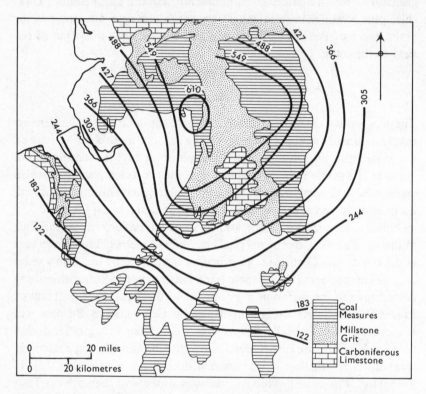

Fig. 46. Pennine coalfields and neighbouring areas. The isopachs of the *modiolaris* and Lower *similis-pulchra* zones show the region of greatest thickness to be near Manchester; originally contoured in feet, here in metres. (Adapted from several sources.)

the Barnsley or Top Hard of Yorkshire reach three metres. Compilations and isopachs of the productive measures (Figs. 45, 46) show that the beds are thickest 1500 metres, in Lancashire, somewhat less in Yorkshire and fall to below 1000 metres in Nottinghamshire. The centre of the subsiding basin therefore probably lay in east Lancashire.

Sedimentary cycles have been described especially from the Middle Coal Measures of Nottinghamshire and Derbyshire, where nine cyclothems are found in some 200 metres of strata. The ideal cyclothem is as follows:

coal
seatearth
sandstone
siltstone or silty mudstone
pale grey mudstone with shells
dark grey to black shales with marine fossils
(coal)

This sequence represents the inundation of the coastal flats by the sea, and after an initial deepening a gradual silting-up until the forests recolonized the land surface. It is, of course, only a theoretical example; other versions have been recorded in South Wales (p. 222). In the East Pennines it is common to find the coal or the marine episode missing, or the upper part expanded by thick lenticular sandstones. The Mansfield Marine Band is the thickest, reaching 10 metres or a third of the total cyclothem, and it contains a large variety of marine invertebrates, both swimming and benthonic forms.

Upper Coal Measures: upper Westphalian C and Westphalian D

More varied conditions and sediments are represented in these later beds, which include red and grey measures with thin coals only in the latter. The fullest succession is in North Staffordshire where four lithological divisions, or successive facies, have a maximum thickness of 780 metres. Stratigraphical fossils become increasingly rare upwards but the lower two divisions belong to Westphalian C and the upper two probably cover much of Westphalian D:

Keele Formation, brown and red sandstones and mudstones
Newcastle Formation, grey sandstones and shales with occasional thin limestones
Etruria Marl Formation, red and purple marls or mudstones
Black Band Formation, grey measures with thin coals and ironstones

The lowest division is only definable locally but there is a broad alternation between grey measures with plants and bivalves and those that introduce red or brownish muds. This gradual change to oxidizing conditions affected all the northern coal basin but the onset was irregular; even within this coalfield the facies boundaries are almost certainly diachronous.

In Lancashire the situation and facies are comparable: the Ardwick Formation of the Upper Coal Measures consists of primary red measures similar to the Etruria Marl, but the exposures are poor and the beds dip steeply under the Permian cover. In addition secondary reddening is conspicuous in some parts of the coalfield, affecting all levels even down to the Millstone Grit. The thickness of red beds, including both types, locally exceeds 300 metres. East of the Pennines the Upper Coal Measures are largely hidden under the Permian cover and probably do not rise above Westphalian C. Coals are few and sandstones prominent; the concealed uppermost beds include some primary red sediments.

COALFIELDS OF THE NORTH-EAST AND NORTH-WEST

The two major fields in this region, of Whitehaven in Cumbria and of Northumberland and Durham, have a good deal in common but do not add very much to the broader picture already presented. In both the upper and lowermost measures are poorly developed.

In Durham and Northumberland an irregular syncline trends north-north-east, but its eastern limb is largely under the sea or, in the south, under a Permian cover. The 600 metres of productive measures are largely in Westphalian B, the lower and higher beds containing more sandstones. A few major marine bands can be correlated with the Pennine sequence (Harvey = Clay Cross; Ryhope = Mansfield) but the Upper Coal Measures are small in area. The main seams are now increasingly worked in the south-east, chiefly under the Permian (beneath which there is an extensive zone of reddening) and also beneath the North Sea. Exploratory boreholes have been sunk several kilometres offshore.

The exposed Whitehaven Coalfield forms a strongly faulted strip on the coast north of St Bees Head and is part of the Carboniferous rim to the Lake District (Fig. 25). Again the Lower and Middle Coal Measures are thin (450 m) but normal; however the great bulk of the Upper Coal Measures, up to 300 metres, includes red and purple mudstones with rare freshwater shells and thick sandy beds with secondary reddening. Round the northern outcrops this colouration falls in level until it affects all the Westphalian strata and the coals are destroyed. In the north-east even Dinantian rocks are affected.

The Ingleton Coalfield of west Yorkshire, on the south side of the Craven faults, is small in extent and economic value but helps to fill in the long gap between the Pennine and northern coalfields. The productive measures are thin (400 m) and are overlain unconformably by the familiar red Upper Coal Measures, somewhat thicker and extending up into Westphalian D. The latter are dominantly sandy and locally coarse, but also include shales with plants, rootlet beds and bivalves.

SOUTH AND WEST MIDLANDS AND NORTH WALES

Considerably greater variety is found in the several coalfields (Fig. 47) that lie in a broad arc along the northern margin of the Midland Barrier and St

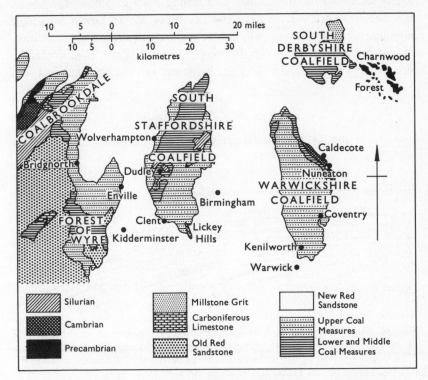

Fig. 47. Geological sketch map of the Midland coalfields and neighbouring areas.

George's Land, from Leicester to North Wales. They are separated, and some still partly covered, by Permian and Trias and reflect the rather irregular foundations that underlie this southern edge of the great northern coal basin. As in North Staffordshire, Upper Coal Measures are thick and there are additional problems in defining the base of the Permian above.

Structure

In the Midlands the coalfields tend to protrude through the Trias as some kind of complex horst on which the Coal Measures lie with their own folding and faulting. In a very general way the dominant strike swings round with the arc of the outcrops. In the east it is north-westerly, or Charnoid, affecting the joint coalfields of Leicestershire and South Derbyshire. North–south trends are conspicuous in Warwickshire and South Staffordshire, the former conforming most closely to the classic 'syncline on a horst'. In the latter the measures rest on a Lower Palaeozoic basement, chiefly Silurian, which is brought up in sharp local anticlines. On the margins of the Welsh Borders (Forest of Wyre, Coalbrookdale, Shrewsbury) Caledonoid strikes predominate, as would be expected.

Lower and Middle Coal Measures

The succession in the Midlands is broadly similar to that in North

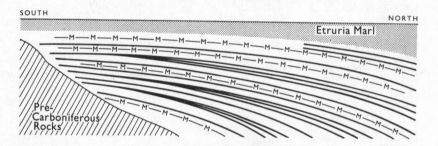

Fig. 48. South Staffordshire Coalfield. Diagram illustrating the splitting of coal seams to the north and general thinning to the south. In that direction the Etruria Marl facies appeared earlier while peat swamps persisted in the north, forming the two top seams. (From Hains and Horton, 1969, p. 40.)

Staffordshire but the productive measures are always thinner because subsidence was less near the basin margin. Minor variations on this pattern can be seen in the influence of local structures. For instance there is a marked easterly thinning from the South Derbyshire to the Leicestershire field in the direction of the Precambrian of Charnwood Forest. In Warwickshire the lowest zone recognizable is *communis*, lying on Cambrian rocks. Irregularities are particularly apparent in South Staffordshire where the Silurian ridges coincide with thin or even absent seams and the measures are banked up against them.

The famous Thick Coal of South Staffordshire (Westphalian B) is equivalent to over 60 metres of strata to the north, with three named seams, and itself becomes thinner to the south. A similar pattern of splitting and thinning affects the Thick Coal of Warwickshire. The top and bottom of the productive measures show this diminution most clearly, the lowest seams failing southwards and some of the highest being replaced by the diachronous base of the red beds (Etruria Marl) above (Fig. 48). In North Wales (Clwyd) the much faulted Coal Measures plunge steeply under the Trias to the east. They have links with both Lancashire and North Staffordshire and are probably continuous with the measures there under the thick cover of the Cheshire basin. Similarly there are deduced to be concealed measures between the south Midland coalfields, eastwards into Leicestershire and southwards to north Oxfordshire.

Upper Coal Measures

These also resemble those of North Staffordshire in being alternations of red and grey beds, but the diachronous boundaries already foreshadowed in that coalfield become rampant farther south. Although the same facies names have been used – Etruria Marl Formation, Newcastle (or here Halesowen) Formation and Keele Formation – with the rarity or absence of reliable stratigraphical fossils it is doubtful if they have any correlative value. In Warwickshire thick red beds with conglomerates (the Enville Formation) come in above and underlie the presumed Lower Permian breccias, but even here no firm indication of Carboniferous horizons higher than Westphalian D has been found.

As in the lower measures the total thickness diminishes towards the margins and red facies tend to replace grey; the seams disappear though sometimes their seatearths can be traced. Another widespread effect is the strongly transgressive base of the Upper Coal Measures, seen for instance in the Shrewsbury and Coalbrookdale fields and farther south in the Forest of Dean and Oxfordshire (p. 222). Thus the long-inherited boundaries, or modest uplands, had become much reduced and ceased to be effective barriers; late Westphalian sediments overrode them. As the Hercynian orogeny and uplift proceeded some were rejuvenated but in a different environment and producing different detritus.

SOUTH WALES, FOREST OF DEAN, BRISTOL AND RADSTOCK

As in earlier Carboniferous times the south-western outcrops were part of a distinct province, which probably extended eastwards to link up with the underground Kent Coalfield and, in Upper Westphalian times, with underground measures in Oxfordshire (p. 228). The southern margin is obscure; the very different Lower Westphalian facies in Devon has already been described and in later times there is some suggestion that a ridge or upland developed on the site of the Bristol Channel.

Structure

The South Wales Coalfield is divided by Carmarthen Bay into two unequal portions, the main basin and that of Pembrokeshire (south Dyfed, Fig. 22). The highly incompetent measures of the latter are overfolded and overthrust as a result of pressure from the south. Most of the rocks exposed are lower Westphalian and, with many fault-gaps, can generally be correlated with those of the main basin.

The main coalfield is an east–west syncline (Fig. 21) with low dips on the north and higher ones, up to 45°, on the south. Superimposed on this structure is a complex system of folds and faults or fault belts. The most important of the last is the Neath disturbance which cuts the coalfield in two, extending north-eastwards from Swansea Bay, and is one of several that follow a Caledonoid trend. East–west and particularly north-westerly faults are also prevalent, the latter commonly transcurrent; one result of this structural complexity is that the Upper Coal Measures occupy a large but very irregular outcrop in the centre and southern sections. In detail the Lower and Middle Coal Measures are intricately deformed, much more so than in the northern coal basin.

Farther east lies the Usk Anticline with its Silurian core, and beyond that the Forest of Dean Coalfield is a relatively simple syncline. Both folds have a north–south trend in line with the Malvern structures farther north. The Bristol Coalfield (Fig. 24) is similar, but that of Radstock (partly concealed under Trias and Lower Jurassic) is much more complex. It is dominated by great east–west fractures and the southern limb is overturned and overthrust, as the block of the Mendips has been pushed northwards against the less competent Coal Measures. These structures form part of the Hercynian front in South-West England.

Succession

The most fundamental character of this province lies in the upper measures, which contain in the lower part thick arenaceous incursions (the 'Pennant Sandstones' of an earlier classification) and in the upper part a return to productive measures with coal seams. Swelled by the sandstone groups the upper Westphalian is exceptionally thick in the south-west though the output and value of the seams as a whole is less than in the north. Some 2400 metres of Westphalian strata are present in the Swansea region of South Wales.

The Lower and Middle Coal Measures contain the majority of workable seams, which reach 2.5 metres in the Radstock Coalfield and considerably more in South Wales, where in the east thicker seams result from the amalgamation of thinner ones in the west. The rock types are similar to those elsewhere, though the *lenisulcata* Zone is sandy like the uppermost Namurian below; various versions of cyclic sequences resemble those of the northern basin.

Marine bands are plentiful in South Wales and their distribution and fauna give some support to the postulated direction of more open sea to the west. In the *lenisulcata* Zone there may be as many as five marine levels in 30 metres of strata; the Subcrenatum Marine Band at the base reaches 15 metres, whereas it is only a metre or two in the Pennines. The faunas are also unusually varied, particularly in the south; a 15-centimetre shale in the Cefn Coed Band yielded some eighty marine forms including such rarities (for the Coal Measures) as crinoid ossicles and a zaphrentid coral. The Lower and Middle Coal Measures are thickest towards the south in Gower and at Swansea and thinnest in the north and east, in Gwent; there is a comparable northerly thinning in the Bristol Coalfield. These differences recall those of earlier Carboniferous times when subsidence was also greater in the south and the Usk Anticline exercised a positive effect in the east.

The Pennant sandstones of Wales are typically coarse, feldspathic, often massive and cross-bedded, thick wedges of ill-sorted fluviatile sands; they include pebbles of shale, coal and ironstone, indicating lithification and erosion of the beds below. However, in contrast to the lower measures in which the current directions suggest a northerly source, as might be expected, the Pennant detritus was apparently derived mainly from the south, and the mineral content supports this. Hence the deduction of a ridge somewhere on the site of the Bristol Channel. This change in geography more or less coincides with others – the cessation of marine incursions, the widespread red facies and extension of deposits in the Midlands.

Like many other Westphalian facies the Pennant sandstones are not uniform in their character or onset. Even where typical and thickest (over 1200 m near Swansea) the measures still include thin coal seams. A

'Pennant' cyclothem can be detected, in which the median sandstone component is enormously thicker than the remainder. Eastwards there is a gradual change, both the sandstones and total thickness decreasing. At one point on the eastern margin the total thickness of measures is probably less than 150 metres. Much of this decrease may be attributed to the persistent elevation, or non-subsidence, of the Usk axis so that very little permanent sediment remained, but minor breaks contributed also. After the Pennant sandy incursions, forest swamp conditions were re-established so that productive measures reappear with several workable coal seams. Red beds are much less common than in the Midlands but occur locally at more than one level in the Upper Coal Measures.

In the Bristol and Radstock coalfields the sedimentary pattern is similar, with some thinning to both north and west; red beds are prevalent in the uppermost beds at Bristol. The thickest and most complete sequence is at Radstock, comparable to that near Swansea. However in the former region there is the unconformable Mesozoic cover and in the latter the upper surface is one of erosion so there is no direct information on how long the coal swamps lasted. The floras suggest that the uppermost beds in South Wales and the Forest of Dean may just extend up into the lowest Stephanian Stage.

The Forest of Dean is in part like those fields just described, but being farther north also has links with the Welsh Borders. Lower and Middle Coal Measures are largely lacking, but the Upper Coal Measures consist of a lower sandy sequence ('Pennant') and an upper ('Supra Pennant') of productive measures. The former lies with conspicuous unconformity on everything from Lower Old Red Sandstone in the east to Drybrook Sandstone (Dinantian) in the west – yet another transgressive base to the upper measures.

COAL MEASURES IN SCOTLAND

The Scottish coalfields comprise several separate largely synclinal areas in the Midland Valley and a few small patches outside the faulted rift. The major regions (Fig. 49) are the Ayrshire Coalfield, large and much faulted in the south-west; the large Central Coalfield in the middle and the much smaller Douglas field near the Southern Uplands; and in the east the deep Midlothian basin and the Fife Coalfield, which join up under the Firth of Forth. None has any cover except the lavas and red sandstones of Mauchline in Ayrshire, both thought to be Lower Permian.

Productive Measures

Owing to the separation and pronounced faulting of these fields, hampering correlation, the Scottish coal seams have been given an exceptionally

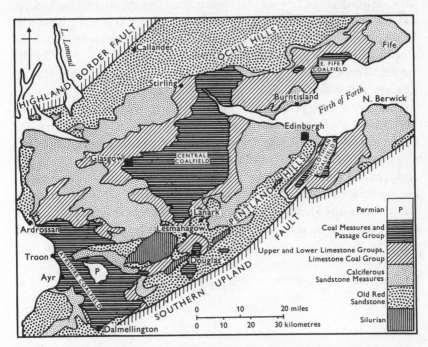

Fig. 49. Geological sketch map of the Midland Valley of Scotland. Igneous rocks are not differentiated but are important among Old Red Sandstone and Lower Carboniferous outcrops (cf. Fig. 42).

large number of local names but some generalizations can be made on the succession as a whole. The coal-bearing strata, sometimes called 'Upper Productive Measures' in contrast to those of the Limestone Coal Group, range through Westphalian A and B. Alone among the major coalfields the Ayrshire output from Westphalian seams is greater than that from the Namurian. Bivalves and plants are fairly common but marine faunas less so and the bands contain *Lingula* more often than goniatites. *Gastrioceras subcrenatum* is among the absentees so that lacking a palaeontological datum line the base of the Coal Measures in Scotland has to be taken arbitrarily in each coalfield.

At the base of Westphalian B (and also of the Middle Coal Measures) the Queenslie Marine Band corresponds to the Vanderbeckei; it is normally a

Lingula assemblage but a small area with goniatites and pectinoids in the
Central Coalfield suggests a temporary deepening there. The upper limit
of the Productive Measures is not wholly satisfactory in that the most
important Scottish incursion is the last – Skipseys Marine Band – which
has a varied fauna of *Lingula*, calcareous brachiopods, nautiloids, bivalves
and goniatites. It is not, however, equivalent to the Top Marine Band of
Yorkshire but to the Mansfield or Aegiranum. This is a level of general
lithological change in Scotland and consequently the base of the locally
defined Upper Coal Measures is drawn at a slightly lower level than in
England and Wales.

There is a general tendency for marine strata to be more abundant in the
west than in the east; for example, the topmost Productive Measures, the
Lower *similis-pulchra* Zone, contains five marine bands in Ayrshire and
only two in Fife and Midlothian. Regional variations in thickness are also
apparent and each of the major fields seems to have been an independent
gently subsiding basin. The maxima include 430 metres in south Ayrshire,
520 in the nearby Douglas Coalfield, less in the centre and Midlothian but
an expansion to 580 metres in Fife. Thicknesses also vary abruptly across
major faults, so that although the controls seem to be inherited from
earlier Carboniferous times the differences were less, and the tendency to
levelling up in Westphalian sedimentation is demonstrated in Scotland as
elsewhere.

Upper Coal Measures

Above the Skipseys Marine Band the coals are few and thin and the plants
and shells much less plentiful, although species of the *phillipsii* and *tenuis*
zones have been found. There are regional changes in thickness, partly
governed by differences in subsidence and partly by later erosion, so that
only an incomplete and variable amount remains. The maximum of 450
metres is again in south Ayrshire, possibly extending up into Westphalian D.

The term 'Barren Red Measures' has commonly been used for this
highest Scottish division, red sandstones and mudstones being typical
though not ubiquitous. However there is a considerable body of evidence
to suggest that much, and perhaps all, of this colour is secondary despite
the great thickness of strata affected; grey beds with poor coals may occur
above the Skipseys Marine Band, as in the Douglas Coalfield, or red beds
well below it. These variations have been studied particularly in Ayrshire,
where in addition to reddening (essentially the oxidation of ferrous iron

compounds with hematite as an end-product) there are other changes attributed to the downward percolation of waters beneath a late Carboniferous or early Permian land surface. Of these the most spectacular is the replacement of coal by carbonate rocks, usually dolomite. Such effects are known elsewhere in Scotland and it was seen how reddening destroyed certain coal seams in the English Midlands.

Outlying coalfields

The principal outcrop on the Southern Uplands is the Sanquhar basin, where nearly all the rocks preserved are some kind of Coal Measures. The Westphalian beds overlap the small amount of Passage Group and limestones (? Limestone Coal Group) onto steeply folded Ordovician. There was thus progressive overlap in Upper Carboniferous times. Essentially the same relationship is present farther south at Thornhill, the other Carboniferous outlier preserved in this depression across the Southern Uplands (Fig. 26), but here the limestones are thicker, probably including some upper Dinantian strata, and there is a partial cover of New Red Sandstone. Elsewhere it is doubtful if the final spread of the Carboniferous swamps ever submerged these uplands completely.

A westward extension is indicated by the Productive Measures on Arran, where parts of Westphalian A and B are known, lying on lavas that resemble those of Namurian age in Ayrshire. All the measures are reddened, lack coals and show signs of overlap among themselves, so that they may be local marginal deposits. In the Machrihanish Coalfield of Kintyre several seams survive in grey measures, though less productive than those in the Limestone Coal Group below; here also there are Namurian lavas.

Two small outcrops much farther to the north-west have been taken as evidence of a wide highland embayment – at the Bridge of Awe, Argyll, and in Morvern, north of the Sound of Mull; the latter outcrops include sandstones, plant-bearing shales and a thin coal, but with no proof of age. At some period in the Carboniferous, however, it seems probable that the peat swamps invaded a low-lying portion of the now reduced Caledonian uplands.

Upper Carboniferous vulcanism: Scotland and elsewhere

To set these volcanic episodes in their wider setting we must hark back to the preceding chapter. There minor igneous activity was noted in northern

England, the Midlands and central Ireland, together with the major outbursts in Scotland. Despite the variations in scale there are some common characters. This Dinantian magmatism was overwhelmingly alkali basaltic and followed mainly contemporaneous fold axes, lines of fracture or the hinges of blocks; it is noticeably absent from the deeper basins. In Scotland and northern England such trends were commonly Caledonoid.

In Upper Carboniferous times this same alkaline activity continued, for instance in the West and East Midlands and as before more widely in Scotland. However although somewhere in the Midland Valley there were eruptions during most of the period the scale was somewhat lessened. During the Namurian the main products were tuffs, as in Ayrshire and particularly in Fife; here many small vents built up impersistent cones, later to be eroded and the volcanic debris spread among the sediments. By Westphalian times vulcanicity diminished further but lavas are known from Ayrshire and under the Firth of Forth, where basalts, tuffs and agglomerates replace much or all of the Lower Coal Measures. A large number of minor intrusions such as plugs and dykes were also formed. Hundreds of volcanic necks are known and with the intrusions tend to be concentrated around the Forth and in the south-west, with a more isolated group of necks north of Glasgow.

Towards the end of the Carboniferous this alkaline activity seems to have been abruptly replaced by a tholeiitic suite that had no extrusive phase. Very many quartz dolerite dykes were intruded along a conspicuously new east–west trend. These are not only widespread in the centre and north of the Midland Valley but (unlike the previous group) extend well into the margins of the Highlands. Akin to these, and locally fed from them, is a major sill, or sill complex, cropping out on either side of the Forth. Both types of intrusion are essentially similar to the Whin Sill complex of northern England, though that covers a larger area and dykes are less conspicuous. The two are probably co-magmatic and the Whin Sill has been dated at 301 ± 5 million years; it seems probable that all these tholeiitic intrusions represent a single short igneous episode with a magmatic source and stress system sharply distinct from the very much longer alkali basalt vulcanism.

The last phase, which just overlaps into the early Permian, is a return to alkali basalt vulcanism. The only major outcrop is the Mauchline Lavas of Ayrshire, associated with minor flows in the Southern Uplands, but intrusions of probable Permian age have also been found in Fife and in the west English Midlands. The Fife example is anomalous but the others are distinguished by their associated trend; this follows neither Caledonoid structures nor the east–west tholeiite dykes but new north to north-westerly lines, which were to become an important factor in the development of the Permian and Triassic sedimentary basins.

COAL MEASURES IN IRELAND

These measures are restricted in space and time and none rises above Westphalian A. The largest coalfield is that of Leinster or Castlecomer, close to the western margin of the Leinster Massif; the smaller Slieve-ardagh field, 32 kilometres to the south-west, is very similar. Both are synclinal and, as in all coalfields in southern Ireland, there is no cover of later rocks. Over 300 metres of Coal Measures are known in the Leinster field, their correlation being clearest at the bottom, with the Subcrenatum Marine Band and the *lenisulcata* Zone. Anthracites and bituminous coals are known and the 60-metre Clay Gall Sandstone is a conspicuous bed, named after its rounded mudstone pebbles. There are some lithological resemblances to both Lancashire and South Wales and the faunas suggest free migration around a reduced St George's Land. The large Namurian outcrop of Limerick and northern Kerry carries two diminutive outliers of Coal Measures, mostly taken up with the Clay Gall Sandstone and beds below.

Much farther north, the Kingscourt outlier (Fig. 31) includes a small amount of Coal Measures, all in the *lenisulcata* Zone; as in the Namurian strata there are faunal associations with the Pennine coalfields. At Dungannon, in Tyrone, the Coal Measures of Coalisland amount to about 300 metres, in lower Westphalian A. There are twelve seams, some reaching nearly three metres. This is a considerably richer sequence than is found elsewhere but the area is small, faulted and lies under a cover of Trias and drift. Its general position, in relation to the Longford–Down Massif, links Tyrone geographically with the Scottish Midland Valley, but the stratigraphical succession more resembles that of the Pennines (with the Subcrenatum Marine Band and the *lenisulcata* Zone), another minor instance of the way that structural and stratigraphical affinities do not always go hand in hand across the Irish Sea.

These few paragraphs on the Irish coalfields form a somewhat melan-choly postscript to the earlier sections describing the extensive and still very valuable Coal Measures of England, Wales and Scotland. Strati-graphically we are grateful for the glimpse we get of a wide westward extension to the tropical forest swamps, coastal flats and shallow inundations of the Silesian lowlands. Economically they are just enough to show what Ireland has lost through the later erosion that stripped those measures all too thoroughly.

CONCEALED CARBONIFEROUS ROCKS OF ENGLAND AND OF THE SEA AREAS

It is convenient at this point to review both Dinantian and Silesian rocks that are known underground in the south and east of England and the small amount of information from the North Sea. As before, the descriptions can be related to the London and East Anglian Platform (the western part of the London–Brabant Massif) and the Precambrian ridge from Leicester to Norfolk.

Around or outside the margins of the London Platform Dinantian rocks have been recorded in a few boreholes in Kent, Surrey and Sussex, and also near Cambridge and Northampton. They do not, as far as is known, make up much of the centre of the platform; the facies is calcareous and resembles that of the South-West Province. Arenaceous facies of Namurian age have not been recognized. Coal Measures are naturally much better known. A few named coalfields are really only underground extensions from old-established ones – such as Cannock Chase (from South Staffordshire) or Selby (from Yorkshire). The Vale of Belvoir is an underground field near the south-eastern end of the Nottinghamshire–Derbyshire Coalfield; the sequence is also similar, though thinner.

The Kent Coalfield is more distant and forms a link with the Silesian in northern France. The lowest Coal Measures rest on Carboniferous Limestone or, to the north, overlap onto the Silurian edge of the London Platform; the structure conforms well to the theoretical buried syncline (Fig. 50), the axis of which plunges eastwards towards the Straits of Dover. Both facies and fauna of the 900 metres of Coal Measures resemble those of Bristol, but there are some absences and in addition to the Namurian unconformity there is probably a gap in Westphalian C. The upper measures include a Pennant-like sandy group, raising an unanswered query about the source of detritus here.

There is no other quite comparable concealed coalfield but boreholes have proved a large area of Upper Coal Measures underlying parts of Oxfordshire, certainly extending as far north as Banbury and probably joining up with the southern part of the Warwickshire Coalfield. The measures overlie a Silurian or Devonian basement and in the south are related to the upper beds at Bristol; an arenaceous group, again like the Pennant sequence, is overlain by mudstones with several workable seams of over 1.5 metres. The total thickness penetrated is about 400 metres.

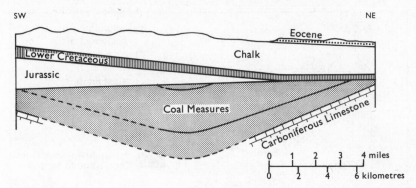

Fig. 50. Diagrammatic section through the Kent Coalfield, showing the synclinal structure and the overstepping Mesozoic cover; vertical scale not given. (After Trueman, 1954, p. 156.)

The rather thinner continuation to the north consists mainly of grey beds overlain by red and has hence been compared with the Halesowen and Keele facies (p. 219) of Warwickshire; thin coals are present in both. These Oxford measures have only low dips and little deformation and must therefore lie north of any Hercynian front.

It thus appears that in later Westphalian times (upper C and D) the increasing spread of grey and red measures with their forests and coal accumulation stretched widely across the Midland Barrier; or to put it in another way this became the westerly termination of the London–Brabant Massif, now finally separated from the rather more shadowy Welsh Massif or residual St George's Land. The break was emphasized in Permo-Triassic times by the deep north–south depression of Worcestershire and the lower Severn – this possibly initiated by early Permian faulting.

In eastern England (Fig. 51) the large concealed portion of the East Pennine Coalfield extends beyond the coasts of Lincolnshire and York-shire, south of Flamborough Head, and more irregularly to the north Yorkshire coast; the valuable Selby field lies between York and the outcrop. Probably Namurian and Dinantian underlie the Westphalian but only a few boreholes have reached them; where they rise from beneath the Coal Measures to underlie Permian rocks the pattern suggests gentle folds, comparable to those in the exposed outcrops. Lower Carboniferous beds are present in an anticlinal belt from Middlesborough to south of the Tees. They are largely of basin facies, with much shale and thin limestone and it seems likely that the eastern margin of the Askrigg Block is not far east of

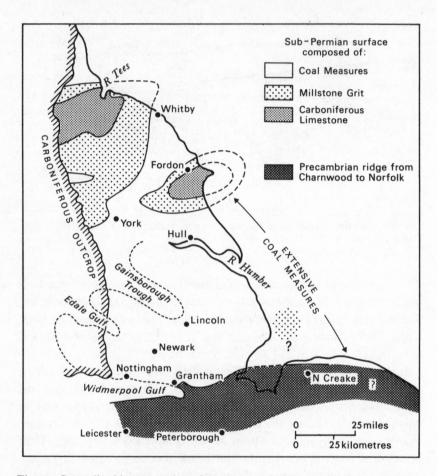

Fig. 51. Generalized interpretation of the rocks underlying the Permian and Trias from Teesside to Norfolk, together with three subsurface 'troughs' in the East Midlands. (Adapted from several sources.)

the present outcrop. A faint suggestion of yet another easterly block is found in a small amount of Yoredale-like beds in a borehole inland from Flamborough.

Farther south there are two components to the subsurface stratigraphy – the concealed Coal Measures themselves and an earlier pattern of blocks and troughs (or gulfs) that has much in common with the structures seen in the exposed outcrops of Carboniferous Limestone and Millstone Grit. These two components are partly related, however, and where the earlier groups expand into the gulfs there is commonly a thickening in the Coal Measures as well.

The three gulfs shown in Fig. 51 are chiefly defined by an expansion of the Millstone Grit, which is over 600 metres in each, compared with much thinner sequences on the intervening shelves. The Gainsborough Trough is a shallower and narrower continuation of the Craven Lowland Basin; similarly the Edale Gulf extends south-eastwards from the basin facies north of the Derbyshire Block. The Widmerpool Gulf is the most extensive, lying south and south-east of that block, and illustrates the thickness changes particularly well. From 90 metres on the northern flank (resting on D limestones), the Millstone Grit expands to over 800 metres in the gulf and is underlain by thick shales. Few boreholes have penetrated the Lower Carboniferous completely, but at Eakring, near Newark, over 900 metres of limestones (C_1–D_2) with sandstones and conglomerates are underlain by over 520 metres of red conglomerates; the latter are probably Carboniferous basement beds, but might be Old Red Sandstone.

Igneous rocks have been found, especially along the east–west fracture belt that borders the Widmerpool Gulf on the north; here 250 metres of altered basalts replace part of the Millstone Grit and most of the Lower Coal Measures. It was probably from similar sources that the thin ash beds (bentonites) were derived that have been found among Namurian and Westphalian sediments in the Midlands. Some of the Lower Carboniferous eruptions seem to have been submarine but later on they became subaerial. At least in the East Midlands the thick tuff beds known underground suggest that numerous small ash cones were built up above the shallow waters or wet swamps of the coastal lowlands. The rock types and eruptive mechanisms are very similar to the cognate activity in Scotland, but in England it was on a smaller scale and the extrusive products are more scattered.

Apart from a small number of coal seams worked close inshore, Carboniferous rocks under British seas have little economic value and hence are only known in certain situations, or somewhat by chance. There are some simple continuations from land outcrops, such as those in the northern part of the Irish Sea (including some Coal Measures) and in the southern part, and off the western Irish coasts. Similarly the central Devon synclinorium, with its folded Carboniferous, is known to extend westwards from the Cornish and Devon coasts. Such submarine data, however, do not add much that is fundamentally new to the land-based picture.

In the North Sea the situation is rather different, since Silesian and much more occasionally Dinantian strata have been penetrated below the important Lower Permian sands. The strong similarity to the continental coalfields, in the Coal Measures especially, had long ago suggested that there was continuity from England to Belgium, the Netherlands and Germany and this has proved to be correct. On the British side the Westphalian is thick (900 m or so penetrated east of Yarmouth) and some Stephanian comes in to the east, in Dutch and German waters. This spread of coal-bearing strata is presumably the source of the gas reserves now trapped in the sands above.

Carboniferous rocks are thought to overlie the Mid North Sea High (p. 270, Fig. 58) only here and there, perhaps comparable to those of the Southern Uplands, but some are known in the Forth Approaches Basin, related to the eastern end of the Scottish Midland Valley. Few records penetrate so deep in the northern North Sea but it is likely from wider considerations that there was still a Caledonian upland in much of this area, part of the source-lands of the great deltaic spreads of Millstone Grit and Coal Measures in northern Europe.

THE HERCYNIAN OROGENY: EFFECTS AND PROBLEMS

In Britain Hercynian effects have been noted especially in South-West England and, in the more stable regions to the north, in the structures of the coalfields. The accompanying uplift resulted in a long period of erosion so that the latest Carboniferous rocks are but doubtfully present and the earliest Permian decidedly rare.

The full orogenic panoply is only present in the south-west, where the stratigraphical break extends from the latest recognizable measures (probably Westphalian C) to the basal post-orogenic breccias and the Exeter Volcanic Group (p. 259). The latter lies approximately at the Carboniferous–Permian boundary, so that within this gap the complex folding and thrusting took place (? late Westphalian), with some cleavage in the finer grained rocks. Then followed the intrusion of the granites; from Dartmoor south-westwards these crop out as far as Land's End and make up the Scilly Isles. A submarine outcrop, Haig Fras, about 100 kilometres west-north-west of the latter, belongs to the same group. Geophysical evidence strongly suggests that the mainland intrusions are upstanding domes, or cupolas, of a single elongate granite batholith.

The granite emplacement was preceded by some minor basic intrusions and succeeded by acid bodies, such as aplites and pegmatites. There is much contact metamorphism and an important belt of mineralization, the veins cutting both granite and country rock. An average age determination for the various granites is about 300 million years, or early Stephanian. Consequently in this, the type area for Britain, the Hercynian orogeny was almost or entirely completed with the Carboniferous Period.

It would be logical at this point to analyse the Hercynian orogenic belt in terms of plate tectonics in the same way that an attempt could be made for the Caledonian belt. But although the orogeny is younger, the Upper Palaeozoic rocks commonly well known and metamorphism not ubiquitous, a satisfactory synthesis at the present time has not really been achieved. The remnants of the belt extend from Poland through middle Europe to Brittany and southern Ireland, but on the American side there is no obvious continuation comparable to Newfoundland and the northeastern Appalachian counterparts to the Caledonian structures. Much farther away to the south-west the Alleghenian orogeny of the Southern or Newer Appalachians is of about the same age (Carboniferous–Permian), but the tectonic situation and degree of deformation are different.

In Europe some of the problems are visible even on a simplified map (Fig. 6): the Hercynian massifs are widely separated by a thick later cover and in the south are overrun by the Alpine folds; among them the structural trends are markedly divergent or arcuate. Though compression was dominantly from the south and more pronounced metamorphism is apparent in that direction (e.g. southern Brittany, the Central Plateau of France and the Vosges), nappes are not on an Alpine or Caledonian scale whereas high-angle faulting is important. Other characters include a sparsity of ophiolites and andesitic vulcanism but extensive granitization.

In this complex situation what of the small British sector (essentially from Cornwall to mid-Devon) and especially the implications of the thin deep water shales, the spilitic lavas and the basic to ultrabasic intrusions? Two principal views have emerged. In the first these are interpreted as part of an ophiolitic suite with the deduction that a Hercynian subduction zone lay across south Cornwall, or between there and Brittany. In the alternative view the igneous rocks are considered to be much nearer alkali basalts than abyssal or oceanic tholeiites and so present no evidence for nearby oceanic crust or a subduction zone; it is further postulated that the latter lay far to the south, possibly within the Alpine belt, at the margin of the Tethys ocean, so that the Hercynian belt and northern Europe were essentially part of the same plate. There are variants on both themes and the disrupted heterogeneous Hercynian massifs, with their diverse structural trends, may have developed as microplates, variously rotated in the larger northward movement of Africa.

Such major issues are not much clarified by looking eastwards, towards central Europe. As in Dinantian stratigraphy there are strong similarities from England to Belgium and north Germany (the Rhineland and the Ruhr). They are perhaps less pronounced in the Namurian, though that tends to be a barren clastic division in the continental outcrops also; but the Westphalian Coal Measures all along this belt are closely comparable, down to the individual rock types, floras and faunas, and particularly in the detailed equivalence of the marine bands.

Many more complexities enter into the major central European massifs comprising especially Czechoslovakia and its borders, but with cognate features from the Vosges and Black Forest to Thuringia, the Harz Mountains and the Polish frontier. Here it has been maintained that throughout the late Precambrian and Palaeozoic the geological history and abundant magmatism have been wholly intracontinental, with no evidence for remnants of oceanic crust or a Hercynian subduction zone. The case is detailed and well argued but many other geologists would prefer to interpret the belt along more orthodox plate tectonic lines, obscure or controversial though they may be in several regions.

In this temporary impasse we may leave the Hercynian problem, which may prove more tractable in the future, and of which the British sector is only a small part. To the north, with the close of the orogeny, the British Isles and north-western Europe became a relatively stable region and in many ways a simpler one. Carboniferous and early Permian vulcanicity was continental, or cratonic, and the new sedimentary history begins with the sparse coarse clastics of early Permian times.

REFERENCES

Ramsbottom *et al.* 1978 (*Special Report* No. 10, Silesian).
General: Dodson and Rex, 1971; Fitch *et al.* 1970; Floyd, 1972; Francis, 1965*a, b,* 1978; Johnson, 1973; Kelling, 1974; Kent, 1966; Mitchell *et al.* 1978; Ramsbottom *et al.* 1974.
Namurian (Millstone Grit, etc.): Allen, 1960; Collinson, 1969, 1970; Francis, 1961; Jones, 1974; Ramsbottom, 1969, 1977; Walker, 1966.
Westphalian and coalfields: Calver, 1968, 1969; Dunham and Poole, 1974; Edwards, 1951; Francis, 1979; Murchison and Westoll, 1968; Mykura, 1960; Scott, 1979; Thomas, 1974; Trueman, 1954.

British Regional Geology: Bristol and Gloucester District, Central England, The Midland Valley of Scotland, Northern England, North Wales, Pennines and Adjacent Areas, South Wales, South-West England.

The Coal Measures in particular appear in many Memoirs of the Geological Survey. Views on Hercynian tectonics are summarized in Ager, 1975 and Francis, 1978.

8

PERMIAN AND TRIASSIC SYSTEMS

These two systems are linked together in the British Isles by their lithological and environmental characters and in many outcrops by difficulties in determining the boundary between them. Being the second great group of continental strata, they were early called the New Red Sandstone, a name which still has useful applications. The rocks include red, buff and occasionally grey-green sandstones, siltstones and marls, with local breccias and conglomerates and important marine carbonates and evaporites. Although macrofossils are known from some sediments, chiefly marine invertebrates in the carbonates, there are still great thicknesses that are virtually unfossiliferous. Nevertheless the last decade has seen marked advances in sedimentology, stratigraphy and correlation, the latter particularly through palynology (the study of spores which were wind-distributed) and other plant microfossils. These are found principally in the grey or greenish siltstones and mudstones; in the red beds they are rare, probably being destroyed by oxidation.

Lithological marker formations have also been employed, especially in the cyclic sequences of the Zechstein and other saline seas. West of that

Table 32. *Permian and Triassic classification in Central Europe*

TRIAS	RHAETIC or RHÄT KEUPER MUSCHELKALK BUNTER	Relation to Alpine stages is given in Table 36.
PERMIAN	ZECHSTEIN ROTLIEGENDES	

marine facies thick groups of red dune-bedded sandstone may be more or less contemporaneous with it. Pebble beds form correlative markers in the lower Trias of the Midlands and associated with them certain stratigraphical breaks, unconformities or disconformities, appear to have a useful time-significance. Thus although the New Red Sandstone has been a classic instance of few or unreliable correlative characters, a good deal has been achieved by using a combination of many incomplete methods. As a result the major British divisions can be broadly correlated with the type sequence of this continental facies, in Germany (Table 32). The standard marine successions – chiefly Russian for the Permian and East Alpine for the Trias – can be only indirectly compared. Two important marine groups (the large fusulinid foraminifera and the ammonoids) are lacking in this country; the shelly faunas of the marine carbonates do provide some help but are limited to parts of the Zechstein.

After the Hercynian movements there was uplift over Britain and the area became one of varied relief and locally pronounced erosion – part of an arid belt in the great continent of Laurasia in low latitudes, around 10° to 15° N. Even so there was some plant cover, especially in early to middle Permian times, as shown by sporadic plant remains. In addition to the major uplands, especially in Scotland, a low ridge is deduced along much of the Pennines. The London and East Anglian Platform remained a positive element for much of Mesozoic history. The major rock groups become thinner as the platform is approached and the later overlap the earlier, the centre of the platform being finally covered in mid-Cretaceous times. From the platform north-westwards to the southern Pennines there is a belt in the East Midlands where Permian and Triassic deposits are thinner and more variable than in the major basins. This may be related to the irregular and relatively shallow Precambrian and Caledonian foundations which protrude, for instance, in Charnwood Forest and on the margins of some Midland coalfields.

Following the main Hercynian folding a new pattern of sedimentary basins developed, many being tensional downwarps, partly fault-bounded. New trends also appeared, especially in the west. Thus from Avon and the Malverns a northerly system extends into the West Midlands and the Birmingham region; Caledonoid structures stretch across from the Welsh Borders to north Staffordshire, but the very deep Cheshire basin is partly bounded by northerly and north-westerly faults. The latter trend extends into the northern Irish Sea and is evident on the margins of the Lake District in the Pennine Faults, in south-west Scotland and north-east Ireland. Although these structures were probably initiated in early Permian times the sagging effects were prolonged, so that the infilling often includes thick Trias and occasionally later Mesozoic rocks. It is likely that the tensional rifting in the North Sea was a contemporary development.

In the following sections the Permian and Triassic formations are grouped partly by age and partly by outcrop: those of eastern England come first because they are nearest to the type sequences of Germany and most easily correlated with them. From Permian times onward the sea areas become a major part of British geology; accordingly their structures are summarized at the end of this chapter, together with their Permo-Triassic stratigraphy.

THE ZECHSTEIN SEA: UPPER PERMIAN

This sea, in an arid setting and periodically highly saline, occupied much of northern Germany, the Netherlands, south Denmark and Poland, and stretched across the North Sea into north-east England (Fig. 52). In this country its south-western border was the London–Brabant Massif and probably there was a westerly shore somewhere in the region of the Pennines; the outlet to the more open sea was almost certainly to the north.

East of the Pennines the Zechstein sediments comprise a carbonate facies, the Magnesian Limestone (dolomites and limestones), evaporites and subordinate clastic beds, particularly red marls or siltstones. The carbonate–evaporite cycles can be correlated broadly with their German counterparts and the Marl Slate, at the base, has plants and fishes similar to those of the Kupferschiefer. The underlying Lower Permian or Rotliegendes is represented in this region only by impersistent breccias and dune sands, the Yellow Sands. The visible discordance beneath the Permian here is inconspicuous, but along the 200-kilometre outcrop the basal beds rest on anything from Dinantian limestones to Upper Coal Measures and the same diversity is found underground. Early exploration in the southern North Sea was particularly concerned with Permian beds and structures since the Rotliegendes sand is the principal gas reservoir.

There was a marked lateral variation in the Permian deposits as they were laid down in the outcrop areas or farther to the east where they are known from boreholes; there are thus differences in the proportion of clastic components and in the development of carbonates and evaporites. To these original differences have been added a variety of secondary effects. Evaporites, being the precipitates of the more concentrated brines, are highly soluble; consequently in a humid climate they are not found in surface exposures, where both the composition and structure of the rocks may be altered by their removal. Then there are diagenetic changes in the evaporites themselves, whereby one mineral is replaced by another.

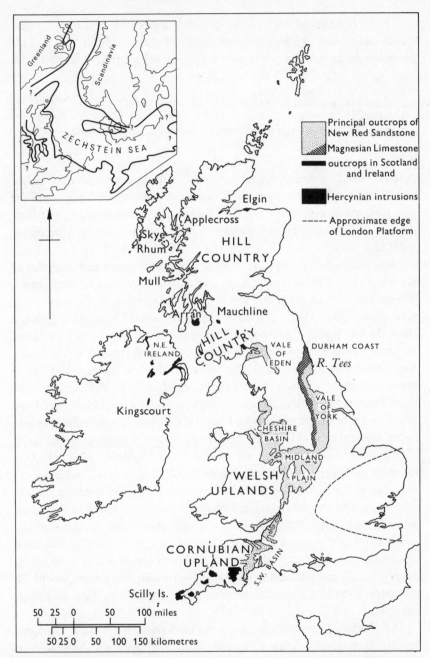

Fig. 52. Outcrops of Permian and Trias (or New Red Sandstone) with Magnesian Limestone (Zechstein) differentiated. Inset: the Zechstein Sea and contemporary neighbouring lands, including Greenland.

In a body of normal sea water evaporating under arid conditions the less soluble salts are precipitated first and the more soluble later. This produces the primary mineral phases of the ideal marine evaporite cycle:

1. calcite or aragonite, $CaCO_3$
2. sulphates, gypsum, $CaSO_4 . 2H_2O$ or anhydrite, $CaSO_4$
3. halite, NaCl
4. potash salts (especially potassium and magnesium chloride) and also polyphalite

A land-locked basin evaporated to dryness will, however, produce only a small thickness of evaporites and this process cannot possibly account for the 1000 metres or more of the Zechstein or other major evaporite deposits. Consequently replenishment from the open sea is a necessary corollary.

The classic concept was the 'barred basin' into which new supplies of sea water flowed from time to time over a lip or bar. Periodic subsidence, or possibly rises in sea-level, would result in a return to the earlier less soluble precipitates and the evaporation cycle would begin again. In some cases the late highly soluble phases were protected by a covering of marl before the reversal took place. In recent years an alternative environment has gained much support – that of very shallow lagoons and saline supratidal flats (sabkhas), such as those bordering the Persian Gulf and other coasts in arid climates. Carbonates, especially aragonite, have been found forming in the lagoons and intertidal channels, and sulphates and halite interstitially among the sediments above tidal level or in small surface pools. Even the more soluble polyhalite $Ca_2MgK_2(SO_4)_4 . 2H_2O$ is known as a diagenetic mineral. The replenishment here is from sea water in periodic floods or blown inland by onshore winds. Algal 'mats' are associated with the carbonates and blue-green algae are found in the highly saline waters and among the halite crystals. Nevertheless, although the components and textures of sabkha deposits have been matched in several fossil evaporite sequences, the former as they are known at present are on a very much smaller scale, and long-continued subsidence would be required to build up several hundreds of metres from such near-tidal precipitates.

The English Zechstein can be exemplified from three areas, Durham, central to north-east Yorkshire, and south Yorkshire and Nottingham-shire. Of these the second is the fullest, especially in the borehole sequences near the coast which compare well with those of the southern North Sea and Germany.

Zechstein cycles in central and north-east Yorkshire

On this western shore of the Zechstein basin the first inundation covered either a bare rocky shore, or a thin layer of scree above the rock pediment (the basal breccias), or yellow dune sands, which became largely redistributed by water: these are the main residues of Lower Permian age. Thereafter four evaporite cycles developed (Table 33) with the initiation of

Table 33. *Zechstein cycles in north-east England*

County Durham	North-east Yorkshire	Central Yorkshire
(Lower Mottled Sandstone or Bunter Sandstone)		
	Saliferous Marls	
Permian	Cycle 5	Upper Permian Marls
Upper Marls	Cycle 4	
	Cycle 3	Upper Magnesian
Upper Magnesian Limestone		Limestone
Middle and Lower	Cycle 2	Middle Permian
Magnesian Limestone		Marls etc.
(with reef facies	Cycle 1	Lower Magnesian
and anhydrite)		Limestone
Marl Slate	Marl Slate	Yellow Sands
Yellow Sands and breccias		
(various Carboniferous strata)		

a fifth which was not completed. Above the first carbonate phase of each the evaporite salts thicken enormously eastwards, and subsurface, into the basin (Fig. 53). In addition to their absence in the outcrops to the west through subsequent solution, there was also an original facies change towards the contemporary shore. At outcrop, therefore, the Upper Permian of mid-Yorkshire has traditionally been referred to as:

Upper Permian Marls
Upper Magnesian Limestone
Middle Permian Marls
Lower Magnesian Limestone (with locally Lower Marls)

Of these the Lower Magnesian Limestone belongs to the first cycle and the Upper to the third; reefs and reefy patches occur in the former, similar to the more extensive examples in Durham.

The evaporites come in eastwards, replacing the marls in particular, and the Upper Permian Marls pass into the very detailed sequence of cycles 3, 4 and 5. Of the various formations of the basin, the thickest are the Hayton Anhydrite (145 m) of the first cycle and the Fordon Evaporites

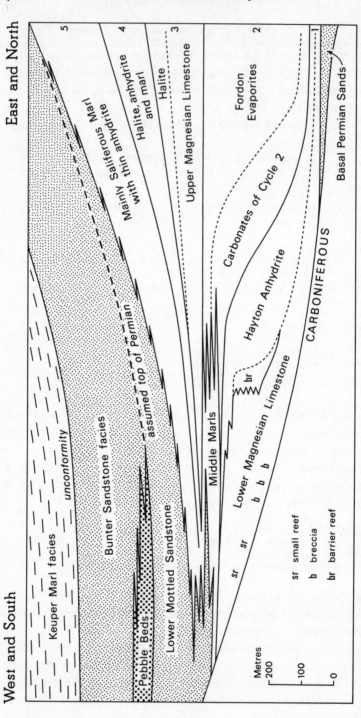

Fig. 53. Generalized cross-section through the Permian and Trias of Yorkshire. The Permian sequence to the north-east is only known underground and shows the great expansion into the basin. The English Zechstein cycles 1–5 are numbered. (Adapted from Smith, 1974, p. 122.)

(over 300 m) of the second – the latter including anhydrite and the late phase polyhalite as well as abundant halite. Units in the later cycles are thinner but some are very persistent, for instance the 'Upper Anhydrite' of cycle 4 remains remarkably consistent across the North Sea to Germany.

The maximum known thickness of the Zechstein within the British Isles is around 600 metres, with more offshore. In the last stages the basin had probably become very shallow and then fine red silts and muds (the Upper Marls) spread far and wide, with interstitial evaporite salts and signs of emergence. There followed a dominantly sandy sequence, with intermittent fluviatile deposition on a vast coastal plain; this has variously been called the Lower Mottled, or 'Bunter', Sandstone (or Sherwood Sandstone Group of Table 36) and is believed to span the Permian–Triassic boundary.

Facies change towards the shore, north and south

In south and east Durham the exposures are very largely of Magnesian Limestone, as in Yorkshire overlying breccias and yellow sands. Above the latter the first Zechstein marine formation is the Marl Slate – a fissile silty dolomitic rock, with plants, plant debris and fishes. Together with the abundant shelly fauna of the Lower Magnesian Limestone it shows that at this stage the incoming Zechstein Sea was not highly saline. The Lower and Middle Magnesian Limestones (of Durham terminology) both belong to the first cycle and the Middle is outstanding for its barrier reef, here seen exposed, though also known subsurface in Yorkshire. Much of the reef consists of a shelly dolomite with brachiopods, gastropods and crinoids held in a framework of bryozoans; towards the top algal growths appear, and the whole terminated in a broad reef-flat, possibly intertidal. There was a gradual migration of reef growth towards the basin and an offshore facies, found in boreholes, includes 160 metres of anhydrite.

The Upper Magnesian Limestone barely extends into the base of the third cycle. It includes oolites and a variety of concretionary and collapse structures – the last two probably representing the solution of dolomite, and possibly of sulphates, in the complex beds below. Algal growths (stromatolites) are abundant at the top, with evidence of sabkha conditions, intertidal or supratidal. The western shore was not far away.

A traverse from central Yorkshire southwards shows another shoreward approach but rather differently, since the facies change is accompanied by a marked thinning of all the beds traditionally ascribed to the Permian (Fig. 54). Naturally the evaporites are the first to go. In the Hayton

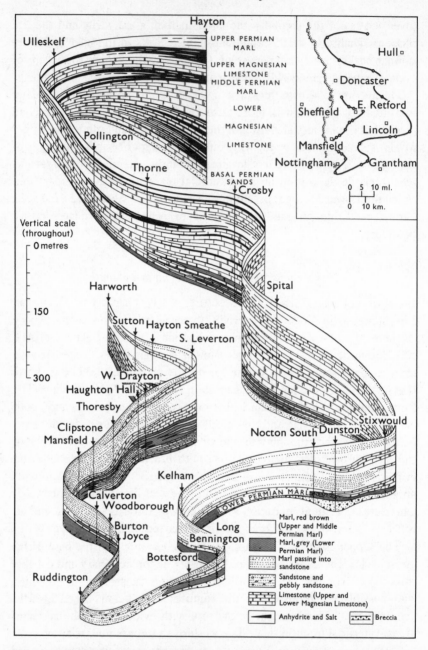

Fig. 54. Ribbon diagram of the Permian rocks of south Yorkshire, Lincolnshire and Nottinghamshire, to show the facies change and the lateral passage into their sandy beds in the south. (From Edwards, 1951, p. 97.)

borehole between York and Hull there are substantial beds of anhydrite in cycles 1 and 2. These have largely disappeared in south Yorkshire, where a lesser amount of marls replaces the evaporites. The Upper Magnesian Limestone tapers out in Nottinghamshire and only a very thin lower bed is present at Nottingham itself, to disappear a little farther south. All that remains is a thin sandstone of 'Bunter' type, the base of the Trias here being formally taken at the base of the pebble beds above.

PERMIAN RED BEDS IN THE MIDLANDS

Within and around the Midland coalfields there is a variety of red beds, commonly with breccias or sandstones, that are overlain by the widely transgressive Bunter Pebble Beds and in places themselves overlie those formally adopted as uppermost Carboniferous (p. 219). The uplands here were subjected to arid or semi-arid erosion and in the depressions between the spurs there accumulated angular screes, pebbles and sands, the finer components spreading out onto the lower ground to the north. Occasional sheets of water evaporated to give impure carbonate rocks. To the south rose the irregular hill country of the Mercian Highlands – a rejuvenated version of the earlier Midland Barrier.

Beds formed in such conditions naturally show much local variation in thickness and in the composition of the breccias. Fossils are almost absent, for life must have been sparse and conditions of preservation exceedingly poor. These red beds have received many local names around the coalfields, correlation between them is by no means certain and it is likely that there are gaps in some sequences.

The Clent, Enville and Kenilworth Breccias

Several variable formations have some similarities here, though they may not be strictly contemporaneous. The thick breccias are characterized in particular by a dominance of Precambrian volcanic rocks, the angular chips and boulders often being coated with hematite. For their origin it is likely that more Precambrian outcrops existed in Permian times than are seen in the Midlands today. The Clent Breccias, locally very coarse, amount to 90 metres in the Clent Hills near Birmingham.

The two Kenilworth breccias of Warwickshire are much thinner and accompanied by finer beds; probably they are attenuated representatives farther from the main source of detritus. A very few fossils are known from this region, including an amphibian skull and reptilian bones. The conifer *Walchia* (or *Lebachia*) is also known; on balance these remains tend to support a Lower Permian age. Tetrapod footprints are locally not uncommon in the finer-grained sediments of the Midlands as a whole.

The Dune Sandstone and quartzite breccias

In the West Midlands extensive spreads of aeolian desert sands are known as the Dune Sandstone, or Bridgnorth Sandstone Group. On the west side of the boundary fault of the South Staffordshire Coalfield it amounts to about 300 metres, but on the east its place is taken by a series of thin local breccias – the Quartzite Breccia of Birmingham and the Moira Breccia farther east – probably to link up with similar patchy deposits in south Yorkshire. To what extent these screes and the Dune Sandstone are truly contemporaneous is not determinable but in places both overlie the Clent or Enville breccias and underlie the Bunter Pebble Beds.

Other aeolian sands in the west and north have been correlated with the Dune Sandstone on a basis of similar lithology and mode of accumulation. They include the 'Lower Mottled Sandstone' of Cheshire and south-west Lancashire and a similar rock in the Vale of Clwyd, North Wales. Farther afield more tentative correlatives include the Penrith Sandstone and others in southern Scotland (p. 261). The Dune Sandstone may be thus early Zechstein, part of a wide spread of barchan dunes that lay west and south of the developing saline sea. The foreset dips recorded in the several outcrops of this facies are fairly constant and suggest that the British region was at that time in the belt of north-east trade winds.

In some places, notably Warwickshire, the Permian red beds and breccias follow without obvious unconformity on those taken to be Upper Carboniferous, and it may be that in some of the hollows north of the Mercian Highlands sedimentation continued without a major interruption. However the absence of reliable fossils precludes a firm deduction on the age of either group and in semi-arid conditions of varied relief the accumulation of screes, gravels, sands and desert dust is likely to be intermittent. A succession of conglomerates and breccias does not necessarily imply persistent sedimentation or a local unconformity more than an unimportant phase of erosion.

NORTH-WEST ENGLAND AND IRELAND: THE BAKEVELLIA SEA AND ITS SURROUNDINGS

The outcrops in this province lie round the edge of a large basin, the central parts of which are now submerged in the Irish Sea (Fig. 55). In Cumbria the facies tend to be marginal and variable, formed in depressions bordering the rejuvenated hills of the Lake District. Linked with them are the Permian of west Lancashire, red beds at the northern end of the Isle of Man and several outcrops in north-east Ireland. Small amounts

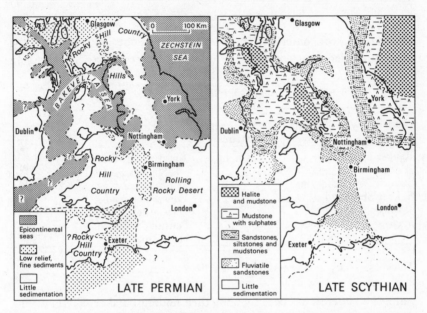

Fig. 55. Permian and Triassic palaeogeography. On the left the relation of the Zechstein and Bakevellia seas, and the neighbouring uplands. On the right the main facies of the upper Scythian (cf. Table 36); these lie above the Hardegsen Disconformity and include beds once grouped as lower Keuper. Many uncertainties in the west and south-west of both maps. (Adapted from Pattison *et al.* 1973, pp. 224, 236.)

of Magnesian Limestone occur in several places and evaporites underground. The source of the marine waters here is problematical. The shoreward approach of the Zechstein Sea along the east side of the Pennines does not suggest a major inflow from that direction, and a northern indirect passage round the north of Scotland is a possibility. Because of this separation the name Bakevellia Sea has been proposed for this north-westerly province, after a bivalve that is common in the carbonate facies.

The mainland outcrops of north-west England include all or most of the formations that are given in Table 34 together with their approximate correlations. Palynology and the cover of Lower Jurassic on the Carlisle plain help to establish the upper beds as Keuper, while the evaporites of the St Bees Shale and the Magnesian Limestone are similarly Zechstein. The Permian–Triassic boundary lacks fossil evidence and has been arbitrarily taken at the base of the St Bees Sandstone.

West Cumbria and Furness

Relatively narrow outcrops, chiefly of red sandstones, extend down the west Cumbrian coast southwards into the Furness district, but more

Table 34. *Permian and Trias of north-west England*

South and West Cumbria	Thickness (metres)	Vale of Eden	Thickness (metres)
Keuper Marl facies	400 m	Keuper Marl facies or Stanwix Shales	*c.* 300 m
Kirklinton Sandstone and St Bees Sandstone	750 m	St Bees Sandstone	*c.* 300 m
St Bees Shale and St Bees Evaporites	137 m	Eden Shales and evaporites	75–180 m
Magnesian Limestone and breccias (local)	1–2 m	Penrith Sandstone and Brockram	0–460 m

Thicknesses vary greatly and only generalized amounts are given.

variety and interest has been found in their underground exploitation. The northern area may be taken as an example. Below the St Bees Sandstone, which forms the headland of that name, and the underlying St Bees Shales, surface outcrops show only a small thickness of Magnesian Limestone overlying basal breccias, but in boreholes to the south evaporites and further carbonates appear, thickening westwards towards the Irish Sea. Above a thin mantle of breccias on the Carboniferous floor three cycles developed in which carbonates and quartz siltstones are succeeded by sulphates – gypsum and especially anhydrite. Traces of algal mats are abundant in the dolomites and anhydrite, and other sedimentary and textural characters strongly resemble those of modern sabkhas. The dolomites of the two lower cycles contain shelly faunas and are tentatively correlated with the two lower cycles of north-eastern England. Although the St Bees Evaporites only amount to about 50 metres, and are thus thin compared with their Zechstein contemporaries, this accords with their position, well inshore. Their shallow water and supratidal characters are impressive in that they are found throughout much of the cyclic components.

Less detail is published on the south Cumbrian, or Furness, sequences but among a variable group of shales with breccias and gypsum, borehole sections have recorded up to 20 metres of carbonates or anhydrite; here also the latter fade away towards the contemporary land, to the north. West of the Cumbrian Boundary Fault, Keuper Marl underlies drift on Walney Island and contains substantial beds of salt. Thinner salt sequences interbedded with mudstones have been found in boreholes on the northern tip of the Isle of Man.

South Lancashire: Permian beds

From underneath the thick Trias of the Cheshire plain (which is best considered with that of the Midlands, p. 252) two Permian formations appear along the southern margin of the Lancashire Coalfield – the Collyhurst Sandstone, overlain by the Manchester Marls. Farther west the outcrops are supplemented by a group of boreholes at Formby, north of Liverpool.

The Collyhurst Sandstone lies with pronounced unconformity on several divisions of the Carboniferous and is marked by abrupt changes in thickness; in one of the Formby boreholes there is over 700 metres but none in another a few kilometres away. Often such changes coincide with faults, as if these Permian sands accumulated in an arid landscape of Hercynian fault-scarps. Lithologically the Collyhurst Sandstone resembles the Dune Sandstone of the Midlands but it also includes some thin bands of breccia like the residual desert screes of the basal Upper Permian on the east side of the Pennines.

Such a position is reinforced by the rare fossils of the Manchester Marls, which near the base include a specialized fauna of marine bivalves and foraminifera. The most probable equivalence is with the first Zechstein cycle; very small amounts of yellow dolomite are also known. This facies of minor carbonates and red muds thus forms part of the irregular eastern margin of the Bakevellia Sea.

The Vale of Eden

At present this valley lies between the Pennines and the Lake District hills and it was similarly a low-lying tract in Permo-Triassic times, intermittently invaded by a gulf of the Bakevellia Sea. The Triassic rocks resemble those elsewhere in the north-west but the Permian facies are rather different (Table 34), with some local developments particularly towards the gulf margins. The 'Brockram' of southern Edenside is especially conspicuous; great wedges of breccia were deposited at the foot of the uplands, up to 350 metres thick. Typical brockram refers to a coarse rock in which angular fragments of Carboniferous Limestone are embedded in a red sandy matrix, but there are other rock types. In Edenside it also contains very rare fragments of quartz dolerite similar to the Whin Sill suite so that this sill, or a related dyke, was within reach of erosion by Lower Permian times. At this southern end of the gulf the overlying beds are thin grey silts and sands with some carbonates and sulphates; they are particularly noted for plant remains, accumulating in an inshore lagoon, and the age is approximately that of the Marl Slate or basal dolomites of Durham.

In central Edenside the upper brockram passes laterally into a different facies – the massive Penrith Sandstone, over 400 metres thick, aeolian in the lower part but water-distributed in the upper. In the former the very well rounded and sorted grains are coloured red by a surface film of hematite and cemented with secondary quartz. The Eden Shales above correspond approximately to the St Bees evaporites and shales of west Cumbria; they include three thin anhydrite beds, the uppermost being underlain by a thin dolomite. In its higher position and modest fauna it is probably equivalent to the carbonate of the third Zechstein cycle. It is possible, but not certain, that this marine invasion did not come in from the Bakevellia Sea but was a small late overspill from the Zechstein basin, perhaps across the Stainmore depression.

North-eastern Ireland

Permian and Triassic beds here occur in two main tectonic and topographic situations. The most important outcrops are in the downwarp that continues across from the Scottish Midland Valley, for instance around Belfast, inland along the Laggan valley and in Tyrone. Farther north they form part of the rim of Mesozoic strata that appears round the edge of the Tertiary Antrim Basalts (Fig. 56). Essentially the sediments accumulated on the westerly margins of the major Bakevellia basin and as such have much in common with those of west Cumbria. Tyrone provides a representative sequence (Table 35) with its sporadic breccias, the carbonates of the first Bakevellia marine invasion and farther up (? second cycle) thin limestones and anhydrite, and then the 'Permian Marls' with more thin evaporites. Here also the Permian–Triassic boundary has to be taken arbitarily.

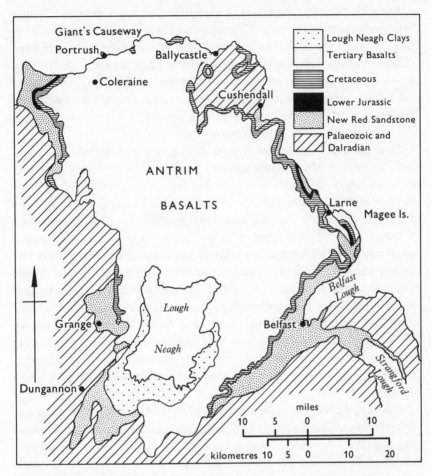

Fig. 56. Geological sketch map of north-east Ireland, showing the main outcrops of Mesozoic and Tertiary rocks; Palaeozoic and Dalradian not differentiated.

Table 35. *Permian and Trias of Tyrone*

Succession and lithology	Thickness (metres)
'Keuper Marl' facies passing down gradually into	350
'Bunter Sandstone' facies	410
'Upper Permian Marl', with gypsum	150
Magnesian Limestone	23
Basal Sands	10

The major Triassic sequences retain a lithological division into a Bunter Sandstone facies below and a Keuper Marl facies above; as in northern England there are no Bunter Pebble Beds. The sandstones include some aeolian deposits but much was deposited in shallow water, with thicknesses between 300 and 400 metres. There are desiccation cracks, ripplemarks and very rare fossils, especially the small crustacean, *Euestheria*. 'Bunter Sandstone' was apparently the source of a rich fish bed, of *Palaeoniscus*, found in Tyrone in the 1830s.

As in parts of England the Keuper Marl facies spans several stages and includes beds of Muschelkalk age (see p. 253); it is exceptionally thick in the Port More borehole in north Antrim, as is all the Permo-Trias penetrated there, confirming the view of a deep basin off this coast. The dominant sediments are the usual red mudstones, with disseminated dolomite and anhydrite, but it is the halite on the east coast that is outstanding. In all over 500 metres of salt-bearing strata are known at Larne, where the salt is exploited. As in the typical English sequences the whole is topped by a small amount of Tea Green Marls (p. 256). Farther south in the Kingscourt outlier (Fig. 31) there is a small strip of red rocks along the western faulted margin. It was earlier divided into Permian Marls below and Trias (mainly Bunter facies) above, and the microfloras show that these titles are approximately correct. The lower division includes massive gypsum and anhydrite and grey shales with plant remains.

It is very difficult to draw a broader picture of Ireland in Permian and Triassic times. The similarity with many sequences in north-west England suggests that the various facies, marine, fluviatile, saline or aeolian, stretched far beyond the present outcrops. But the evidence has been removed, and the sedimentary history of Ireland in Mesozoic and Tertiary times is almost entirely a blank, except in the north-east where limited information can be gained from the incomplete border of Mesozoic rocks that were protected from erosion by a cover of Tertiary basalts.

RED TRIAS IN THE MIDLANDS

Here we are reviewing the largest of the English outcrops, and also some more general stratigraphical problems. The Alpine stages Scythian to Norian, as recognized in the German sequence, can be applied to the English Trias in a broad way (Table 36), principally through palynology. Where the spore samples are inadequate more reliance has to be placed on lithological characters and correlation becomes more uncertain. However in practice such formations as pebble beds, major evaporites and minor sandstones in the thick marls do seem to be fairly effective marker horizons within a single depositional basin.

For a century or so the German terms Bunter and Keuper were used in this country entirely on a lithological basis and it is not surprising that some of the results were confusing; it was also thought that the Muschelkalk Sea did not reach Britain. In recent years it has turned out that

Table 36. *Triassic lithological groups and equivalents*

Alpine stages	Germany, lithological		English Midlands, traditional grouping	English, revised
RHAETIAN	Rhät		Rhaetic	{ Penarth Group
NORIAN				{ Mercia
CARNIAN	Keuper		Keuper Marl (facies)	{ Mudstone
LADINIAN				{ Group
ANISIAN	Muschelkalk		Sandy facies	}
		Upper		Sherwood
SCYTHIAN	Bunter	{ Middle	Bunter Pebble Bed*	Sandstone
		Lower	Bunter (or Lower Mottled) Sandstone	Group }

* Also called Kidderminster Conglomerate Group.
The equivalents set out are only approximate and the boundaries between some lithological units are recognized as diachronous.

whereas the Bunter of the Midlands was approximately correct, though only representing part of the German Bunter, the English 'Keuper' extended far lower than the German type series and thus included beds of Muschelkalk age. 'Keuper Marl' or 'Keuper Marl facies' for the upper part remains a common term, but 'Keuper Sandstone' for the lower is basically unsatisfactory and has largely disappeared. Modern lithostratigraphic replacements are given in Table 36.

In spite of these problems, if we do not attempt too rigid a time-relationship, some common facies-progressions can be recognized through the Triassic period. Coarse fluviatile beds, with some sedimentary breaks, are widespread near the base and finer siltstones and mudstones, locally with substantial evaporites, towards the top; saline waters (? ultimately marine) dominate the later scenes. Much more restricted cyclic sequences, on a medium or small scale, also show a decrease in grain-size upwards. As erosion and sedimentation progressed the old Hercynian uplands were reduced, the basins filled and subsided and there was progressive overlap so that the later formations commonly conceal the earlier.

Bunter Pebble Beds and Sandstones: lower and middle Scythian

These two facies form a group of sandy beds that in certain parts contain scattered or lenticular seams of pebbles (especially towards the base) or, more rarely, thick banks of shingle with very little sandy matrix. In the West Midlands the distinction between Bunter Pebble Beds below and Bunter Sandstone ('Upper Mottled Sandstone') above is fairly clear, each amounting to about 120 metres; there is an unconformable junction with the Dune Sandstone or Permian breccias below. However to the north-east the pebbles diminish and are inconspicuous in a slightly pebbly sandstone and a similar dwindling takes place northwards into Lancashire.

Compared with the earlier breccias these pebbles are not only much better rounded, particularly the smaller ones, but more varied. Quartzite and vein quartz are abundant with lesser amounts of chert, limestone and volcanic rocks. In some areas local types predominate, such as Cambrian quartzites, Silurian sandstones and Carboniferous Limestone, but there are others, especially among the quartzites, that have no known origin. Equally enigmatic are rare pebbles whose nearest visible sources are far to the south, including those bearing tourmaline and other minerals of Devon and Cornwall, Devonian sandstones and large cobbles containing Ordovician brachiopods like those cropping out at present in Normandy.

A southerly source is thus suggested for some of the Bunter gravels but not necessarily as far away as France; it is at least possible that outcrops of Ordovician or Devonian strata were available in areas now buried in southern England or on the site of the western English Channel. A deep through valley from this direction is deduced from the exceptionally thick Trias found in boreholes in south Worcestershire and Gloucestershire. This north–south depression is aligned with the lower Severn Valley and has been called the Worcestershire Graben, the western margin being the faulted edge of the Malvern hills and the eastern possibly marked by faults also. The whole is part of the tensional effects mentioned earlier in the chapter.

The rounded pebbles among fluviatile sands suggest that the rare heavy rainfall in the uplands caused flash-floods to sweep down the valleys, losing themselves and spreading their detritus on the sandy plains to the north. The succeeding Bunter Sandstone is a fine-grained red rock, cross-bedded but not aeolian. From Shropshire it thins eastwards till in the East Midlands it is not known. To the north-west in Cheshire it continues to be a recognizable division and, as the pebbles below fade out, the whole has been called 'Bunter Sandstone', amounting to some 320 metres.

Upper Scythian, Anisian and Ladinian: sandy facies

Throughout much (perhaps all) of the Midlands there is some kind of break within the upper part of the Scythian; the beds above the break vary in age and some are conglomeratic. This unconformity appears to be one of the better marker horizons in this difficult complex of Triassic facies and probably represents the same period of erosion that is known in north-west Germany – the Hardegsen Disconformity. If the two are truly contemporaneous it suggests a remarkable uniformity either side of the southern North Sea basin.

In the west and central Midlands (Fig. 55) sedimentation was resumed with fluviatile sands (the earlier 'Keuper Sandstones') in which sedimentary cycles show an upward progression from gravels to sands, silts and muds. This facies has also provided more fossils than most of the Trias – a variety of plants, crustaceans, fish and amphibian remains, indicating a climate no more than semi-arid. Presumably the preceding slight uplift had rejuvenated the ridge or sill linking the southern Pennines with the London Platform, in Permian times the edge of the Zechstein Sea, because here the fluviatile sands are absent. On the far side (Nottinghamshire and Lincolnshire) boreholes have shown thin green mudstones with evaporites.

The next phase was the influx of a different facies, formerly called the Waterstones, transitional in type between the sandy beds below and the marls above. It comprises thin sandstones intercalated with finer beds, with many signs of very shallow water: desiccation cracks, footprints and salt pseudomorphs. There was thus a far-reaching marine invasion, or perhaps very shallow oscillating conditions on wide tidal flats. Fish and marine microfossils have been found and a *Lingula* bed in Nottingham-shire. The last suggested several years ago what recent research has confirmed – that the Waterstones are in part equivalent to the Muschel-kalk, which is much more obviously marine in Germany and in the North Sea boreholes includes substantial halite. In the English Midlands the Waterstones facies is probably diachronous, in that it appears earlier in the north-east and north-west and is later, for instance, in Worcestershire.

The Keuper Marl facies

This is the thickest, most extensive and most conspicuously overlapping of all the Triassic lithological divisions; in the basin areas, notably of Cheshire and north-east Ireland, it is vastly expanded by salt-bearing strata. The base of the facies naturally varies in age according to the development of the underlying sandy formations, but the Norian and Carnian stages in the Midlands are universally represented by some type of 'Keuper Marl' and in some regions the Ladinian also.

Typically the rock is a reddish siltstone; the colour is due to minute granules of hematite and the bulk consists of fine angular quartz grains and the evaporitic minerals, dolomite or gypsum. The grey-green layers and patches are probably caused by the reduction of the iron oxide. Some thin but widespread beds at the top are more uniformly green than the rest and have been separated off as the Tea Green Marls, but even this colour may be secondary. Interspersed among the marls the greyish lenticular sandstones with rhombs of dolomite are called skerries; the more massive beds show cross-bedding, slumped horizons and salt pseudomorphs. An exceptional example is the more extensive Arden Sandstone of Warwickshire and Worcestershire, unusual also in its fossils. In addition to fish spines and scales, reptilian and amphibian bones, there are poorly preserved shells that may be marine. If so the debris may have been washed together in very shallow or intertidal waters.

The two main economic precipitates of the Keuper Marl are gypsum and halite. The former is widespread in thin bands and patches and a more massive bed, up to 20 metres thick, has been worked in Nottinghamshire. The halite is outstanding, particularly in the basins; some salt-bearing sequences around the Irish Sea have been mentioned and there are others in the Midlands and Somerset. All these are outstripped, however, by the Cheshire basin. The 1500 metres of marl facies here includes two salt-bearing divisions (200 and 400 m approximately) making up one of the richest and most concentrated salt-producing fields in the world.

There is some uncertainty about the origin of the halite and also how far the beds are useful in correlation. Palynology suggests that the lower and upper saliferous divisions of Cheshire are Anisian and Carnian respectively; these positions would agree with the halite formations in the North Sea, while the lesser beds in the Midlands would correspond to the upper division. Apart from the salt itself many sedimentary features of the marls are characteristic of shallow waters that must have been highly saline, but this is not a typical marine evaporite sequence: the halite is not underlain by massive sulphates and carbonates as in the Zechstein basin. Nevertheless the enormous volume (290 km³ in the Cheshire basin and the probability of further basins in the Irish Sea) does seem to demand a marine source, but whether from the north-west or south-west is not clear. Moreover some authorities consider deep rather than shallow waters to be more probable, with precipitation from the bottom saturated brines.

The gradual extension of Keuper deposition is shown by the widespread overlap of the marl facies around the basin margins. The best known is the partial cover on the Precambrian of Charnwood Forest and the neighbouring Mountsorrel Granite, a small Caledonian intrusion. The marls can be seen banked up against steep slopes and filling valleys between Precambrian spurs. Here there are glimpses of a late Triassic landscape, as the red muds and silts lapped round and finally buried the modest hills and ridges.

While still upstanding such hills were subject to the rigorous erosion of an arid climate. The granite, for instance, is remarkably fresh immediately under the marl cover, with the feldspars bright and unweathered. There are signs of insolation (weathering under pronounced temperature variations) and some of the surfaces are polished and grooved by the wind. Although certain features have been claimed as Pleistocene the total effect of Triassic erosion and burial is well supported and at present forms a striking exhumation. Finally at the end of this long period of much reduced uplands, saline seas and wide salty flats there entered the more normal waters of the Rhaetic invasion.

PERMIAN AND TRIAS OF THE SOUTH-WEST

Around the Severn estuary and into central Somerset (Figs. 24, 57) the Keuper Marl facies is the main lithological type in surface outcrops, overlapping onto the surrounding ridges and uplands, these being chiefly of Carboniferous Limestone. On the Welsh side there is a similar relationship, marls overlapping an irregular floor of Carboniferous Limestone and Old Red Sandstone. A former westward extension is suggested by patches of red conglomerate on the Gower Peninsula and the gash breccias of south Dyfed. These are large swallow holes or caverns filled with angular blocks in a red matrix; the surface Trias has long since been eroded but this unusual type of subterranean evidence remains.

Where Triassic and Jurassic strata are banked up against the Carboniferous Limestone in the Mendips, around Bristol or in Glamorgan, there is a local marginal facies called the Dolomitic Conglomerate – though in

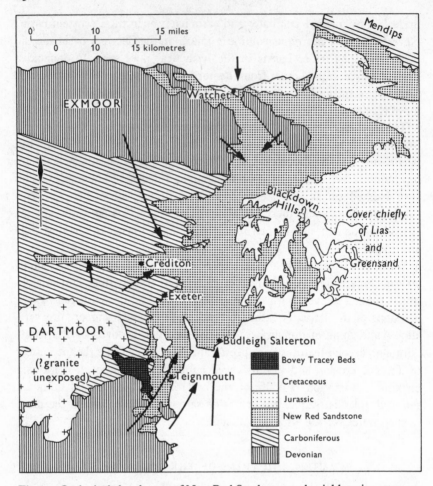

Fig. 57. Geological sketch map of New Red Sandstone and neighbouring outcrops in South-West England, showing sources of detritus.

fact boulders may be angular and formed of limestone as well as dolomite, set in a red sandy matrix. This fringe of debris accumulated against the Carboniferous outcrop and locally may be as much as 13 kilometres wide, as in places north of the Mendips. In Glamorgan and the southern Mendips burrows and solution hollows penetrate the limestones and in them have been found traces of small terrestrial animals. Their bones were washed into underground water-courses and are now embedded in the muddy or silty infilling. Some of the remains probably date from the time when the limestone hills were islands in the Rhaetic or Lower Jurassic sea, but others are earlier and include several small carnivorous dinosaurs, lizard-like reptiles and forms near the reptile–mammal boundary. From these rather unusual sources we get more information about Triassic quadrupeds than from any other region in Britain except the sandstones of northern Scotland.

South-East Devon and Somerset

Here, in the main south-westerly basin, the rocks are best exposed along the coast from the Torquay district to Teignmouth, Budleigh Salterton and into the westernmost cliffs of Dorset. A long deep trough of Permian beds follows the Crediton valley in the central Carboniferous synclinorium. The general trend of the Permian and Trias is north–south, the outcrop truncating the pronounced Hercynian structures on which the basal rocks lie with conspicuous unconformity. Table 37 is a composite

Table 37. *Permian and Trias of south-east Devon*

Succession and lithology	Thickness (metres)
Keuper Marl facies	200+
Red Sandstones	118
Budleigh Salterton Pebble Beds	26–30
Red Marls	275
Red Sandstones, with breccias	255
Breccias, sandstones and local clays { Exeter lavas at or near base	c. 300
(Devonian and Carboniferous rocks)	

sequence with few correlative data. The best are the isotopic ages from the lavas, and there are also very sparse vertebrate fragments of Triassic age and miospores from the upper red sandstones. The Permian–Triassic junction remains arbitrary; it has commonly been taken at the base of the pebble beds, but with a possible alternative position somewhat lower.

At the base the marginal depressions of the Cornubian upland were filled with a great variety of breccias and conglomerates which crop out along the Crediton valley and in the southern coastal district. Local Devonian limestones and Carboniferous rocks contribute to the lower beds, lavas and tourmaline rocks to some of the upper ones; however there is little indication of the erosion and exposure of the metamorphic aureole or of the Dartmoor Granite itself. The breccias represent desert fans derived from the nearby uplands and pass eastwards and upwards into the red sandstones and siltstones, mostly water-laid but with some aeolian lenses.

Near Exeter and along the Crediton valley there are several small outcrops of potash-rich lavas, mostly trachytic but some basaltic. Some lie directly on the Carboniferous floor but others succeed the lowest breccias. Isotopic dating from two outcrops (about 285 m.y.) suggests that this minor vulcanicity occurred around the Carboniferous–Permian boundary. Accordingly some of the earliest southern breccias are probably very late Stephanian – a time-sequence that accords with a late Carboniferous age (possibly late Westphalian) for the major Hercynian deformation in this province and with early Stephanian for the post-tectonic granites.

The Budleigh Salterton Pebble Beds have some resemblances to the Bunter Pebble Beds of the Midlands. Quartzites and sandstones are common, some rounded and some subangular. Devonian rocks with fossils are known, tourmaline rocks and also the quartzites with orthids already noted as rare components in the Bunter pebbles; Carboniferous Limestone appears in more northerly outcrops. Along the Devon cliffs the Keuper thickness is small because of the prolonged erosion preceding the deposition of Cretaceous greensand but a full sequence of over 400 metres was penetrated in a borehole in the north Somerset plain, including a substantial thickness of halite. To the east incomplete borehole evidence, reinforced by seismic surveys, suggest that another major Mesozoic downwarp, the Wessex basin, was initiated in Triassic times, between the Cornubian uplands and the margin of the London Platform. The south-western outcrops are thus the fringing edge of that major basin, which extended well out into the Channel and to the north linked up with the Worcestershire through valley.

BASINS AND OUTLIERS IN SCOTLAND

With the exception of the Elgin sandstones the red beds in Scotland present little firm evidence of Permian or Triassic age and the term New Red Sandstone is often used here.

The South of Scotland

The Southern Uplands are traversed by several elongate depressions or basins sculptured into the Lower Palaeozoic greywackes, mudstones and shales, and for the most part trending between north and north-west (Fig. 26). Some were through valleys across the uplands in Carboniferous times and are still marked by outliers of that age. The Thornhill Coalfield retains an incomplete cover of New Red Sandstone and farther south-east the River Annan runs through the parallel basin of Lochmaben. The third of these cross depressions has been partly eroded by the sea and is seen as a low neck of land at Stranraer, between Loch Ryan on the north and Luce Bay on the south. The isthmus between the Belfast and Strangford loughs is analogous on the Irish side and a deep basin occupies the North Channel between Scotland and Ireland.

The Scottish hill-girt depressions received a variety of red detritus – sandstones, breccias and conglomerates – and are surprisingly deep for their area. That at Dumfries is less than 12 kilometres across, and at Stranraer less than eight; but over 1500 metres of sandstones and breccias have been deduced, by gravity surveys, in the former and even more in the latter. It is probable that such basins subsided as they were filled and may have had a contemporary fault margin on one side, the other acting as a hinge. The sandstones are commonly aeolian-bedded and the breccias reflect the vigorous erosion of the upland rocks on either side.

The red rocks of the Thornhill and Sanquhar coalfields form a link between Dumfries and Ayrshire where the Mauchline Lavas and Mauchline Sandstone form an isolated outlier (Fig. 49) above Westphalian D Coal Measures. The red sandy sediments at the base of the lavas suggest that 'New Red Sandstone' conditions had already set in and this agrees with a flora of probable Lower Permian age interbedded with the flows. The Mauchline Sandstone above is a noted dune deposit. Some 460 metres thick, it is remarkably consistent in the lack of pebbles and water sorting, and in the rounded grains and great wedges of aeolian bedding. It is logical to see this facies as a Scottish outpost of the major dune formations, linked with the Penrith Sandstone and that in the West Midlands.

Arran and the West of Scotland

It was seen how, during Carboniferous times, there was some kind of westerly extension to the Midland Valley of Scotland, for although the evidence is scanty, sediments extend beyond the rift margin to lie on Dalradian schists. A lowland in this direction, west of the Scottish Highlands and linked with north-east Ireland, becomes clearer in the later systems. Deduced Permo-Trias basins occur offshore (p. 272) and Mesozoic sediments occur patchily in the Tertiary Volcanic Province.

The southern half of Arran is composed of the central Tertiary ring complex and a broad outcrop of New Red Sandstone, nearly 1000 metres thick. Three lithological groups have been distinguished:

3. { marls with salt pseudomorphs and calcareous beds
   { red and green marls
2. red and variegated sandstones
1. red partly aeolian sandstone, interdigitating with thick breccias

The breccias of the lowest group contain many basalt fragments and, with the aeolian sands, resemble the Mauchline beds and are deduced to be similarly Lower Permian. Microspores characteristic of early to middle Trias have been found in the lower part of group 3 and the upper part has a Keuper Marl facies. The New Red Sandstone here, therefore, may be complete or nearly so.

In western Scotland the small remnants of red beds can be mentioned more briefly. From south to north they extend from Kintyre and Islay to Mull and the nearby mainland, and thence to slightly larger outcrops on Skye and at Applecross. The Stornaway Sandstone on Lewis (Outer Hebrides) probably belongs here also. In general coarse deposits are common, derived locally and uncertain in age; but in a few places (e.g. Mull and Applecross) the stratigraphic position, closely underlying Rhaetic or Jurassic, allows a firmer ascription to the Trias.

Elgin, eastern Scotland

On the opposite side of Scotland where the Grampian Highlands slope down to the Moray Firth, there is a poorly exposed outlier of New Red Sandstone – part of a much larger Mesozoic basin that underlies the firth. These red rocks of Elgin are highly unusual in providing a number of fossil reptiles.

Three lithological divisions have been distinguished: the lower and upper include aeolian sandstones whereas the middle is fluviatile with pebbly beds. The outcrops are sparse in drift-covered country and the relations between them rather obscure. However in the lower group the sandstones of Cutties Hillock have yielded a small fauna that is tentatively assigned to the lowest Trias (uppermost Permian is a possibility). The Lossiemouth Sandstones in the upper group have angular grains, locally a cement of barite and a much more prolific fauna, the age being upper Keuper (probably lower Norian). These sediments are also unusual in their lenses of aeolian bedding; elsewhere the Keuper is largely or entirely water-laid, much of the waters being saline. Reptiles are not fitted for a fully desert environment and it may be deduced that they lived near the small water-courses, with some vegetation, and that the dunes formed in the truly arid areas in between.

Being so isolated and exceptional the Elgin sandstones do not add much to the regional picture of the period. Nevertheless, with the Keuper fossils of the English Midlands, the reptilian burrow-dwellers of the Mendips and Glamorgan and the more widely spread footprints where no actual bones remain, they are a reminder that animals could survive, at least in some places and at some times, among the inhospitable conditions of Permian and Triassic Britain. Further, it may not be accidental that the fossils seem to be more related to the uplands than to the plains, saline flats and shallow seas, and most of our sediments and precipitates inevitably come from the latter.

THE RHAETIC

The especial problems of these beds in Britain reflect the contrast, or even clash, between their depositional environment and their formal stratigraphical grouping. They are the first representatives of the open shelf seas that replaced the Keuper saline facies and which were to dominate sedimentation in north-western Europe for the remainder of the Mesozoic Era. As such they have an obvious link with the Jurassic beds above. But conditions of deposition, however important, are not an international arbiter of stratigraphical grouping, which in Mesozoic stratigraphy is based primarily on the ammonite faunas. These are lacking in the British facies but in the marine Trias of the Eastern Alps they are present; although the Rhaetic species are not abundant, they are nearer those of the Norian Stage below than the Hettangian (lowest Jurassic) above so that the Rhaetic becomes the uppermost Triassic division.

The main English outcrop is comparable with that of the Jurassic (Fig. 60) in that it extends across country from the south-west to the north-east. Only in the former region are the beds well exposed in cliff sections, especially along the Bristol Channel, in the north being hidden by the alluvium of the River Tees. The divisions are mainly lithological with some assistance from the commoner bivalves; as such their definition is not always easy. Very shallow marine conditions are typical and it is remarkable that the principal units are as extensive and dependable as they are.

South-West England and South Wales

The type section is on the north Somerset coast near Watchet (Fig. 57) and there are others farther up the Severn estuary and on the Glamorgan and Dorset coasts. The beds that at one time or another have been called Rhaetic are shown in Table 38, but with some reservations at the bottom

Table 38. *Rhaetic succession of Somerset*

	Formations and beds	Thickness (metres)
LILSTOCK FORMATION	Pre-planorbis Beds	4
	White Lias, and Watchet Beds	10
	Cotham Beds	6
WESTBURY FORMATION	Westbury Beds	15
	Grey Marls	3.5
	(Blue Anchor Formation or Tea Green Marls)	

and top. The Grey Marls of Somerset and Glamorgan comprise shales, siltstones and calcareous mudstones; they are overlain non-sequentially by the distinctive Westbury Beds and were it not for their marine shells (which include *Rhaetavicula contorta* typical of later horizons) would not be easy to separate from the Tea Green Marls below. However they presumably reflect an early south-westerly marine invasion.

The succeeding Westbury Beds are much more extensive and over the rest of the outcrop form the recognizable base of the series. Black pyritous shales with impure limestones and sandy beds are typical, laid down in shallow but partly stagnant waters, probably of varying salinity, comparable with some modern estuarine or lagoonal black muds. Shells, especially *Rhaetavicula*, are locally common in the more calcareous, better aerated sediments. Where these shales overlie Keuper Marl, for instance at Aust Cliff on the Severn, the basal conglomerate is also a bone bed; here are rolled and eroded bones and teeth of lung fishes, probably derived from fresh waters, and rare remains of marine reptiles. Bone beds, often in a shallow water sandy facies, are characteristic of the lower Westbury Beds, though not always at the base.

The much thinner Cotham Beds are also widespread – grey, silty and calcareous mudstones with rather rare fossils, *Euestheria, Pseudomonotis fallax* and pectens. North of the Mendips there is a freshwater intercalation with liverworts, and Cotham Marble is a local name for a calcilutite with algal structures. All this suggests a gently shelving shore near the limestone islets of the Mendips and Bristol region.

The Westbury and Cotham beds make up the Rhaetic in its most restricted sense but commonly the White Lias is also included. This is known chiefly in Somerset and Dorset as an impure calcilutite with marly partings. Contemporary conglomerates, eroded and bored surfaces and mudcracks again suggest very shallow and temporarily emergent conditions. The Watchet Beds are a local, south-western, marly facies of no regional significance, but larger problems have centred on the Pre-planorbis Beds, not in themselves but because they affect the formal Triassic–Jurassic boundary. As a thin sequence of limestones and shales they resemble the Lower Jurassic above; however they lack the ammonites of the lowest stage there, and since palaeontological evidence is the governing factor this small unit is retained in the Rhaetic.

The Midlands and East England, Ireland and Scotland

The Westbury and Cotham facies with their bivalve faunas continue northwards along the main outcrop, poorly exposed at the foot of the Jurassic scarp. In the East Midlands the irregular foundations that affected Triassic thicknesses and sedimentation had modest reflections on the Rhaetic and underground to the south-east sandy accretions come in, probably from the nearby landmass of the London Platform. Northwards into Lincolnshire and Yorkshire the double facies sequence is maintained but redder marls appear in the upper part, so that there is a partial return to a Keuper-like facies.

The Prees outlier on the Cheshire plain shows a normal Midland development, but thereafter there is a long gap to north-east Ireland. Near Belfast the Keuper is overlain by about 16 metres of Rhaetic – typical black shales containing *Rhaetavicula contorta* overlain by black and grey shales with *Protocardium rhaeticum*. A bone bed like those of the south-west occurs at the base with bones and teeth of lung fishes and reptiles. On Arran similar shales and thin limestones are only found as jumbled and downfaulted fragments in the central ring complex, but the bivalves are a sufficient indication of age. That the Rhaetic seas also invaded the western lowlands of Scotland is shown by a section in western Mull where some 25 metres, mainly of dark calcareous sandstones with bivalves, are overlain by grey-green shales and impure limestones. The former are probably the Westbury Beds, to which, as in Glamorgan and the south-east Midlands, local erosion has added a sandy component.

In terms of volume or thickness the British Rhaetic beds are only a very thin skin on top of the far greater amount of continental Trias below. As an extreme example, a borehole section in the Cheshire basin shows 20 metres of typical Rhaetic overlying 1350 metres of Keuper Marl and salt. The significance of this thin division, however, is multiple. It is the most reliable lithological marker in the Trias of this country, from which sections are commonly estimated downwards. In environment the beds represent a transition from saline continental conditions to those of the open sea at a crucial stage in Mesozoic history. The quite remarkable cross-country maintenance of the Westbury and Cotham facies shows how the discordant influences on Triassic sedimentation had been lessened, with the uplands much reduced and the basins filled, their part-bounding faults becoming inoperative, at least for a time. Variations there are in the Rhaetic beds, but on a small scale and especially caused by littoral incursions. Such was the relative simplicity achieved by the end of the post-Hercynian continental phase.

THE SEAS AROUND THE BRITISH ISLES

A summary of the neighbouring seas is interpolated here since more is known of Permian and later rocks than the somewhat isolated data noted in previous chapters. Generalized maps and sections can now be drawn (e.g. Figs. 1A, 1B, 1C), though where Mesozoic and Tertiary basins are under active exploration some information is still unpublished. The sources include geophysical surveys, especially seismic; shallow boreholes and grab samples, especially where there is little superficial cover; and in the regions of potential hydrocarbon accumulations (oil and gas), the most valuable, deep boreholes. Since Carboniferous and earlier rocks appear to be largely without such potential reservoirs, boreholes rarely penetrate far below Permian beds. A 'basement' may refer simply to pre-Permian or pre-Mesozoic rocks, but often is much older, such as the Precambrian or Lower Palaeozoic basement deduced or penetrated off western and northern coasts. This review does not go outside the edge of the continental shelf (or approximately the 200-metre contour) but beyond it some shoal regions rise from the ocean floor (Fig. 6): Porcupine Bank (*c.* 53° 30′ N, 14° W), Hatton Bank (*c.* 59° N, 18° W) and Rockall Bank with Rockall Islet emergent; two deep ocean boreholes have been sunk on Rockall Bank. Though not considered further, these 'microcontinents' are related to the geology of north-western Europe as portions of the continent detached in the later stages of Atlantic rifting.

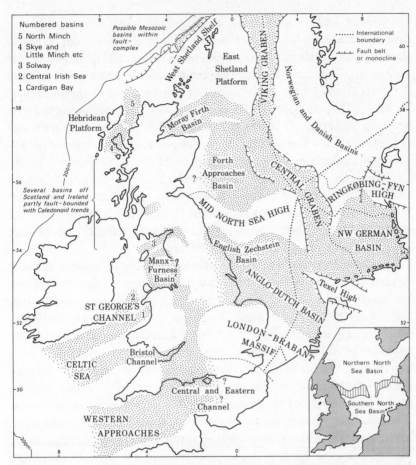

Fig. 58. Regions and structures around the British Isles. Basins known or deduced to have Permian or Mesozoic infilling, stippled. All boundaries are approximate or tentative; some names also are informal or provisional. Inset: major divisions in the North Sea. (Adapted from Kent, 1975*b* and other sources.)

Our summarized tour of the seas (Fig. 58) will be taken clockwise from the English Channel. In some areas there is a simple correspondence to the geology of the neighbouring lands; in others some surprising differences are revealed. Also from now on the stratigraphy of the sea floor is added to that of the land outcrops, system by system, towards the end of each chapter.

The English Channel

The London–Brabant Massif separates the southern North Sea basin from the Channel and consists largely of Palaeozoic rocks with a Cretaceous cover. Immediately on the south-west side of the massif the Wealden structures swing round to the south-east and cross the Channel into the Boulonnais. In general this small eastern section has been a stable area and the strata known (Jurassic to Eocene) are very like those on the neighbouring lands. The junction with the central section (Fig. 1C) is marked by a line from Bembridge to St Valéry-en-Caux, some 30 kilometres south-west of Dieppe. On the English side this line begins as a continuation of the Isle of Wight monocline, but soon, with reduced throw, turns into a south-easterly trend – a structural direction continued by faults in northern France. Two other monoclinal, east–west, structures characterize this part of the Channel, the rocks ranging from Permian to Lower Tertiary; the visible effects here and in the eastern section are presumably mid-Tertiary (cf. p. 364) but related to deeper structures.

The western section is by far the largest and, merging with the Western Approaches, extends from the continental margin to the Start (South Devon)–Cotentin line. This buried basement ridge or persistent positive feature separates the central structures from those of the west, which trend dominantly west-south-west. Several of the latter are concentrated in the Alderney–Ouessant fault zone, stretching just north of those two islands. On the south side the wide shallow basement off Brittany contains much metamorphic and folded Precambrian, overlain by some Chalk and Eocene. The analogue off the Devon and Cornish coasts (Devonian and Carboniferous) is much narrower but spreads out westwards to surround the Scilly Islands. A major Permo-Triassic basin on the seaward side turns north-east to come onshore in south-east Devon. From here to the Ouessant fault zone the main structure is a broad belt of irregular subsidence, with Jurassic, Cretaceous and Tertiary rocks known. It is possible that the land areas on either side have remained more uplifted through the relative buoyancy of their Hercynian granites; the cumulative downthrow along the Alderney–Ouessant line amounts to several thousand metres.

The Western Seas

From the Western Approaches northwards, off the coasts of Ireland, Wales and western Scotland, there is rather less information available. The Devonian and Carboniferous fringes continue to the west of Cornwall, at least as far as Haig Fras (a submarine Hercynian granite outcrop, 50° 10′ N, 8° W) and possibly farther. The outer part of the Celtic Sea, like the western Channel, contains much Chalk and Tertiary and some

Jurassic. To the east and north a narrow basin, of Trias and Jurassic, extends into the Bristol Channel and another expands into the rather complex structures of St George's Channel; in between a divide, presumably Lower Palaeozoic, extends westwards from South Wales.

Caledonian structures dominate St George's Channel and the southern Irish Sea with a complicated history of accumulation and subsidence from Carboniferous times. A central ridge, approximately from the Lleyn Peninsula to Wexford, separates a lesser north-westerly basin, partly obscured with some Carboniferous filling, from a larger one with Mesozoic and Tertiary on the south-east. The latter links up with the Celtic Sea in one direction and in the other with Tremadoc Bay and the Mochras borehole (cf. p. 310). The south-eastern margin is faulted against the Welsh Lower Palaeozoic across Cardigan Bay, perhaps along a continuation of the Bala Fault.

The northern Irish Sea has a Lower Palaeozoic fringe on the Irish side and some Precambrian extends north-westwards from Anglesey. A basin existed in Carboniferous times and these rocks, under drift, occupy much of the western part, with the addition of the Mesozoic and Tertiary outcrop of Kish Bank offshore from Dublin. Two conspicuous Permo-Triassic regions link up with the shore outcrops: the large Manx–Furness basin in the south-east and the smaller Solway basin in the north-east. They are partly separated by a ridge (possibly of Manx and Skiddaw slates) from the north of the Isle of Man to the northern Lake District.

From the latitude of Belfast northwards, up the south-west coast of Scotland, structures are complex and the information more patchy. Caledonoid faulting governs the distribution of many, perhaps all, basins. A tongue of Permo-Trias extends well up into the Firth of Clyde and a second, connected with the deep basin off the north Antrim coast, reaches up west of the Mull of Kintyre. In the Hebridean region at least three major fractures affect both land and sea structures: the Great Glen Fault swinging south-westwards from southern Mull; the Camasunary Fault from southern Skye, passing between Rhum and Eigg to the Skerryvore lighthouse; and the Minch Fault close against the eastern shore of the Outer Hebrides. In their Mesozoic expression these faults throw down to the east and the infillings tend to be fault-controlled sags, thickening against their westerly faulted margins; the two largest basins lie in the North Minch and inside the Minch Fault – the latter including Skye and the Sea of the Hebrides. Permo-Trias and Jurassic rocks have chiefly been recognized with obvious relations to the onshore outcrops; Cretaceous is less in evidence.

Some partly analogous structures are known west of Orkney and Shetland and there is a wide stretch of Permo-Trias and Jurassic strata west and north of Cape Wrath. The sea around the Orkneys and north to the west Shetlands is floored by a pre-Mesozoic basement including stretches of Old Red Sandstone, and there is an extensive shelf west of the Shetlands. This is bounded by the West Shetland Boundary Fault and farther out lies the complex downfaulted edge of the continental shelf. Between them the West Shetland Basin is another sagging structure, with probably a Mesozoic and Tertiary infilling. Farther south along the continental margin where there are large cumulative throws towards the ocean floor (2000 m estimated) there are other similar inward-tilted basins. On the eastern side the Great Glen Fault is a convenient boundary with the northern North Sea. It is commonly supposed to run through the Shetlands but the link is uncertain and it may splay out east of these islands. The amount and direction of post-Devonian movement, as deduced from its submarine course, is also controversial.

The North Sea

The particular problems of the North Sea arise from its size (substantially a greater area than the British Isles); the concentration of deep exploration in certain areas with extensive information still unpublished; and the great depths to which the older rocks have been depressed by persistent, but not wholly regular, subsidence since Permian times and possibly earlier. A pair of linked northern and southern basins (Fig. 58, inset) developed early. Faulted structures, especially grabens, are important in both and they are separated by two positive regions – the Mid North Sea High on the west and the Ringkøbing–Fyn on the east; the former is roughly a continuation of the Southern Uplands but swings round somewhat to the south-east. An isolated Silurian find is reported from beneath the Dogger Bank and on the other side there are records of Precambrian and Lower Palaeozoic rocks at moderate depths on the Danish mainland. The Central Graben cuts through this east–west barrier; as with other structures the margins may not be simple faults but fault belts or monoclines.

Throughout northerly to north-westerly trends are common, and Caledonoid directions, which might have been expected in the north between Scotland and Scandinavia, are not prevalent. Here the Caledonian basement must be very deeply buried and the older rocks are known only occasionally or at depth. Thus the East Shetland Platform includes metamorphic rocks and these and Lower Palaeozoic sediments have very occasionally been met in the floor of the Viking Graben. Old Red Sandstone is known from a few boreholes and is thought to be widespread, related to the Scottish outcrops; the enigmatic marine Middle Devonian at the edge of the Central Graben has already been noted (p. 165).

So far, either the records are only tentative pointers to the very deep rocks and structures, or they correspond to the land outcrops without adding much that is fundamentally distinct. In the southern basin the first goal of hydrocarbon exploration was the gas-bearing Lower Permian sands and pre-Permian rocks have not been much explored. However, Coal Measures are known to be continuous across the southern basin and join up with those of the Netherlands and Germany.

From Permian times onwards the scale of the offshore data is rather different and in the south at least the basin presents a record that substantially amplifies that of the land areas. To anticipate for a moment, the marginal or inshore character of British land-based Mesozoic and Tertiary stratigraphy is continually emphasized: in its shallow waters, facies variations, non-sequences and some major unconformities. Some of these operated on the raised portions of the North Sea also but in the grabens the sedimentation was more continuous, and increasingly from late Cretaceous times onwards the North Sea was *the* major basin of north-western Europe.

Two types of structure, long-continued in their effects, influence the history of the southern basin. The first comprises the halokinetic effects, caused by the salt structures of the Zechstein or Upper Permian. These, in the form of salt pillows and plugs, were active from the late Triassic, doming the strata above or even piercing them. They affected stratigraphy, sedimentation and erosion well into Tertiary times. Then much of the basin is dominated by south-easterly trends, which are known underground in the Netherlands and north Germany and have been noted in eastern England (p. 99).

'Inversion structures' are a particular variant here, though apparently not in the northern basin. Typically a subsiding trough is later inverted by being transformed into a rising ridge or a more complex anticline, which is then eroded along its crest. Such features characterize the Dutch sector and in British waters the Sole Pit inversion, 100 kilometres east of the Humber, is an example. Here the Chalk has been removed from the crest, and in general some of the thickest Chalk is developed in 'subsequent' or marginal downwarps beside these structures. In age, inversions appear to range from late Cretaceous to mid-Tertiary and the Weald may be regarded as a late inversion though an unusually large one. The governing

movements were probably vertical, though with some subsidiary compressive effects; it is uncertain how far they may be related either to Alpine
stresses in the south or to those of Atlantic rifting, though the latter
presumably influenced the European margins during much of Mesozoic
history.

PERMIAN AND TRIAS OF THE SEA AREAS

The submarine stratigraphy of these systems can be treated in three
regions, related in varying degrees to that of the land outcrops. Little more
needs to be added about the western and northern seas. The many faulted
basins on the west and north-west of the Hebrides (p. 269, Fig. 1A) are
usually deduced or known to have a Permo-Trias infilling comparable to
the basins of north-west England and Scotland. Aeolian sands and a
'Keuper Marl' facies are typical and no Zechstein carbonates or evaporites
are known as yet, though if thin they might not be detected by geophysical
methods. The Irish Sea certainly contains Keuper salt as part of the
Manx–Furness basin, analogous to that in the Larne boreholes. The small
amount of Trias penetrated at the base of the Mochras borehole (Tremadoc Bay) is noted with the much more extensive Jurassic (p. 310).

In the English Channel these red rocks are not known in the eastern
section but they become important in the centre. Here there is the offshore
continuation of the Devon outcrops, leading eastwards into the Wessex
basin where the Trias is probably thick. In the central and western section
as a whole the Permian Period probably begins a new stage in the
geological history of the Channel, with continental deposits lying on an
eroded and possibly faulted basement. At this time, before the opening of
this part of the Atlantic and the associated opening of the Bay of Biscay,
the Western Approaches were partly blocked on the south-west by the
Iberian massif of Spain.

No clear palaeogeography of the Channel or the outer Celtic Sea
emerges, however, partly because of the very wide continental shelf which
is now submerged and on which any Permo-Trias is deeply buried.
However there are a few pointers to an upland of some relief in which
thick basic lavas, essentially like those of the Exeter group, were a component; the western Channel includes a pronounced positive gravity
anomaly which would not result simply from a thick sedimentary basin.
Such an upland, at least till middle or later Triassic times, could have
provided the large cobbles of the Budleigh Salterton and Midland pebble
beds. Until the Rhaetic invasion it is probable that any waters would have
been restricted and possibly saline.

The previous section demonstrated how different were the data on the North Sea and it is in this direction, across the southern basin, that the major continental comparisons are found, in the Netherlands and especially Germany. The latter country is the source of the lithostratigraphic terms (Table 32) and the systems are very extensive. Permian outcrops occur chiefly on the borders of the Hercynian massifs and the Trias spreads out in between, but both are well known underground. The Rotliegendes is a widespread sandy formation and the Zechstein salt forms major halokinetic structures in north Germany and the Netherlands. The Muschelkalk Sea entered from the south-east, bringing a variety of marine and lagoonal facies and faunas; it then spread into the Triassic lowlands of eastern France, the North Sea (and just into England) and into Denmark and Poland.

Permian and Triassic strata stretch right across the southern basin, from eastern England to northern Germany, but the facies tend to resemble the latter. The Rotliegendes aeolian sands are concentrated in the western and southern portions and here are the principal gas wells; farther out in the centre there is a more shaly facies with thin evaporites – products of ephemeral lakes. The Zechstein evaporites thicken greatly into the basin, especially the halite, and the halokinetic structures, which are absent from the English mainland, are widespread right across to the eastern shores. The Zechstein cycles, which can be traced from England to Germany, have already been mentioned (p. 243) and the Trias of the basin follows conformably. It contains fewer sandy facies than in Britain, more evaporites and the Muschelkalk is well developed; substantial halite is found in the upper Bunter, Muschelkalk and Keuper divisions.

Permian and Trias are either incomplete or thin over the paired barriers between southern and northern basins but the Zechstein is continuous through the Central Graben. In general the rifted pattern was developing during the Permian as it did in Britain, to govern the interlocking sedimentary and tectonic history for much of Mesozoic times. In the north Permo-Trias is known more sporadically but marginal dune sands are recorded in the south-west, as they are in the southern basin. Possibly they accumulated on this side after being blown by the prevailing winds across the desert depression. They reach the margins of the Moray Firth, which is an exceptionally deep basin, faulted both along its margins and in the basin centre. In the west, close to the Scottish mainland, the trend is Caledonoid but farther east it is more east–west. The Mesozoic infilling has direct relations to the onshore outcrops, for instance the Permo-Trias of Elgin and the Jurassic of Brora in Sutherland.

A Zechstein basin probably extended from the outer Forth Approaches to the northern Danish sector and dolomites are known from the Moray Firth and around 60° N in the Viking Graben, where also the Trias is exceptionally thick. This is a possible, but not certain, route for the entry of oceanic water to supply the Zechstein evaporites.

Various volcanic rocks of early Permian age have been mentioned previously and further examples are known from the North Sea – on the flanks of the Ringkøbing–Fyn High and lesser examples on the western edge of the Central Graben and in the Viking Graben. There is thus a link with the much more extensive vulcanism of the Oslo Graben, which includes mildly alkaline basalts and rhomb-porphyries. The graben is a north–south structure and among the complex massifs of central Europe there are some analogous rifts, though with more varied trends, which contain Permian sediments and volcanic rocks and were possibly initiated at the same time.

REFERENCES

Smith *et al.* 1974 (*Special Report* No. 5, Permian).

Warrington *et al.* 1980 (*Special Report* No. 13, Trias).

General and regional: Arthurton *et al.* 1978; Audley-Charles, 1970*a*, *b*; Craig, 1965; Elliott, 1961; Evans, 1970; McLean, 1978; Pattison *et al.* 1973; Smith, 1973, 1974; Steel, 1974, 1978; Sylvester-Bradley and Ford, 1968, Chs. 9, 10; Warrington, 1970, 1974.

Sea areas, general and Permo-Trias: Curry and Smith, 1975; Evans and Thompson, 1979; Kent, 1975*a*, *b*; Naylor and Mounteney, 1975; Pegrum *et al.* 1975; Riddihough and Max, 1976; Smith and Curry, 1975; Woodland, 1975; Ziegler, 1978. Several Reports of the Institute of Geological Sciences include data on offshore geology.

British Regional Geology: Central England, Northern England, Northern Ireland, South-West England.

9

THE JURASSIC SYSTEM

Although the name of this system was derived in the last years of the eighteenth century from the Jura Mountains, the coast of England, and particularly that of Dorset, provides some of the best sections. Indeed several of the stages and traditional divisions (Table 39) are named after

Table 39. *Jurassic stages and divisions in southern England*

	Stages	Lithological divisions
UPPER JURASSIC	PORTLANDIAN	Lulworth Beds (lower Purbeck) Portland Beds
	KIMMERIDGIAN	Kimmeridge Clay
	OXFORDIAN	Corallian Beds Oxford Clay (upper part)
MIDDLE JURASSIC	CALLOVIAN	Oxford Clay (lower part) Kellaways Beds, etc.
	BATHONIAN	Great Oolite, limestones, etc.
	BAJOCIAN	Inferior Oolite,
	AALENIAN	limestones, etc.
LOWER JURASSIC	TOARCIAN	Upper Lias
	PLIENSBACHIAN	Middle Lias
	SINEMURIAN	Lower Lias
	HETTANGIAN	

The equivalence set out is in some cases only approximate; zones are given in the Appendix (pp. 422–3).

English towns and districts. Moreover, of all the systems the Jurassic exhibits most clearly the principles and practice of stratigraphical palaeontology. It cannot compare with the Carboniferous System in variety of rock type or with the Ordovician in the complex interplay of vulcanism, tectonics and stratigraphy; but the relationship between sedimentary strata and their faunas and the identification and correlation of beds by means of their contained fossils were first and most clearly established in Jurassic rocks.

Not all the stages in Britain are well provided with ammonites but most of them are, and these are then incomparably the best zone fossils. Their value was already appreciated in the middle of the last century and many of the zones and stages defined then still survive, with some changes in nomenclature, as working units to the present day. It is thus not accidental that discussions on the principles and practice of stratigraphical palaeontology include persistent references to Jurassic rocks and faunas, and the precision that is possible in their study may be taken as a kind of standard against which correlations and methods of other ages may be measured. The long history of Jurassic research does not mean that there are no problems left in the classification of these strata. Although abundant in many formations the ammonites are not ubiquitous; even among marine rocks they are rare or absent from most of the bioclastic limestones and particularly from the reef facies.

The period is not one of the longest nor are the strata outstanding in thickness in this country. The measured maximum is over 1500 metres in boreholes in southern England though more may originally have been deposited in north-west Britain. Fine-grained clastic sediments predominate – marine shales and mudstones; sandstones are relatively unimportant though they include deltaic facies in the Middle Jurassic and are economically significant in the North Sea. The limestones, though varied and conspicuous, are much less in total thickness than the shales.

Most of the structures affecting the Jurassic rocks are relatively simple. In north-east England and the Midlands there is a gentle regional dip to the east or south-east, the many minor divergencies having little effect on the broad crescentic sweep of the outcrop. In southern England the mid-Cretaceous and Tertiary folds are more acute and there is some Cretaceous overstep; in particular the latter conceals the Upper Jurassic over much of north Dorset and Wiltshire. The erosion of two anticlines on the western chalk scarp produces the vales of Wardour and Pewsey (Fig. 24) which help to fill the gap. In south Dorset the overstep and folding is complicated by the acuteness of the anticlines, which locally have an overturned northern limb associated with reversed faulting. The general effect, seen for instance around Weymouth and Lulworth (Fig. 65), is to form small asymmetric anticlines bringing up Jurassic beds while the overstepping Cretaceous is conspicuous on the flanks. Although intra-Jurassic movements are rarely seen there was probably mild tectonic warping. Some local influences affected the Lower Jurassic of the Cotswolds and, more conspicuously, the Upper Jurassic of north-east Scotland. Vulcanicity is an important factor in the northern North Sea and traces of ashfalls have been detected in the land outcrops of various ages, as in the Lower Cretaceous (p. 338).

In its wider setting Jurassic history inherits the scene already set by the Rhaetic invasion. An epicontinental sea developed widely over northern Europe (Fig. 59) and was connected to the Tethys in the south. No full

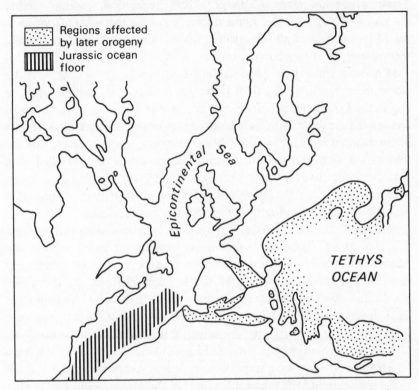

Fig. 59. Generalized interpretation of the palaeogeography in mid-Jurassic times. The area around the British Isles is still occupied by epicontinental seas but a strip of ocean floor has appeared in the Atlantic rift farther south.

analogues are known at present but the waters were probably shallow and near the margins merged with wide tidal flats and marshes. Within the British area two landmasses can be traced with some certainty – the low-lying London Platform, as in earlier periods, and another probably rather more pronounced in the Scottish Highlands. Arenaceous incursions suggest a low northern Pennine ridge, probably linked to southern Scotland and thence to the Mid North Sea High (Fig. 58). In the west, with very few outcrops, the scene remains tenuous; Cornubia may well have been emergent but any major Welsh island has become much more doubtful with the discovery of thick Lower Jurassic bordering the Irish Sea.

Although still in low latitudes (possibly around 30° to 35° N during much of the period) the marine influence and slight northward shift had greatly ameliorated the climate after the persistent partial deserts of the Trias. There are many traces of a lush vegetation, conifers, ferns, horsetails, Ginkgoales and Bennettitales, clothing the lowlands where a varied reptilian fauna also flourished. Invertebrates, including reef corals, were abundant in the shallow seas.

In another context the Jurassic Period also opens a new door: it is the oldest from which oceanic floor is known, of about 180 million years, or within the Lower Jurassic; from now on there is increasing opportunity to add sea-floor history, with its successive palaeomagnetic anomalies, to that of the continents. It is probable, though not very precisely dated, that the southern section of the North Atlantic became a true oceanic rift during Jurassic times. Away from the immediate rifting edges, which do not abut very closely on the British outcrops, the seas flooded the continental margins and these conditions continued into the Cretaceous.

Within the area of the major English outcrop from Dorset to Yorkshire (Fig. 60), there is an obvious distinction between the broad sedimentary basins, with thicker and more complete sequences, and the intervening divides. South and south-west of the London Platform lay the major Weald and Wessex basins, partly linked and both continuing into the English Channel. The western and northern margins of the platform were more irregular; in particular there was a belt of shallows, centred on Oxfordshire, which perhaps extended across to the Midlands in areas from which Jurassic strata have since been eroded. A somewhat similar belt lay to the west, around the limestone islands of the Mendips and northwards across the 'Radstock Shelf' to Bath and Bristol. Between here and Oxford there was the deep basin of the lower Severn and the Cotswolds, also a Triassic inheritance.

North from the London Platform a much shallower basin appears in Northamptonshire and Lincolnshire, perhaps better seen as the relatively stable East Midlands Shelf. This and the deep Yorkshire basin, which has some affinities with the southern North Sea in its Mesozoic history, are separated by the Market Weighton structure or block. This was an unusually persistent element that caused thinning or absence of all formations from the Rhaetic to the Lower Chalk.

The divides between the basins have been given various names, of which the non-committal 'structure' or 'swell' are common – the latter

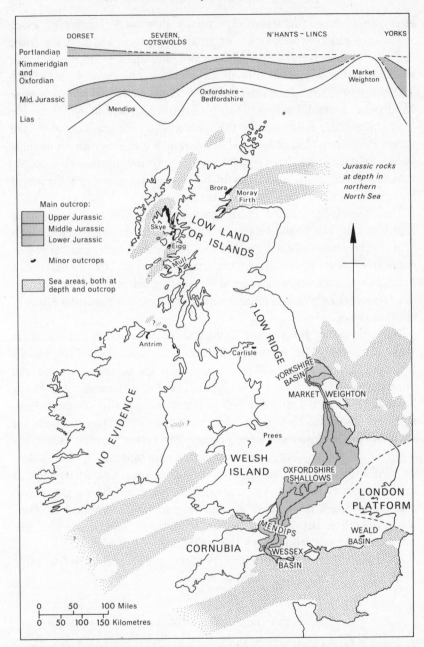

DORSET   SEVERN, COTSWOLDS   N'HANTS – LINCS   YORKS

Portlandian

Kimmeridgian and Oxfordian

Market Weighton

Mid. Jurassic

Mendips   Oxfordshire – Bedfordshire

Lias

Jurassic rocks at depth in northern North Sea

Brora   Moray Firth

Main outcrop:

Upper Jurassic
Middle Jurassic
Lower Jurassic

Minor outcrops

Sea areas, both at depth and outcrop

Skye

LOW LAND OR ISLANDS

Eigg

Mull

? LOW RIDGE

Antrim

Carlisle

YORKSHIRE BASIN

MARKET WEIGHTON

NO EVIDENCE

Prees

WELSH ISLAND
?

OXFORDSHIRE SHALLOWS

LONDON PLATFORM

MENDIPS

WEALD BASIN

CORNUBIA

WESSEX BASIN

?

?

0   50   100 Miles
0   50   100   150 Kilometres

Fig. 60. Distribution of Jurassic rocks showing the main outcrop, minor outliers or inliers and generalized areas in the seas. At the top, outline section showing relative thicknesses across basins and swells, not to scale.

after the German 'Schwelle'. Essentially they were regions of lesser subsidence and only occasionally of actual emergence, and since the basin and swell pattern is influential in the major outcrops it seems possible that similar structures affected Jurassic sedimentation to the west and north. The histories of the various swells varied considerably and probably their underlying causes did as well. Thus the Mendips are within the Hercynian belt, Oxfordshire is adjacent to the London Platform and may well have varied Palaeozoic foundations, while Market Weighton (with no obvious basement effects) has been ascribed on geophysical grounds to a deeply buried granitic mass, analogous to the intrusions below the Pennines, though this can only be a tentative suggestion.

Comparisons in sedimentary influence have been drawn with the Lower Carboniferous blocks and basins but the Mesozoic topography was much gentler; the effects of even the Market Weighton structure are only just detectable on a section drawn to true scale. That the swells had a substantial influence is a measure of the delicate balances that operated on the Jurassic sea floor, the foundations that lay below it and the effect of changes in ocean-level.

During much of Jurassic and Cretaceous times the marine faunas show some distinction into a Boreal Province and a southern (Tethyan or Mediterranean) Province. The former includes outcrops around the present Arctic Ocean, in Russia, Poland, northern Germany, the North Sea and northern Britain. The belt of intermingling, with elements from both provinces, shifted from time to time but frequently included southern England and northern France. The Lower Jurassic faunas are fairly cosmopolitan but provincial differences appear in the middle division and become more acute in the upper one. The current list of British zones (Appendix, pp. 422–3) is based primarily on the boreal succession, since there were enough migrants to the south to effect correlation; it is also an advantage in the later Portlandian Stage when ammonites are not found in southern England.

As in other chapters the stratigraphy of the land areas is taken first and a summary of the sea floors added at the end.

LIAS OR LOWER JURASSIC

The term Lias has long been used for the lowest of the three major Jurassic divisions, which is also the most uniform in lithology. Even so there is more variation than appears at first sight or in poor exposures, and what seems to be a monotonous series of shales proves on detailed examination to have many minor variations. There are additions of coarser detritus, silt or fine sand; the precipitated components include ironstones and impure limestones, the latter especially towards the base. Intermittently local stagnant conditions gave rise to dark grey or black bituminous shales, sometimes laminated or accompanied by much iron pyrites. In these, benthonic faunas are sparse or absent but floating or swimming forms are well preserved and include ammonites, fishes and marine reptiles. In the other rock types, formed in well aerated shallow waters, there is commonly a rich fauna. In addition to the ammonites, bivalves are abundant (especially oysters and gryphaeas), brachiopods, belemnites and many other invertebrates; trace-fossils are locally common and varied, especially in the inshore or subtidal sediments.

The base of the Jurassic is drawn where the ammonites of the lowest Hettangian zone enter and the 'Pre-planorbis Beds' are referred to the Rhaetic below (p. 265). Formally the Lias is a lithostratigraphic term but it corresponds closely to the following stages:

Upper Lias	Toarcian Stage (Whitbian and Yeovilian Substages)
Middle Lias	Pliensbachian Stage, upper part
Lower Lias	{ Pliensbachian Stage, lower part, Sinemurian and Hettangian Stages

Variations in the main outcrop

Of the three Jurassic divisions the Lias is the most affected by changes in thickness in the various subsiding basins and across the intervening swells. In addition each basin had its own variants in sedimentation and the way in which certain zones increase or decrease in thickness, are cut out entirely or change in facies is often extremely complicated. Such changes, reflecting sedimentary conditions, can only be traced in a system like the Jurassic, where contemporaneity can be checked from outcrop to outcrop against the detailed zonal succession.

In the south the succession in the Wessex basin – the type area for Britain – is well seen in the Dorset cliffs, where the beds dip gently eastwards along the sweep of Lyme Regis bay. Most of the 280 metres is composed of Lower and Middle Lias. The latter tends to be sandy with conspicuous beds of calcareous sandstone jutting out; in the former much attention has been paid to the lowest beds, the Blue Lias. The facies here consists of four principal rock types: marls, limestones (calcilutites), very finely laminated limestones and bituminous shales. The last two represent the stagnant conditions already mentioned but the first two are rich in fossils, especially ammonites and bivalves. The repetitions of the four variants, signifying aerated and stagnant conditions, form one of the many types of repetitive or cyclic sedimentation found in Jurassic rocks; here the distinction is accentuated by diagenetic changes as well. Similar conditions and sediments extended across into South Wales, though only part of the Lower Lias is known here. As in the Rhaetic period abnormal littoral facies abut against Carboniferous islands, but the Liassic types consist of massive limestones with chert, sandy and conglomeratic beds. Some of the limestones are rich in corals, which elsewhere are rare as might be expected in the muddy type of Lias.

In the main outcrop the Wessex basin stretched northwards through Somerset to the Mendips, which also formed limestone islands or shoals. Beyond lay the shallows of the Radstock Shelf where sedimentation was slow and intermittent; the lowest zones are thin with layers of phosphatic pebbles and locally the Middle Lias is missing altogether. North of Bristol there is a marked expansion into the lower Severn basin where the Lias is argillaceous and thick (514 m), as found in a borehole in the northern Cotswolds. Within this major downwarp smaller synclinal sags developed during the Lias and became faulted on their eastern margins so that there is local thinning over the subsidiary anticlines. The dominant trends are north–south, as in the Trias, and similarly they may result from tectonic influences in underlying rocks.

The Oxfordshire shallows produced a general reduction to 168 metres, but in an uneven manner. Thus at its extreme attenuation the Upper Lias is no more than a couple of metres, including a basal limestone with eroded ammonites of several zones. In the irregular expansion that succeeds northwards through the East Midlands this upper division continues to be affected by condensed sequences and internal overlap within its zones, reflecting very shallow shelf conditions with only intermittent sediment accumulation. Northwards from Lincolnshire all the Jurassic formations become thinner but the Lower Lias is least affected. At Market Weighton 30 metres survive and it is the only component at outcrop, being overlain directly by the Albian (uppermost stage of the Lower Cretaceous). The swell was not a complete barrier, however, and the seas and their faunas had free passage round it, certainly to the east and possibly to the west also.

Where the full Liassic sequence reappears in the Yorkshire basin it amounts to over 370 metres, well exposed along the northern cliffs and only the lowest zones hidden under the Tees alluvium. Compared with their southern equivalents the beds are more argillaceous and less variable but include many of the typical subsidiary components – beds rich in silt, impure limestone, ironstones and organic matter. Similarly minor variations in bottom conditions and depth of water produced small-scale repetitions or cyclic sequences, concurrently with variations in the bottom faunas, chiefly bivalves, and burrowing forms.

Ironstones

Until recently by far the largest source of iron ore in this country was the bedded ironstones of the Lias and the Aalenian Stage above. Although little is being worked now the iron-bearing facies is important in the Lias. In composition the various ironstones have much in common. The iron is present as siderite, chamosite and limonite while the non-ferrous components include calcite, clay minerals and an opaline form of silica. In texture the rocks consist of ooliths, fragmental and fossil debris and matrix. Siderite, chamosite and calcite are important in the matrix and the ooliths were originally of chamosite, locally altered to siderite or limonite.

All the ironstones show lateral variation from what has been a rich workable rock to clays or sandy limestones that are only locally iron-bearing. Thus in south Humberside the 10 metres of the Frodingham ore at Scunthorpe consist very largely of a calcareous oolitic ironstone, rich in belemnites and bivalves, which expands southwards into 50 metres of clays that are only partly ferruginous. Similarly the Middle Lias Marlstone is locally enriched in Leicestershire and north Oxfordshire and at the scarp of Edge Hill it contains six metres of ironstone; but as the Oxfordshire shallows are approached terrigenous detritus from the London Platform enters and the iron content falls. The most valuable of these bedded ores was the Cleveland Ironstone Formation. In the north of the Yorkshire basin several seams reach a combined maximum of 24 metres but southwards an internal non-sequence develops and the seams and intervening shales decrease; as with the rest of the Middle Lias the whole disappears well north of Market Weighton.

Ironstone precipitation must be considered as an integral part of the Liassic sedimentary scene, affecting most of the major provinces except in the south; the details of its origin, however, are still somewhat problematical. The ultimate source of the iron is thought to be the slowly flowing rivers from the warm Jurassic hinterlands. Workable ironstones are commonly thinner than the equivalent shaly or sandy beds, which may also include lesser amounts of the same iron-bearing minerals, and bear signs of condensed or interrupted deposition. Probably they represent shoal areas where the clastic sediments were sparse, yet where gentle currents and aerated waters promoted the formation of ooliths and the flourishing benthonic shelly faunas. Then through subsequent chemical changes both ooliths and matrix were affected by complex diagenesis.

Changes in late Toarcian times

At many places along the main outcrop there are signs of major warping, sandy incursions, erosion or non-sequence towards the top of the Upper Lias, chiefly in the upper Yeovilian Substage. Examples are taken from south and north.

From Gloucester southwards the most conspicuous formation in the Upper Lias is a thick mass of fine-grained yellow sands with a sporadic calcareous cement. They are known by different names in different parts of their outcrop but especially as the Cotswolds Sands in the north and the Bridport Sands on the Dorset coast. In the field they appear to be a single formation but their relation to the zones and subzones of the upper Toarcian and lowest Aalenian (Fig. 61) shows that the boundaries above

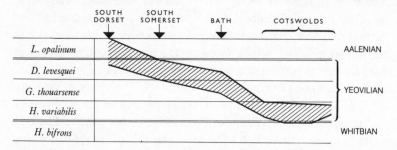

Fig. 61. Diagram showing the distribution and age of the Cotswold–Bridport sands, in relation to the zones of the upper Toarcian and lowest Aalenian. (After Wills, 1929, p. 139; zones from Dean *et al.* 1961.)

and below are diachronous and in all, from the Cotswolds to Dorset, the sands cover four ammonite zones. The 'cephalopod beds' which are found at various places above or below are thin layers of clay or limestone crowded with ammonites from more than one horizon. They accumulated in current-swept conditions where little or no normal sediment was retained. The origin of the sandy detritus entering this shoaling sea is not very clear, but the mineral content suggests a possible south-westerly source including metamorphic rocks, comparable to that supplying Wealden sediments (Fig. 70).

In their transgressive boundaries and false appearance of contemporaneity the Cotswolds Sands are a classic example of a diachronous deposit. They also show how, in order to establish the details of diachronism, some rigorous time-sequence is necessary, which in this case is supplied by the ammonite species.

Evidence of shallowing, emergence and pre-Aalenian warping is presented in the Yorkshire basin in a different way. Over most of the outcrops

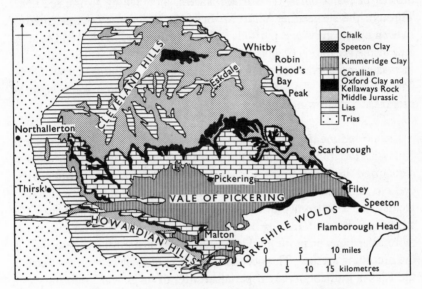

Fig. 62. Geological sketch map of north-east Yorkshire.

the Whitbian Substage alone is present, overlain by a relatively thin Dogger, the lowest Middle Jurassic formation. In a few of the inland valleys however isolated synclines include some Yeovilian strata and on a short section of the coast south of Robin Hood's Bay (Fig. 62) there is a full Upper Lias complement, up to the top Yeovilian zone, and only a small break below a thicker Dogger. This fuller sequence and the abbreviated one are separated by the Peak Fault – probably a transcurrent fracture (probably Tertiary), which has brought together Toarcian sequences which after gentle warping were differently affected by the erosive phase that preceded the Aalenian invasion, a phase ubiquitous in the Yorkshire basin as a whole and common elsewhere.

Another aspect of the full Upper Lias along this coastal stretch is its upward change in facies, beginning with the Jet Rock in the lower part of the Whitbian Substage; from here upwards there is increasing evidence of aeration and shallowing. The Jet Rock is highly bituminous, the ammonite chambers sometimes contain oil, and pyrite is abundant in the dark grey shales – a type of facies widespread outside Britain at this time, for instance in Germany. Bottom faunas are rare or absent, the bottom waters being stagnant and the muds laid down below wave action. Fine laminations are typical of the shales and have been attributed to the annual growth of plankton in the surface waters, later falling to the sea floor. Through the succeeding Whitbian beds the sediments became progressively a lighter grey, less pyritous and less bituminous. This trend culminates in the grey and, finally, yellow beds of Blea Wyke which contain an increasing proportion of sand and evidence of a thriving bottom fauna. The Middle and Lower Lias in this basin also show some signs of progressive shallowing but they are less pronounced.

Liassic sedimentation exhibits several aspects of wider application in Jurassic stratigraphy. Cyclic sedimentation is one of them and it has been recognized in many areas and on very different scales. Evidence for cyclicity is drawn from lithological and chemical variations in the sediments, and also from the faunas adapted to successive conditions. The scale varies from alternating beds of only a few centimetres, for instance of more calcareous or more muddy layers, to the major effects described in the preceding section. The peculiar profusion of these repetitive effects in the Lias is probably related, paradoxically, to its *general* uniformity as a great argillaceous group. Except in a few marginal outcrops there was little coarse detritus carried in to mask the subtle physical and chemical changes on the floor of the shallow sea.

Some variations in interpretation naturally arise; the origin of ironstone, for instance, has produced many problems, and the depth of water and how it may be assessed is a persistent query. On a large scale there is the question of ultimate causes. Are the major changes due to eustatic alterations in sea-level or to tectonic warping? In Jurassic history as a whole some transgressions and regressions are very widespread and for these the eustatic argument is persuasive; certain major faunal breaks also tend to coincide with such phases. However in a small area like Britain, on heterogeneous foundations, some tectonic effects are likely to be added. Presumably they governed the distribution of swells and basins in the first place and such local events as pre-Aalenian warping or other comparable effects among the later formations.

CARBONATE PLATFORMS OF THE MIDLANDS AND COTSWOLDS: AALENIAN, BAJOCIAN, BATHONIAN

These three Middle Jurassic stages have nearly, but not quite, the same range of strata as the traditional lithological terms Inferior Oolite and Great Oolite (Table 40). The latter are retained here because they aid

Table 40. *Limestones of the Cotswolds*

	(Upper Cornbrash and Kellaways Beds)	
BATHONIAN	{ Lower Cornbrash { Great Oolite, limestones, clays, etc.	} Approximate | lithological
BAJOCIAN	{ Upper Inferior Oolite { Middle Inferior Oolite	} equivalents to the
AALENIAN	Lower Inferior Oolite	} stages
	(Upper Lias)	

description of the complex carbonate belt south of Market Weighton; borehole cores may also have to be logged on a basis of lithology (cf. Fig. 66(*c*)).

In the west of England the Inferior Oolite is the most conspicuous group because it forms the fine scarp and long dip-slopes of the Cotswold hills, from Oxfordshire to Bath. The Lincolnshire Limestone, similar in age, is also a feature-forming division. South of Bath the Inferior Oolite is less important but Great Oolite limestones are found in the relatively high ground of east Somerset and Wiltshire; they disappear southwards and those of the Inferior Oolite become exceedingly thin. Although dominantly calcareous, facies variation is a hall-mark of the three stages, while in the Bajocian in particular there are many local breaks in the sequence. At some levels ammonites are rare or lacking, for instance in the lower part of the Bathonian; bivalves, brachiopods and ostracods provide some substitutes but where there is strong lateral change a large number of local names have been used and some correlation problems remain. The various limestones are linked in environment and aspects of their interpretation are taken together (p. 292).

Inferior Oolite of the Cotswolds and Wessex

This division is thickest in the basin of north Gloucestershire, which had earlier subsided with a full complement of Liassic clays, and like the Lias it also becomes thinner over the Oxfordshire shallows. Even in the basin subsidence was not uniform or continuous but was twice interrupted by slight uplift and very gentle warping. There is thus a division into Lower, Middle and Upper Inferior Oolite (Fig. 63); the second phase was the

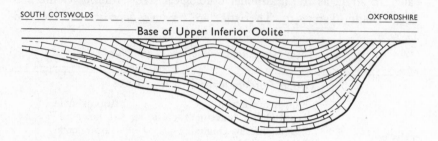

SOUTH COTSWOLDS   OXFORDSHIRE

Base of Upper Inferior Oolite

Fig. 63. Diagram to show the gentle folding and unconformities in the Inferior Oolite of the Cotswolds, the transgressive upper division being taken as horizontal. (After Kellaway and Welch, 1948, p. 68.)

more extensive and after persistent erosion the gentle flexures were planed off and covered by the base of the Upper Inferior Oolite, which here represents only part of the upper Bajocian. A further result is that certain of the underlying beds are confined to the synclines; the fullest succession is at the height of the scarp near Cheltenham but even here the Inferior Oolite is only about 100 metres thick.

Almost all the beds are calcareous with many signs of shallow water and penecontemporaneous erosion. They include sandy and rubbly lime-stones, pisolites and oolitic freestones – the last term referring to the ease with which the rock can be cut in any direction. Erosion levels are shown by planed off and bored surfaces, by non-sequences and in some places the absence of certain lithological units. Shells and banks of shell debris are abundant, together with echinoderms and corals; the last are common in some beds but do not approach a true reef.

When traced from this type area north-eastwards into Oxfordshire the Lower and Middle Inferior Oolite become thinner, with the progressive disappearance of their lithological components; ultimately they disappear and only the transgressive upper division remains. Even this becomes reduced in the north-west of the county to two metres of limestone, the eroded ammonites at the base lying on a similar condensed sequence of the Whitbian Substage, the Yeovilian being absent. This is the very meagre link between the Bajocian outcrops of the Cotswolds and the East Midlands.

Southwards from the Cotswolds erosion was again active over the Mendips and only the upper Bajocian is found, as a thin capping of rubbly limestones which overlie the steeply dipping Carboniferous Limestone. For the rest of the outcrop down to Bridport all three divisions are present but the total is not more than 30 metres and in places very much less, with many signs of erosion and non-sequence. Nevertheless in this south-western belt ammonites are more plentiful than in the Cotswolds; zones and subzones can be recognized in detail even though some may be represented only by thin layers of conglomerate with waterworn fossils.

On the Dorset cliffs the Inferior Oolite forms but a thin layer above the Bridport Sands, the upper part of which extends up into the lowest Aalenian zone (*opalinum*). Above there is a metre or so of sandy limestone forming a transitional facies and then an almost incredibly condensed sequence, in under four metres, of oolitic and conglomeratic limestones, with bivalves, ammonites and sponges. It has been suggested that the Upper Lias sands stood up as a submarine ridge or a belt of intertidal banks, so that very little permanent accumulation was possible in the current-swept waters above and around the shoals. Limestones and oolites found offshore are rather thicker but even so there is little to suggest a major Wessex 'basin' during Inferior Oolite sedimentation.

The Great Oolite of Wessex, Bath and Oxfordshire

This division is nearly, but not quite, equivalent to the Bathonian Stage since the latter includes the thin Lower Cornbrash above (p. 296). It also supplies a good example of the problems that may still attend Jurassic stratigraphy even when the rocks are marine, fossiliferous, conformable and fairly well exposed; many of the beds in the Bath district have been known since the days of William Smith. Difficulties arise not so much from erosions and transgressions like those that complicate the Inferior Oolite, but from the rarity or absence of ammonites and the bewildering variety of local facies, especially among the limestones.

In the west of England two major contrasting facies are found at outcrop and persist underground to the east: the limestones of Bath and the southern Cotswolds and the clays of the Wessex basin, whose subsidence is clearly re-established. Some 300 to 360 metres of Bathonian sediments were penetrated here in boreholes just onshore and offshore from the Dorset coast (Kimmeridge and Lulworth Banks). The facies is very largely the Fuller's Earth Clays, with occasional thin impure limestones and shell beds. As their equivalents are traced northwards along the outcrop towards Bath the whole becomes thinner, to a third or less, and various limestones replace the clays – a typical Jurassic diachronism. A grey limestone with brachiopods, the Fuller's Earth Rock, divides the clay into lower and upper units, and then near Bath the Great Oolite limestones largely replace the upper one.

This is the margin of the Cotswold shelf, or a return to the carbonate platform facies. As before there was a variety of shelly and oolitic rocks with white calcilutites and several coral beds, locally forming small patch-reefs. The Bath Stone, a typical oolitic freestone, was quarried and worked underground near the city; it is one of the great historic building stones and justifiably gives its name to the whole stage. The Forest Marble at the top, named from Wychwood Forest in Oxfordshire, contains more clastic detritus though it is still largely marine – shelly and cross-bedded limestones, sandy beds and clays with occasional plant remains and lignites.

The commercial seam of fuller's earth south of Bath has an unexpected geological interest. The principal clay mineral (montmorillonite) is characteristic of bentonites, or altered volcanic ash. Now that much more is known of Jurassic vulcanism this does not seem so out of place as it once did, since there are the major Middle Jurassic lavas in the North Sea (p. 317) and possibly eruptions elsewhere, for instance along the newly rifted margin of the Atlantic. No certain source is deduced, however.

North-eastwards from the Cotswolds there are also changes in facies and thickness but after the late Bajocian erosion and transgression the concept of the Oxfordshire shallows, between the Cotswold basin and the lesser subsidence in the East Midlands, ceases to be precisely applicable. From now on there is a wider area, including parts of Bedfordshire and Buckinghamshire, that is more obviously part of the western slope of the London Platform. Over this stretch the Great Oolite comprises several thin limestones, with local names and not all easily correlated, the last remnant of the Fuller's Earth having disappeared.

The Stonesfield Slate near the base is a thin unit peculiar to the northern Cotswolds and Oxfordshire – a fissile sandy limestone, locally used as roofing material. However in addition to a normal marine shelly fauna it includes a few plants and a remarkable vertebrate assemblage: land reptiles, flying forms and rare remains of early mammals. The Forest Marble also includes some plants, locally abundant lignite and some freshwater ostracods and shells. Here we see the debris drifted in from the nearby land which was fringed with marshes and tidal flats.

FACIES VARIATION ON THE EAST MIDLANDS SHELF: AALENIAN,
BAJOCIAN, BATHONIAN

Where the three stages expand, in modest degree, from Oxfordshire to the
East Midlands neither the lithological grouping nor the erosion levels
correspond closely with those of the Cotswolds, although limestones are
still important and most of the sequence is marine (Table 41). There was a

Table 41. *Bathonian, Bajocian and Aalenian of Lincolnshire*

Stages	Approximate lithological equivalents
BATHONIAN	Lower Cornbrash Blisworth Clay Blisworth Limestone (Great Oolite Limestone) 'Upper Estuarine Formation'
LOWER BAJOCIAN and AALENIAN	Lincolnshire Limestone Formation Grantham Formation ('Lower Estuarine Series') Northampton Ironstone

marked change towards the end of Upper Lias times and widespread
emergence. The base of the Northampton Ironstone includes a layer of
phosphatic pebbles and lies for the most part on the lower substage, the
Whitbian. The ironstone contains ammonites of the *opalinum* Zone and
much resembles the Liassic ironstones; however the proportion of sand
grains is higher and away from the main orefield around Kettering and
Corby the beds become a ferruginous sandstone of no commercial
importance. Lateral facies change, non-sequences and erosion channels
point to very shallow water and locally the whole ironstone may be absent
through channelling at the base of the beds above.

The Grantham Formation marks a further broad recession of the sea.
The earlier epithet 'estuarine' was misleading, and these beds were formed
in low-lying coastal flats fringing the London Platform on its northern
side, the rootlets of the first colonizing vegetation penetrating the
ironstone or sands beneath. The beds consist chiefly of grey clays, silts and
fine sands, with some resemblance to the Coal Measures in the rootlet beds
and seatearths; nearly all the fossils are fragmentary plants, with a few
shells.

These coastal flats were then submerged by the seas of the Lincolnshire
Limestone. It is thinner than the Cotswold limestones, with a maximum of
41 metres in the south of the county, but otherwise has many features in
common with them. The dominant rock is an oolite but there are also

various bioclastic types and calcilutites; some parts are rich in brachiopods and molluscs but not ammonites. The freestones are a famous building material and have been used widely in eastern England since mediaeval times and also in the House of Commons. The Collyweston Slate at the base is fissile and sandy like the Bathonian Stonesfield Slate, the sand grains resembling those of the beds beneath but embedded in a calcareous matrix during the first flooding of the coastal flats. Towards the south-west the whole becomes thinner, eroded surfaces and channels develop at the base of the upper limestones and finally the whole carbonate facies disappears near Kettering (Northamptonshire), well before the Oxford-shire shallows are reached. Moreover, as on the Cotswold shelf, there were extensive breaks in sedimentation and in the East Midlands the whole of the upper Bajocian appears to be missing.

The Bathonian is even thinner, commonly around 16 metres, but its components are rather similar. The 'Upper Estuarine Formation' of grey and lavender clays also contains several rootlet beds, showing that the oolite and other carbonate banks were colonized during this regressive phase. There is little evidence of channelling but the muds and marshes were locally in contact with marine waters; an oyster bed was established in the middle of the formation and there are mixtures of marine and freshwater ostracods. The Blisworth Limestone is a short-lived return to the clear shoal waters of the carbonate shelf or bank environment, in which accumulated fine-grained bioclastic limestones with oysters but not many oolitic beds. This phase was then terminated by an influx of mud; the Blisworth Clay is variegated and poor in fossils, but it lacks rootlet beds and does contain a few oysters. It may represent muddy lagoonal conditions rather than mud-flats, breached temporarily by the sea which allowed the oysters to settle.

All these Middle Jurassic formations are reduced northwards towards Market Weighton but the Lincolnshire Limestone less than its compan-ions. A small thickness of oolite is found north of the Humber and by some route there was temporary continuity at this level with marine facies in the Yorkshire basin.

The environment of the carbonates

Very shallow waters and incomplete sequences are recurrent themes in the interpretations of the Middle Jurassic carbonate facies. The most usual comparison is with the modern Bahama Banks and Florida Keys. Here there is a range of intertidal and subtidal analogues – oolite shoals and banks (even locally emergent), slightly deeper channels, patches of corals, the seas variously current-swept or quieter and more protected; the fauna varies with depth and type of bottom sediment. The Cotswolds and East Midlands represent only small-scale carbonate platforms of this type, bordering the subdued land of the London Platform. This was elevated enough at times however to supply some terrigenous detritus, chiefly northwards into Northamptonshire and especially in the early stages of the Aalenian transgression. From this source also were derived the transported land plants and animals, freshwater shells and ostracods. In such conditions the submarine shoals, intertidal flats and coastal marshes could be transformed into one another by small changes in relative land- and sea-levels. We should not look for any rigid or persistent geographical entities; the land area of the London Platform merges with the sea areas with fluctuating boundaries.

The extent of the carbonates to the north-west is unknown but the facies is lacking in the Wessex basin and largely in the Bristol Channel; a fringe bordered at least part of the London Platform on the south.

THE DELTAIC FACIES IN YORKSHIRE: AALENIAN, BAJOCIAN, BATHONIAN

The Yorkshire basin is particularly characterized by its deltaic facies which occupies most of the Middle Jurassic. Detailed correlation is rarely possible because although marine waters did invade the delta several times they very rarely brought in ammonites. However it seems fairly clear that the Dogger at the base belongs to the lowest Aalenian zone and the Bathonian is represented by the uppermost deltaic unit (the Scalby Formation) alone. The whole succession is admirably displayed along the cliffs of north Yorkshire and in the inland dales, and the Howardian Hills extend the outcrops farther south towards Market Weighton. Taking marine and non-marine facies together some 250 metres were deposited in the Yorkshire basin.

The marine facies

As would be expected from their geographical setting the marine intercalations tend to thicken to the south where they become substantial components of the delta complex. In rock type they are dominantly sandy, calcareous and ferruginous and are locally rich in a shallow water fauna. The Dogger at the base is the most variable and its changes in thickness are related to the magnitude of the underlying unconformity. Along the short coastal stretch where Yeovilian beds are present these are succeeded by 12 metres of ferruginous sandstones and chamosite oolites with an abundant shelly fauna and occasional corals. The base is marked by an inconspicuous erosion level with pebbles and rolled fossils. Along the remainder of the coastal outcrops north-westwards to Whitby not more than four metres are present, lying with a major unconformity on Whitbian shales. Over north-east Yorkshire as a whole further Dogger facies appear – such as black shales, sideritic sandstones and mudstones.

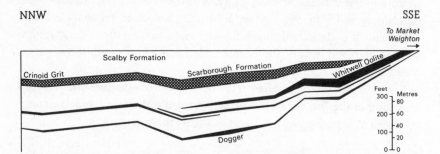

Fig. 64. Diagrammatic section through the Middle Jurassic of Yorkshire to show the various marine formations and general thinning towards the south. The intervening deltaic facies is left blank. (Redrawn from Hemingway, 1974*a*, p. 182.)

Some variants only occupy channels and hollows cut into the beds beneath and contain phosphatic pebbles at the base. It appears that the Upper Lias was uplifted, very gently folded and locally eroded and then the varied but dominantly ferruginous deposits of the Dogger were laid down on an uneven surface in a number of very shallow marine basins or lagoons.

At this point the sands, silts and marsh facies entered the Yorkshire basin, but at times subsidence overtook sedimentation and the seas invaded the delta, though they did not extend uniformly right across the present outcrop area (Fig. 64). Excluding the initial Dogger invasion, four such incursions occurred during the Aalenian and Bajocian, the first and fourth initially from the east, the second and third from the south. Of these the third is clearly an extension northwards of the Lincolnshire Limestone and even the Market Weighton block may have been shallowly covered. Thus in the Howardian Hills to the west a normal Bajocian limestone (the Whitwell Oolite) reaches eight metres while to the east on the coast there is a calcareous and sandy facies and in the north a thinner, more ferruginous equivalent. A different pattern is seen in the last and thickest marine facies – the Scarborough Formation. This reaches 32 metres in the central cliff outcrops, where the sandstones, shales and impure limestones are often richly fossiliferous; the fauna includes bivalves and gastropods, scattered crinoid debris and ostracods, together with a complex range of trace-fossils. In the western outcrops another detrital component was added in the later stages, a coarse sandstone with quartz pebbles and only scattered fossils; this is one of the clearest pointers to the erosion of Carboniferous sandstones somewhere to the west or north, probably the coarser Millstone Grit.

The non-marine facies

The Yorkshire delta was a broad marshy area of silts and clays and more rarely sand. Sluggish streams wandered across it and there were large stretches of open water. The dominant colonizing vegetation was the horsetail thickets of *Equisetites*, which are often represented by rootlet beds when the upper parts have been eroded. Other plants include ferns, ginkgos, cycad-like forms and conifers – a rich flora best seen in the fine-grained plant beds. Animal remains are rare but include the fresh-water bivalve *Unio* and traces of fishes and reptiles, chiefly dinosaur footprints. There are several resemblances to the Coal Measures though the Yorkshire delta is much smaller than the great Westphalian coal swamps. A few thin Jurassic coals add to the parallel, though the quality is usually poor.

This was the dominant marsh facies but there were also periods when the prevailing subsidence was interrupted by slight uplifts; the distributary streams then cut down through the previous marshy clays and silts to produce a series of channels, now usually filled with fluviatile sands and finer detritus collapsing inwards from the channel sides. The general direction of the distributaries seems to have been from the north, with a minority from the west.

After the last marine invasion the delta remained above sea-level, probably for the whole of the Bathonian Stage, here represented by the Scalby Formation and overlying unconformity. Although the grades of the incoming sediments are similar to those of the earlier marsh facies there are some differences. The first influx was of widespread sands, highly quartzose with cross-bedding common – a mature fluviatile deposit. This was succeeded by fine sands, silts and clays, locally with washouts and plant beds, some insect remains and a footprint bed. This last stage of the Yorkshire delta shows further signs of derivation from deeply weathered

Carboniferous source rocks. That provenance, and the known marine transgression from the east and south, has reinforced the view that the northern Pennines were part of a low landmass to the north-west of the basin. The offshore geology also indicates that this land joined up in the north with southern Scotland and then swung round to the south-east as part of an enlarged Mid North Sea High. This arcuate land was covered by Carboniferous rocks to some extent at least as residual patches still remain. The Yorkshire delta thus appears to have been at the head of an elongate subsiding basin, the south-eastern part of which is now under the North Sea (p. 349) – the Sole Pit Trough.

THE CORNBRASH: UPPER BATHONIAN AND LOWER CALLOVIAN

'Cornbrash' was one of William Smith's original terms for a thin rubbly brown-weathering bed of limestone that extends almost from one end of the main outcrop to the other. In western England and the Midlands it commonly forms a flattish belt of land suitable for arable crops.

The name therefore has a respectable antiquity as a rock unit but when the faunas and detailed stratigraphy are studied certain anomalies appear and the uniform outcrop of the small-scale map turns out to be rather misleading. Ammonites reappear in the Cornbrash after their extreme rarity in the beds below but those of the upper and lower divisions are widely dissimilar. *Clydoniceras* of the Lower Cornbrash has Bathonian affinities and a new Callovian genus, *Macrocephalites*, enters in the Upper Cornbrash. As a result the lithological formation, either half of which may be only a few metres thick, overlaps two major Jurassic stages. In its slender way the Cornbrash illustrates neatly the progress of stratigraphical definition from lithological to palaeontological criteria.

A study of variation in the formation across country resolves itself, in more detail than is appropriate here, into minor changes of thickness and facies of the lower and upper divisions. The whole must have been laid down in very shallow water, with contemporary erosion, comparable to the conditions of earlier limestones. It is thickest in the Wessex basin, some 18 metres; from Bedfordshire, where there is only a metre, the outcrop extends to the Humber, nearly all being Upper Cornbrash. Then there is a gap of 80 kilometres over the Market Weighton swell. In Yorkshire only the upper division is present, as a grey hard shelly limestone, which occurring after a period of erosion marks the final submergence of the mid-Jurassic delta.

Although this early Callovian marine invasion is not very dramatic in

Britain it does introduce a new widespread transgression of the European seas, after the irregular but general regressive tendencies of the Bathonian Stage. Thenceforward those seas gradually deepened and widened, with an acme in the Kimmeridgian Stage.

MIDDLE TO UPPER JURASSIC: CALLOVIAN TO KIMMERIDGIAN

As mentioned at the beginning of this chapter the international stages of the Jurassic System are based on the ammonite faunas. Sometimes the boundaries coincide approximately with changes in condition and

Table 42. *Callovian to Portlandian successions*

Stages and components in southern England	Changes to the east and north
PORTLANDIAN:	
Lulworth Beds (lower Purbeck)	Absent except for thin sandy remnants
Portland Stone	in the east
Portland Sand	
KIMMERIDGIAN:	
Kimmeridge Clay facies	Mostly as in the south but locally
throughout	incomplete at the top
OXFORDIAN:	
Corallian, limestones, etc.	Corallian (Yorks.), Ampthill Clay (Midlands)
Oxford Clay, upper part	Oxford Clay, locally thinned
CALLOVIAN:	
Oxford Clay, lower part	Oxford Clay or sands (Yorks.)
Kellaways Beds	Approximately as in southern
Upper Cornbrash	England

lithology, as at the base of the Aalenian Stage and consequently of the Middle Jurassic. The base of the Upper Jurassic is taken at a marked break in the faunas but here there is no convenient lithological correspondence and the boundary, which is also that of the Oxfordian Stage (Table 42), is commonly within the Oxford Clay; the lower part of that lithological unit thus belongs to the Callovian Stage.

The Kellaways Beds and other Callovian sands

After the initial Callovian transgression, in most regions there was a gradual return to deeper quiet waters, with clays and shales. During the transition, however, there were some minor sandy incursions and these now form part of the Kellaways Beds, commonly as sand (the Kellaways

Rock) overlying clays; there may be minor diachronous junctions between the two. Oysters and gryphaeas are fairly common and in some regions ammonites.

This pattern of sedimentation does not continue regularly over the Market Weighton swell. In the Yorkshire basin the stratigraphy differs in several ways. The Callovian as a whole is more varied, thicker and is interrupted by two phases of gentle folding, erosion and unconformity. Sandstones predominate but there are also shales and oolitic beds. The whole amounts to nearly 30 metres on the coast near Scarborough and decreases southwards; in the same direction, towards Market Weighton, the uppermost sandstone (the Hackness Rock) is strongly transgressive over the lower members. Another result, in this flexured basin, is that these varied sandy transgressions correspond to most of the lower and middle Oxford Clay of the Midlands.

The great clay formations

In central and southern England thick clays succeed the Kellaways Beds and dominate the sequence through the Oxfordian and Kimmeridgian stages. In lithological terms they are as follows:

Kimmeridge Clay from Dorset to Yorkshire
Ampthill Clay   Midlands and eastern England
Oxford Clay   Dorset to Yorkshire, but more restricted in the latter

All these divisions are interrupted over the Market Weighton swell while in southern England and much of Yorkshire the Ampthill Clay is replaced by a fossiliferous carbonate shelf facies – the Corallian Beds, considered below. Exposures also vary and where the three clays are superposed they form the very flat lands of the Fens and the Wash, where information is chiefly from boreholes. However the lower part of the Oxford Clay is very well exposed in large brick-pits near Peterborough and most of the Kimmeridge Clay crops out on the Dorset coast.

As with other Jurassic strata thicknesses increase, a little irregularly, southwards. The Oxford Clay amounts to some 150 metres in Dorset and 180 in a Wiltshire borehole, decreasing in the Midlands to about 70 metres. The Ampthill Clay is only well recorded in Fenland boreholes; it is also about 70 metres thick, including some lower more silty and calcareous beds. In the same area there are 127 metres of Kimmeridge Clay – a formation which shows most clearly the expansion into the Wessex and Weald basins; combined cliff and borehole sections in Dorset reach 503 metres, and in a borehole of the central Weald the maximum of 560 metres.

In Bedfordshire and Buckinghamshire the outcrops are approaching the low shelving edge of the London Platform and here the Kellaways Beds are reduced and the upper zones of the Oxford Clay are missing. This influence can be considered as part of the mildly unstable region of the Oxfordshire shallows, but its manifestation inevitably shifts eastwards because the outcrops of the younger formations extend further in that direction.

Faunas, facies and conditions have several resemblances to the Lias, though the clastic influx is less and largely of silt grade and there is very little ironstone. The faunas of the Oxford Clay are especially well known because the brick-pits have afforded exceptional opportunities for collection. Ammonites, bivalves and belemnites are abundant. Marine reptiles are represented by ichthyosaurs and plesiosaurs and a variety of fishes. Terrestrial dinosaur bones have been found, a reminder that even these marine clays were laid down not far from land.

Grey mudstone is a central rock type, well bedded or even finely laminated; calcareous beds are the main variant, some approaching cementstones. At certain levels the mudstones are bituminous and in these the fossils tend to be pyritized. Small-scale rhythms or alternations have been noted between the more bituminous and the more calcareous types, together with the normal grey mudstones. The Kimmeridge Clay is markedly rich in organic components and here the bituminous mudstones can be termed oil shales. A particularly thick seam on the Dorset coast was once known as Kimmeridge Coal and it can be ignited. Such layers are also rich in phytoplankton, dinoflagellates and hystrichospheres; the thin limestones are largely composed of coccoliths (minute calcareous nannoplankton), at some levels by enormous numbers of a single species. The occurrence of these microfossils and the fine laminations of the oil shales suggest periodic plankton 'blooms' in the surface waters, and possibly annual layering in the bituminous facies, similar to that deduced in the Lias.

Round the main outcrop into Wiltshire and the Midlands the normal clay facies persists, but with some diminution and interruption by Cretaceous overstep. Thus in Oxfordshire yellow sands with large concretions replace some of the upper beds and boreholes in Cambridgeshire show the upper zones to be missing. In Lincolnshire there is some expansion but the break at the top persists, this being a major feature of the East Midlands Shelf and the northern margin of the London Platform.

During the same period the Yorkshire basin maintained certain distinctions. Over most of the outcrop the true Oxford Clay facies only occupies a single lower Oxfordian zone (*mariae*), being underlain by the variable incomplete Callovian of the last section and overlain by Corallian sandstones and limestones. As before, however, Market Weighton does not coincide with a precise line of facies change: on its northern flank there are small outcrops with a fuller Oxford Clay sequence, like those of the Midlands, and small amounts of Ampthill Clay are found. The Kimmeridge Clay is very badly exposed, its outcrop being planed down to form the Vale of Pickering. Some 225 metres, chiefly of the lower clays, are known from boreholes; southwards it is truncated by the Lower Cretaceous overstep. As such, however, it is thicker than in the Midlands or eastern England and may well have originally been complete or nearly so.

The dominance of the clays from late Callovian to Kimmeridgian times reflects the gradual deepening of the seas, with sporadic stagnant bottom waters. The great thickness and extent of Kimmeridge Clay, especially the lower and middle parts, marks the acme of the transgression which carried this facies widely over north-western Europe.

Corallian Beds: middle and upper Oxfordian

In certain conditions of shallow and intertidal waters these beds interrupted the dominant clay facies. There are two main outcrops – from Dorset to Oxford and in Yorkshire. The first is well exposed on the Dorset coast, though being only some 65 metres thick the Corallian is only a tenth of the total clays above and below; the varied sequence is locally rich in fossils though as in other carbonates ammonites are only sporadic. Particular interest has been shown in the small-scale cyclic sequences, four cycles in all. Each is presumed to begin with an abrupt transgression and then to consist essentially of a slower regression with shallowing and the entrance of increasing clastic detritus; the fullest sequence is thus limestone–clay–sandstone. The third cycle is almost all limestone, including the Osmington Oolite, following an exceptionally wide transgression; in this cycle very little detritus entered the Wessex basin and reef growth was more prolific than at other times. The individual facies vary according to local conditions and extent of the submergence, but they include offshore shelves (with oolites), coral beds, subtidal, intertidal and occasionally lagoonal sediments. Bivalves are abundant in the shelly beds.

Corallian beds can be traced inland and the cycles identified, with some variation. Like all the succeeding strata they disappear under the Cretaceous in north Dorset. In Wiltshire some new facies appear including the striking Steeple Ashton coral bed, which is part of the third cycle. This grew on a shoal of cross-bedded oolites and includes a specialized echinoderm fauna. At the top of the fourth cycle the Westbury oolitic ironstone represents an unusual lagoonal facies.

The Corallian is again well seen in the horseshoe of low hills that border Oxford on the west, south and east, the north and centre of the city being founded on Oxford Clay. There are three main facies in what is probably a slightly condensed sequence, the arenaceous Lower Calcareous Grit being overlain by Coral Rag and the Wheatley Limestone. Coral Rag is a common term in this group for reef patches, coral and shelly detritus in a loose rubbly limestone; the Wheatley Limestone was also formed from coral and shell debris and contains some ooliths. It may overlie or underlie the reef spreads or occupy channels between them. Both facies have specialized faunas, usually of echinoids and bivalves; as in Wiltshire and Yorkshire the reef corals are principally *Isastraea* and *Thamnasteria*. Ammonites are conspicuously absent, as in other reefs.

Despite its impressive richness and variety the Oxford Corallian corresponds only to the middle and lower parts of the Wiltshire and Dorset succession, the calcareous rocks being equivalent to the Osmington Oolite of the third cycle. The Kimmeridge Clay followed after a period of uplift and erosion.

Oxford is nearly at the margin of the Corallian shelf facies, to the north-east giving way to the Ampthill Clay. Within the clays, however, there is the very small outcrop of reef rock and oolite at Upware, 15 kilometres north-east of Cambridge. It is associated nearby with the calcareous and siltstone basal facies below the normal Ampthill Clay and was not far from the low northern shore of the London Platform. Presumably a few short-lived patch-reefs and oolites were formed in littoral waters before being overwhelmed by the clays.

In the Yorkshire basin the Corallian carbonates are similarly conspicuous and a good deal thicker, commonly about 100 metres. All seem to have accumulated in warm shallow waters of varying turbulence. In contrast to the Callovian below only one minor warping phase is found, the basin having again become stabilized. The age is much the same as in the south except that the first sandy incursions were probably somewhat earlier.

Three main formations are distinguished – a Lower and an Upper Calcareous Grit and a middle Coralline Oolite Formation of limestones interspersed with wedges of sand. Internal boundaries are often diachronous. The Corallian as a whole dips southwards off the North York Moors, under the Kimmeridge Clay and alluvium of the Vale of Pickering, the robust Lower Calcareous Grit forming a conspicuous scarp above the softer Oxford Clay. It is rich in the ovoid siliceous spicules of the sponge *Rhaxella*, which colonized the sea floor; the spicules are locally an important component of the rock and on their replacement by calcite the silica released contributed to the formation of chert.

The limestones of the middle formation are very variable and often rich in fossils. They include oolites (also forming building stones), algal and bioclastic types, patches of reef and reef-debris. The faunas include many species of bivalves, crinoids, echinoids and gastropods. Ammonites have been recorded mainly in the calcareous sandstones. The Coral Rag at the top of this formation is the most conspicuous reef member, with reef patches up to four metres high and colonies surviving in position of growth. It is found almost throughout and a main reef belt has been discerned in the north-west, with associated fore-reef and back-reef facies.

This prolific growth and carbonate accumulation was gradually brought to an end by the Upper Calcareous Grit. It is more ferruginous than the lower formation and becomes more clayey upwards, passing finally into the Kimmeridge Clay.

PORTLANDIAN CHANGES

In this, the last stage, major changes came over the Jurassic scene, more radical than any earlier in the period and persisting into much of the Lower Cretaceous. In particular the minor uplifts that had already caused reduced or intermittent sedimentation in the southern Midlands swelled to become a broader shallow region or even a partial barrier that extended north-westwards from the London Platform. Deposition was concentrated in a southern province, the basins of Wessex and the Weald, and in a north-eastern one from Norfolk to Yorkshire, the latter being essentially the English part of the Boreal Province and the onshore fringe of the southern North Sea basin. Here marine conditions continued but their sedimentary record is incomplete or condensed. In the south the Portlandian seas survived for a time to produce the Portland Beds and then a variable mixed facies supervened, which is now seen as the 'Purbeck Beds' – or in more modern terms, the Lulworth and Durlston beds because they span the Jurassic – Cretaceous boundary (Table 43). Ammonites are conspicuously absent, so in southern Britain that boundary has to be drawn on supplementary groups, such as ostracods and plant spores, and also by lithostratigraphic means.

The Portland Beds of the Dorset coast

All the inland outcrops of Portlandian strata are incomplete and meagre compared with the superb exposures on the Dorset coast. If a map of that county shows a lesser area of Portland Beds than of any other formation, a

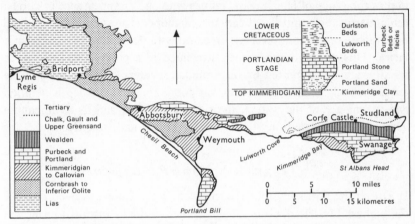

Fig. 65. Geological sketch map of south Dorset; details around Lulworth cannot be shown and the minor amounts of Lower Greensand in the east are included with the Wealden. Inset: lithological components of the uppermost Jurassic Portlandian Stage.

view of the cliffs gives a very different impression. Of all the Jurassic limestones the Portland Stone is the most massive and resistant to wave action. It builds the bastion of Portland Bill, the cliffs of the Isle of Purbeck and the two breakwaters that almost enclose Lulworth Cove. In the first two the beds form part of the gently dipping southern limb of the Weymouth anticline; at Lulworth they are part of the very steep northern limb and are almost vertical.

The greatest thickness is in the Isle of Purbeck, in the gentle basin of east Dorset. Here there are some 75 metres comprising two traditional lithological divisions, the Portland Sand below and the slightly thicker Portland Stone above. On the coast there is a gradual transition in facies from marly beds at the top of the Kimmeridge Clay. The sand proportion increases upwards but mud remains; the Portland Sand when fresh is dark grey and there is a repetition of the thin cementstones. The Portland Stone comprises a range of limestones: shelly rocks, micrites, white limestone with chert seams and large nodules, the Roach (a massive rock made porous by the hollow moulds of molluscs), and oolitic freestone with comminuted shells; only the last, the Portland Freestone Member, is the Portland Stone of commerce. The quarries on Portland Bill in particular have supplied vast quantities of stone for massive masonry since Wren used it in the rebuilding of St Paul's Cathedral.

Although the fauna is locally abundant, with the giant ammonites an impressive component, it is not rich in variety. There is gradual diminution in the range of late Jurassic faunas with the disappearance of corals, echinoderms, belemnites and various brachiopods; as a result the Portland limestones are not coral-bearing, in contrast to the earlier carbonate bank facies. This diminution has been attributed to a progressive increase in salinity, a view supported by the abundant evidence of evaporites in the lower Purbeck facies above. Minor variations are related to depth, aeration and turbulence of the waters, the type of sea floor and its sediments; a gentle swell separated the shallow basins of east and west Dorset.

The Lulworth Beds, or lower Purbeck facies

In Dorset the Portland Stone is succeeded by the Lulworth Beds which is the highest Jurassic division, some 73 metres thick. In broad terms the Purbeck environment seems to have been dominated by a series of saline lagoons, marshes and forest-flats that were affected by small changes in sea-level and local topographic features; most sediments are water-laid but there were emergent areas or phases. The ostracod species in particular reflect oscillations in salinity or small-scale cycles, but true freshwater conditions seem to have been rare. The rocks are all fine-grained, argillaceous, calcareous or evaporitic, the last two showing complex diagenetic changes. An extensive transgression breaks the Purbeck sequence resulting in a bed thick with oysters and some other marine forms; on account of its appearance on weathering it is called, rather absurdly, the 'Cinder Bed' and is taken to mark the base of the Cretaceous (Fig. 65, inset; Table 43).

Most of the Lulworth Beds were formed in lagoons and shallow waters, varying in salinity but commonly hypersaline. Traces of evaporites occur at several levels but are best known from the basal beds, up to 10 metres above the conformable junction with the Portland Stone below. In their present state these beds are largely limestones; the less saline types contain ostracods, foraminifera and sometimes gastropods, whereas the algal sediments were formed under more saline conditions on intertidal flats.

Table 43. *Purbeck succession near Swanage*

	Traditional terms	Lithological summary
Durlston Beds, LOWER CRETACEOUS (60 m)	Viviparus Clays; Marble, shell and Ostracod Beds; Unio Beds; Broken Shell Limestone	Interbedded shales and thin limestones with bivalves, ostracods and some gypsum
	Chief Beef Beds / Corbula Beds	Mainly shales with calcite and gypsum, some bivalves
	Upper Building Stones	Shelly limestones with shale partings
	Cinder Bed (*c.* 1 m) with abundant oysters, some other bivalves and rare echinoids	
Lulworth Beds, PORTLANDIAN (73 m)	Lower Building Stones / Mammal Bed	Limestones with chert, overlying shales, local dirt bed
	Marls with gypsum and insect beds	Variable clays with thin limestones, gypsum near base
	Cypris Freestone and Breccias / Caps and dirt beds	'Basal beds', primary and secondary limestones, breccias and dirt beds

Traditional terms chiefly from Arkell (1947*b*), some being local or ancient in origin. Main stratigraphical fossils are ostracods, plentiful in several beds.

The original evaporite here was gypsum, but through a complex diagenesis the rock is now a porous unfossiliferous limestone, replacing anhydrite. The Broken Beds are breccias, distributed rather irregularly within the basal limestones. They are considered to be tectonic in origin but the effects may have been accentuated in the incompetent or brittle evaporite-bearing strata.

The well known Dirt Beds (carbonaceous shales) are soils, formed during regressive or emergent phases, some including the stumps of trees that grew on the shores. On submergence the forest was replaced by extensive tidal flats and the stumps were coated by algal limestone, and are now called the Fossil Forest. More than one cycle has been recognized, each beginning with a regressive dirt bed and followed by transgression but they differ from those of the typical sabkha (cf. Permian examples, p. 240) and the forests also suggest less arid conditions.

More marls and limestones succeeded the basal Lulworth Beds and the salinity continued to be variable; however there are beds with ostracods, gastropods, some fishes and drifted insects. The Mammal Bed is a lenticular dirt bed near Swanage, famous for its bones of reptiles and early mammals though these are no longer found; the latter were small insectivorous and arboreal forms while crocodiles, turtles and dinosaurs reinforce the picture of a warm climate, but not much more than semi-arid.

THE JURASSIC BASIN OF THE WEALD

The underground structure of the Weald has long been known as the reverse of the surface anticlinorium displayed by the Cretaceous formations; most of the Jurassic strata are synclinal in form (Fig. 66) and

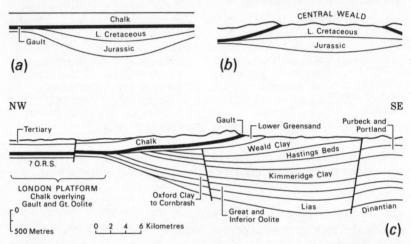

Fig. 66. Sections across the Weald. (*a*) Sketch to show the sagging basin towards the end of the Cretaceous. (*b*) The same uplifted to give the present outline, the Jurassic strata remaining broadly synclinal. (*c*) Cross-section compiled from boreholes from south-west London to east Kent, showing the edge of the London Platform and expansion into the Weald basin. (Adapted from Worssam and Ivimey-Cook, 1971, p. 11.)

all are subsurface except for some minute inliers. The Weald was one of the major subsiding basins; not only do all the Jurassic beds become thinner to the north and north-east as they approach the London Platform but there is some indication of thinning to the south as well, towards some feature near the present coastline. A similar feature, the Portsdown ridge or swell, partly separated the Weald and Wessex basins. Our knowledge of the underground Jurassic beds is drawn from the shafts and boreholes of the Kent Coalfield and more widespread oil exploration. An isopach map for all of southern England is shown in Fig. 67.

In general the facies are similar to those described in the southern outcrops; from the Lias to the Kimmeridge Clay the closest resemblances are with the Cotswolds and we need only note the main differences. The

Lias tends to be more sandy to the east and south suggesting a source of detritus in these directions. A substantial limestone formation in the Inferior Oolite reaches as far as east Sussex but in east Kent a more sandy facies again prevails. From the Great Oolite to the Oxford Clay the rocks resemble those of the southern Cotswolds and the Great Oolite is unusually extensive, overlapping onto the edge of the London Platform and the limestones attaining some 60 metres. The expansion of the Kimmeridge Clay to 560 metres in the central Weald (Ashdown borehole) is one of the clearest contributions to the depression of the basin centre.

The Corallian Beds exhibit thickness and facies changes comparable to

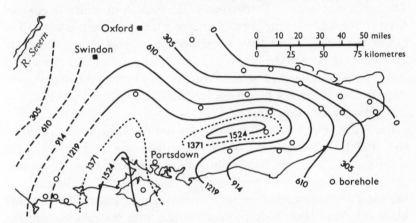

Fig. 67. Isopach map of the Jurassic rocks of the Weald and Wessex basins; original thickness-interval 1000 feet, recalculated in metres. (Adapted from Howitt, 1964, p. 97.)

those of the main outcrop and could probably be interpreted similarly if more detail were available. Along the northern edge of the basin there is a belt of limestone, including the 45 metres of coral-bearing rocks penetrated in the Warlingham borehole south of London; but with the expansion southwards into the central Weald the calcareous facies diminishes and most of the 117 metres at Ashdown is argillaceous. There is a marked reduction over the Portsdown ridge and from late Jurassic to early Cretaceous times thinner sequences are characteristic here, suggesting a positive swell, comparable to the better known examples of the main outcrop.

Changes appear in the Portlandian Stage, for the Portland Beds are thinner and more sandy than in Dorset, with only a thin calcareous bed at the top to compare with the Portland Stone. Thus that lithological unit, so prominent in the type area, may in fact be rather restricted. The Purbeck Beds are thick; in some boreholes they are grouped together but in several the marine faunas of the Cinder Bed can be recognized. The lower (Lulworth) evaporites are prominent with both anhydrite and gypsum and the latter has been exploited in the central Weald. Over the Portsdown ridge all the basal Lulworth Beds of Dorset, with their complex diagenesis, are represented by a thick bed of anhydrite.

EASTERN ENGLAND: THE MARGIN OF THE BOREAL SEA

The uplift that isolated Portlandian sedimentation in southern England has its repercussion in the Boreal Province but in a different manner. The top of the Kimmeridge Clay is marked by a break from Norfolk to Yorkshire but only in the last county is it a major unconformity and here there are no Portlandian strata. Whether any were ever deposited is uncertain but in view of the pronounced subsidence earlier in the Jurassic it is quite possible that some were.

For many years Portlandian beds were not recognized in Lincolnshire and Norfolk but ammonites of that stage have now been found in the lower part of the Spilsby Sandstone and Sandringham Sands, the remainder of both being Lower Cretaceous (Tables 49, 50).* At the base of both there is a disconformity above the Kimmeridge Clay, marked by a bed of black chert and phosphatic nodules together with the rolled remains of ammonites from the highest zones of the Kimmeridgian and other fossil debris. The base of the Lower Spilsby Sandstone also contains some condensed Portlandian horizons.

Thus when the seas of the Boreal Province reinvaded this coastal belt there was a variable period of impersistent sedimentation and winnowing along the shallow margins. Such conditions were not confined to the initial stage because in Norfolk another nodule bed occurs a few metres higher up and yet another is extensive around the base of the Cretaceous. Although these littoral sands are thin, inconspicuous and mostly not well exposed they represent an important part of late Jurassic history in Britain. They form a link with the extensive Boreal Province and the North Sea and the sequence in East Anglia is the basis of the uppermost zones in Britain, where the ammonites of the Portlandian in Dorset fail.

---

* The use of Portlandian here follows the current list of stages and zones (p. 422); the Volgian Stage, based on the Russian sequence, is partly equivalent but extends down into the upper Kimmeridgian; it has commonly been used in the Boreal Province.

The intermediate belt, Wiltshire to Buckinghamshire

It now remains to pick up the traces of Portlandian beds between eastern and southern England. Northwards from Dorset they appear in the vales of Wardour and Pewsey and in a small syncline at Swindon, north Wiltshire; there are further outcrops in a string of outliers between Oxford and central Buckinghamshire. The remnants are much thinner than farther south, as would be expected on the western fringes of the London Platform.

In the Vale of Wardour, Portland and both Purbeck divisions are present, totalling some 55 metres; the main facies are recognizable, including the Cinder Bed incursion. At Swindon the Portland Beds are mainly sands and limestones and rest non-sequentially on the Kimmeridge Clay with a basal nodule bed and derived phosphatic fossils. Thin Purbeck beds (all or mostly lower Purbeck) follow – limestones, marls and more pebble beds, with ostracods and molluscs.

Farther east the sequences are even thinner and more incomplete, with the 'Purbeck' reduced to a few capping outliers. In attempting to sort out these beds much depends on the recognition of the basal Cretaceous marine beds – here taken to be the ferruginous Whitchurch Sands (p. 336). These in places overstep Purbeck onto Portland Beds, and the succeeding Lower Cretaceous formations themselves may be mutually unconformable. Such relations make for detailed difficult stratigraphy but we may see the problems as inherent in the reduced abnormal sedimentation along this stretch between the southern and eastern provinces.

JURASSIC ROCKS BEYOND THE MAIN OUTCROP

To the west and north of the main outcrop further knowledge of Jurassic stratigraphy comes from a few boreholes and minor outcrops as far as north Antrim and Carlisle; more important and more complete sequences occur in north-east Scotland and especially among the Inner Hebrides. In several regions there is a clear link or actual continuity with nearby stratigraphy of the sea floors. However the detail and information available there differs and the land outcrops will be taken first.

North-West England, North Wales and North-East Ireland

The three small English outcrops are all of Lias, in Staffordshire, on the Carlisle plain and the Prees outlier on the Cheshire plain, but the

Hettangian zones in the last are much thicker than in the Midlands. Similarly there are minor strip-like outcrops around the basalts of Antrim and the lower part of the Lower Lias was penetrated in a borehole (Port More) on the north coast; again the individual zones were exceptionally thick. A few derived fossils from the Middle and Upper Lias occur in the basal Cretaceous conglomerates so that later Jurassic sediments were laid down in northern Ireland and comparison with western Scotland also suggests that originally there was a fuller sequence.

The most dramatic of these remnants was discovered in the Mochras borehole, drilled on the shore of Tremadoc Bay in North Wales. The small amount of continental Trias at the base displayed an unusual facies, arenaceous and conglomeratic with much carbonate. The early Hettangian invasion was shown by a slight break and above that 896 metres were penetrated, almost to the top of the Lias – nearly three times as much as the thickest land sequence in the lower Severn valley. The facies of North Wales is also unusually uniform, slightly calcareous mudstones and siltstones; in particularly the Middle Lias ironstones and ferruginous sands are lacking.

This very thick sequence is faulted against the Cambrian of the Harlech Dome, but with no littoral characters evident presumably the seas extended widely across it. A major query thus arises: do all these remnants indicate that the Liassic seas generally submerged north-west England, Wales and (perhaps) Ireland, with increasing depth and uniform subsidence? Or did an irregular arrangement of subsiding basins and intervening swells govern deposition here, as in the main outcrops and south-east England, but only the deeper basins escape subsequent erosion? On balance the second view is held here, and certainly the few outcrops of the sea floors do suggest tectonically complex Caledonoid foundations.

The Inner Hebrides

Northwards from Antrim the Jurassic seas extended widely up the west side of Scotland and faulted fragments of Lias accompany Rhaetic and Trias in the central ring complex of Arran. There are outcrops on several Hebridean Islands (Fig. 9), the largest on Skye and Raasay, and smaller ones on Rhum, Eigg and Muck; a southern group includes smaller strips on Mull and also in Morvern and Ardnamurchan on the mainland. Probably all the Jurassic rocks of the Hebrides owe their preservation to a cover of Tertiary basalts which in places is still present. The succession falls naturally into three unequal divisions: a thick marine group (Lias to

Bajocian), the Great Estuarine Group (approximately Bathonian) and a
thinner incomplete upper group (Bathonian or Callovian to lower Kimmeridgian). A composite succession is given in Table 44 but over the

Table 44. *Composite Jurassic succession of the Inner Hebrides*

Lithological divisions	Thickness (metres)	Stages
Staffin Shale Formation shales	6	KIMMERIDGIAN (lowest beds only)
Staffin Shale Formation shales with sandstones and siltstones	100	OXFORDIAN and CALLOVIAN
Great Estuarine Group shales, thin sandstones and limestones	180	Chiefly BATHONIAN
Bearreraig Sandstone, with clays at the top	180	?LOWER BAJOCIAN and AALENIAN
Upper Lias, shales	24	TOARCIAN,
Middle Lias, Scalpa Sandstone	75	PLIENSBACHIAN,
Lower Lias { Pabba Shales	210	SINEMURIAN and
Lower Lias { Broadford Beds	128	HETTANGIAN
(Rhaetic, Trias or older rocks)		

150-kilometre stretch from northern Skye to Mull facies and thickness
show a good deal of variation.

The Lower Lias is not the largely argillaceous formation familiar
elsewhere. The Broadford Beds include calcareous sandstones and impure
limestones while on Skye there are corals and an oyster bed. Sandy
incursions dominate the Middle Lias but the Upper Lias shows some
resemblance to Yorkshire in the presence of jet, calcareous concretions
and pyrites; the oolitic Raasay ironstone (Whitbian Substage) also resembles the English ores in that it contains chamosite and siderite. The
Bearreraig Sandstone varies in thickness and to some extent in lithology
according to local differences in subsidence and amount of incoming
detritus. It reaches a maximum of 490 metres in southern Skye, thinning
to the north and south, with limestones appearing in Ardnamurchan.
Despite the difference in facies the ammonites, covering most of the
Aalenian and Bajocian stages, are distinctly similar to those of the
Cotswolds.

The Great Estuarine Group (an inappropriate name) is the most
interesting of the Scottish formations. At the base an oil shale denotes a
restriction of the shallow seas and the formation of a temporary stagnant

basin. Above there is much variety: shales, limestones and lesser sandstones, deposited in waters of varying salinity. The more marine forms come in particularly in the upper part – oyster beds, marine algae and one incursion of faunas like those of the Great Oolite in the south. There are also fish and reptile remains and the jaw of an early mammal. Very shallow coastal bays and lagoons are suggested, bordering a low hinterland. This was presumably a subdued Jurassic version of the Scottish Highlands; the mineral assemblage of the sandstones can be matched with the rocks exposed there at present. There is little evidence of full emergence but desiccation cracks are common and calcite pseudomorphs after gypsum have been found in the algal limestones, so that the lagoons became hypersaline from time to time; a partial resemblance to the Purbeck conditions is apparent. A southward thinning affects this group also, seen for instance on Raasay, Eigg and Muck.

The lower Callovian transgression is marked here as elsewhere, and thereafter there was a long period of marine clastic deposition, varying in both place and time from fairly coarse sandstones to siltstones and shales (the Staffin Shale Formation of Table 44). Several Callovian and Oxfordian zones can be recognized, dominantly of boreal type, but the whole is not very thick. Minor cycles show alternations between more condensed sequences, some bituminous shales and more sandy incursions; woody debris was brought in at more than one level. The uppermost shales just extend into the basal Kimmeridgian and here again much more may have been deposited, because the Jurassic deduced to the north (the Minch basin) is exceptionally thick. The Staffin shales contain bentonitic mudstones at more than one level and, as in the Fuller's Earth, their montmorillonite was derived from volcanic ash. There is thus further evidence of contemporary eruptions in west Scotland, chiefly in the upper Callovian and lower Oxfordian stages. As in the Bathonian example there is no clear evidence of the source.

North-eastern Scotland, Brora and the Moray Firth

The close connection between an offshore basin and its onshore margins is especially conspicuous in the Moray Firth. As with the less well known Trias, nearly all the Jurassic rocks are at present submerged (p. 318), and details of the stratigraphy still have to be drawn from the restricted version that crops out on the shores. In the same way that the Jurassic rocks of Yorkshire, Lincolnshire and Norfolk are allied to those of the southern North Sea basin, those of the Moray Firth and its onshore margins are an important outlying part of the northern North Sea basin and that is where the main comparisons lie.

Apart from a small amount of Liassic shale in a borehole in Moray all the outcrops are on the north-west side of the firth, the only one of any size being a 35-kilometre coastal strip on either side of the small town of Brora (Table 45) in south-east Sutherland. Exposures are sporadic and there is

Table 45. *Jurassic succession at Brora, Sutherland*

Formations and lithology		Stages
Shales and sandstones, with boulder beds		KIMMERIDGIAN
Sandstones and limestones	}	OXFORDIAN
Brora Sandstone Formation (part)		
Brora Sandstone Formation (part)		
Brora Shale Formation, including Brora Roof Bed	}	CALLOVIAN
Brora Coal Formation, coal seam, sandstones and shales	}	Probably BATHONIAN
—————fault gap—————		
Shales, sandstones and limestones		Includes some LOWER LIAS

much faulting, the whole lying parallel to the Great Glen Fault offshore and bounded by the Helmsdale Fault on the landward side. Thicknesses are difficult to estimate but the Kimmeridgian occupies all the northern half and is likely to be much the thickest. For all the Jurassic present a total of 1100 metres has been suggested.

The Lias shows a transition from freshwater clays with carbonaceous layers, or even poor coals, through a soft white sandstone to bituminous shales with some ammonites – the latter spanning the boundary between Lower and Middle Lias. The early Jurassic seas thus invaded the lowlands at the edge of the Moray Firth basin. Compared with most British outcrops the non-marine Lower Lias may seem rather unusual but in context of the northern North Sea it is not: a distinctly similar succession is known from south Sweden.

Above a fault-gap, cutting out Middle Lias to Bajocian, Bathonian facies reflect emergence on this side of Scotland also. The base is marked by a massive 17-metre sandstone and above that there are increasingly finer beds which have yielded several plant species also found in the deltaic facies of Yorkshire. The succeeding bituminous shales have microfossils with some marine affinities and the whole small sequence is not unlike the Coal Measures. That likeness is enhanced by the Brora Coal itself; although small in extent it is a metre thick and has been worth working in a region naturally poor in fuel.

The Roof Bed above witnesses the now familiar Callovian transgression and thereafter Callovian and Oxfordian sediments resemble those on the west of Scotland, though the main arenaceous incursions were not simultaneous on each side. After a basal sandy formation the Kimmeridgian conforms to the typical shales (albeit with more drifted plants) and most of the ammonite zones have been recognized. However the stage is chiefly famous for the several boulder beds. The most impressive consist of large angular blocks of Middle Old Red Sandstone, some containing fishes. There is also a minority of metamorphic or Moine rocks and a few masses of Jurassic cross-bedded sands from a different deltaic or onshore environment. At present the Jurassic outcrop is bounded on the north-west by the Helmsdale Fault and the jumbled boulders are thought to have been dislodged from the landward side by contemporaneous movement along that fault.

The influence of nearby land or actual emergence is even more marked here than on the west of Scotland, not only in the Bathonian Stage – so widely a time of regression – but also in the lowest Lias and the numerous well preserved plants. The Kimmeridgian faulting points to another distinction. Judged by land-based stratigraphy that stage appears to be a placid period, mostly of deeper water though with gentle warping and uplift towards the end. In their more vigorous tectonics the Moray Firth outcrops belong to the more active regions of the seas rather than the lands.

JURASSIC ROCKS OF THE SEA AREAS

The Jurassic rocks in the seas of north-west Europe are outstanding for their economic possibilities, especially where deltaic or other sandy facies form potential oil reservoirs; organic-rich shales (notably Kimmeridgian) are probable source rocks. Accordingly much research has centred on the system, with a corresponding proportion unpublished. While all accounts of the sea areas are liable to major revisions this applies particularly to the following summary. The maps of Figs. 1A, 1B, 1C illustrate the sea floors near the British Isles.

Relationships around the North Sea are essentially similar in Triassic and Jurassic stratigraphy but in the latter more is known across the English Channel. North Germany still continues to be a major province; on land there are various Jurassic outcrops near Hanover and a much greater spread subsurface, sloping down to the southern North Sea; together they form the Lower Saxony Basin. Similar rocks occur below parts of the Netherlands and Denmark and small amounts of Lias, partly non-marine, crop out in south Sweden. Much farther north off the Norwegian coast the Lofoten Islands include minor Jurassic outcrops and there are larger areas in East Greenland (cf. Fig. 59). All this is part of the Boreal Province.

Across the English Channel the Paris Basin became a definite entity in Jurassic times. At present it is a very gentle structural basin, on rather irregular foundations; during various periods in geological history it has been called the Anglo-Parisian Basin or with a Belgian component added. Since Cretaceous and Tertiary strata are extensive, Jurassic outcrops appear as a discontinuous rim. Those nearest Britain include two small ones, in the Boulonnais, a structural continuation of the Weald, and the Pays de Bray, a faulted anticline inland from Dieppe; the largest and most important is in Calvados (Normandy), abutting on the east side of the Armorican Massif and with fine cliff sections along the coast. With southern England these French outcrops are part of the intermediate belt where the faunas at different times show both Boreal and Tethyan affinities. The London–Brabant Massif remained a partial geographical barrier.

From the English Channel to the Irish Sea

On the French side neither the Boulonnais nor the Calvados outcrops provide a full succession but together they can be compared with all the English formations. Bathonian to Portlandian strata are exposed in the Boulonnais but are thinner than in the Weald with interrupted or condensed sequences and locally coarser detritus, all reflecting the nearby London–Brabant Massif. The resemblance between Calvados and Dorset is close from the Toarcian to the Kimmeridgian Stage, with the same distribution of clays and limestones; the famous Caen Stone of mediaeval buildings was quarried from Bathonian freestones. It follows that the Jurassic in the English Channel is also very similar to that on either side. It is extensive beneath a later cover and the outcrops include those off the Boulonnais coast, off the Dorset and Normandy coasts (the latter in the Baie de le Seine) and east–west faulted strips in the central Channel. The main tendency is for rather more argillaceous sequences, for instance in the Bathonian, and the Corallian of Dorset passes into a more clayey facies south-west of Portland Bill.

In the western Channel Jurassic outcrops are small but more is reported at depth, mainly of middle and lower divisions (the latter thick) in an argillaceous and impure calcareous facies. This character is also found subsurface in the Celtic Sea – an area of much unpublished information. However in the northern part the sequence seems to be virtually complete with block-faulting developing in Middle or Upper Jurassic times.

The Bristol Channel is a major westerly outcrop adjoining the Lias of Glamorgan and Somerset and linking up with a further faulted area north-west of Lundy. The sequence resembles that of western England up to the Kimmeridge Clay except that, in the east at least, the Middle Jurassic limestones are lacking or incomplete. Portlandian rocks are only present locally in the west but appear to include an upper non-marine 'Purbeck' section.

From the northern Celtic Sea there is an extension into the southern Irish Sea and into Tremadoc Bay where the Lias was penetrated in the Mochras borehole (p. 310). Farther out there are Bajocian and Bathonian rocks as well, including some limestones. The former transgress unconformably onto beds as far down as the Lower Lias, so despite the exceptional Liassic thickness the central part of the Irish Sea was uplifted and eroded in Middle Jurassic times – an interesting sequence of subsidence and upheaval, possibly on deep-seated Caledonian foundations.

Western and northern Scottish seas

From the Antrim coast northwards up through the Hebridean seas Jurassic rocks contribute to the various Mesozoic basin infillings. They can be related in some degree to the land outcrops but the sea floor is complicated by several major Caledonoid faults and Precambrian basement ridges, together with some covering by Tertiary lavas. Only small amounts of Lias are known offshore from north Antrim, related to that found in boreholes onshore. Farther north there are two major Hebridean basins – one stretching west and south-west from Mull and a larger one north-west and south-west from Skye (cf. Fig. 1A). Taking the two together the Jurassic seems to be largely complete and, as would be expected, it is essentially similar to the land outcrops of the Hebrides.

The North Minch basin is a more separate entity on which there is much less information available. However the Jurassic is reported to be exceptionally thick, possibly from 1000 metres in the south to 2000 metres in the north. Even as rough estimates such figures confirm the suggestion of very deep sagging basins in this north-westerly Caledonoid province.

The basin west of Orkney possibly contains some Jurassic, while that in the elongate faulted West Shetland Basin (Fig. 58) is more securely known and is thought to include Upper and Middle Jurassic in a clastic facies.

The North Sea

In addition to the economic sandy facies, Jurassic geology here is notable for large-scale vulcanicity and an associated Middle Jurassic uplift, with more uniform marine facies before and after.

The Lias is best known in the southern basin where there are similarities with northern Germany and eastern England; some of the southern outcrops lie offshore from Yorkshire and Lincolnshire. Less is known in the northern basin but in general the division was more restricted than the later ones with deposition mainly in the Moray Firth, Central and Viking grabens. The sediments include thick sandy spreads in the far north, probably derived from the Vestland Ridge on the Norwegian side and forming an upward continuation from similar Triassic beds beneath. Old Red Sandstone is a likely source for much of the Jurassic detritus in the north.

Major tectonic and geographical changes took place early in Middle Jurassic times; the Mid North Sea High and the Ringkøbing–Fyn High were uplifted to form a wide barrier and in the south continuous sedimentation was restricted to some more depressed area, such as the Central Graben in Dutch waters and the Sole Pit Trough off eastern England. The nearest parallel onshore is with the Yorkshire basin, at this time bounded by higher land to north-west and north-east. Similarly arenaceous detritus was shed into depressions in the north, such as the northern Viking Graben and offshore from south-west Norway.

A dramatic discovery of this time (though not an economic one) is the very large mass of lavas and tuffs where the eastern margin of the Viking Graben intersects with faults extending eastwards from the Moray Firth (c. 58° N, 1° E, Fig. 72). Here there was a great centre of alkaline basalts, some 1000 metres thick, built up above sea-level; a wide range of isotopic dates has been obtained but the stratigraphical evidence supports a Bathonian age. Some local block-faulting also characterized the Middle Jurassic, for instance on the margins of the Central Graben.

From the upper Oxfordian onwards there was a gradual extension and deepening of the seas, with some basal transgressive sands, while in British waters the northern and southern basins were again connected. The Kimmeridgian Stage is exceptionally widespread and uniform in the sea areas as it is on land, with bituminous shales as an important component. Where Portlandian rocks are found, as in the far north, they are also in an argillaceous facies but in many areas the late Jurassic and early Cretaceous movements have left little of this stage; minor vulcanicity continued in the Upper Jurassic, with examples in the Netherlands and ashfalls in Skye and the Viking Graben (cf. Fig. 72).

Jurassic tectonics in the North Sea combine gentle elevations and depressions, often with halokinetic effects (p. 271) and the more episodic movements, especially of fault blocks in the graben areas. The term Kimmerian has been used for such intermittent effects, with 'late Kimmerian' for the most vigorous around the end of the Jurassic. Locally the structures are pronounced but not orogenic in the normal sense, and thus the terms are perhaps a little misleading.

The Moray Firth basin is somewhat aside from the northern North Sea and is an exceptionally deep faulted structure, and contains a more complete Mesozoic succession than appears in the small land outcrops on the margins. However the Jurassic partakes of the same general pattern; the Lias is not well known but includes some marine strata and the Middle Jurassic some non-marine, with thickness changes over contemporary faults. The Upper Jurassic is the thickest, the Kimmeridgian black shales being again conspicuous. All three divisions are thickest in the west, against the line of the Great Glen Fault; over 2000 metres is a generalized estimate.

In its widest setting the North Sea of this period can be seen in two ways. It complements the history of the land areas, being most similar in the most uniform argillaceous facies and least similar in the shallow water and non-marine sediments. It is also a major tectonic segment of north-west Europe, part of the continental block which was gradually rifting in the west as the North Atlantic opened. The ubiquitous tensional structures can be seen as parallels to the rifting process, though the latter is not very precisely dated. The alkali basalts, with oceanic rather than continental affinities, fit into such a tensional scheme, but though the crust may have been thinned under the North Sea no full rift developed.

The opening paragraphs of this chapter set the Jurassic scene and stressed the importance of stratigraphical palaeontology. At the close does any extended pattern emerge? In terms of bulk sediments a few generalizations can be made. Unexpected though it may be when admiring the mellow freestone of the Cotswolds or the cliffs of Portland the dominant deposit was mud, fine-grained detritus from the warm forested hinterlands, which far exceeds any other type; the same holds for much of north-western Europe. Moreover in most of the basins the thicker central sequences tend to be the most argillaceous.

Superimposed on this first character the lesser subsiding banks and swells represent more agitated, current-swept waters where permanent sediments were thinner or none survived. Some carbonate shelves developed here, locally elevated into systems of intertidal shoals and channels or temporary supratidal marshes and flats. Where saline waters were isolated some evaporites were precipitated, on a modest scale. Within the land-based stratigraphy arenaceous incursions were relatively uncommon and largely related to the Aalenian–Bajocian regression and gentle uplift, producing greater fluviatile detritus.

Although much detail has been added to the English and Scottish scene (Ireland remains all too empty except in the north-east) with more studies of the sediments or refinements among the ammonite zones, the broad outline was essentially drawn many years ago. Jurassic stratigraphy and research had settled into an expert but peaceful, stable condition, with no acute problems of tectonics or orogeny to unravel and not even much change in the well known nomenclature – not, in review, a system liable to upheaval and revolution; even the revolution of plate tectonics held only rather distant repercussions. Yet a revolution of sorts there has been and the picture is substantially altered, the cause being primarily economic. Within the past 15 years the Jurassic rocks below the sea areas have become vastly more important than those of the land, where in Britain exploitation is only a minor facet.

As the exploration of the seas has expanded it has become clearer that the rocks and structures of the land and sea floors are not altogether alike, or that there is a difference in emphasis. Despite some modest faulting, and with exceptions in Scotland, tectonic effects in the land outcrops are slight. The more insistent questions are those of sedimentary environment, of palaeogeography (is Wales to be drawn as an island or not?), or the causes of transgressions and regressions. What constitutes a 'cycle' and what causes cycles?

In the seas there was the major block-faulting and jostling of existing fault blocks; major structures were influenced or initiated through halokinetic movements, whereas the absence of salt structures on land is striking. A great volcanic pile was extruded, up to 1000 metres thick and half the size of Wales. Side by side with all these dramatic features the great spreads of delta complexes and sands hold the main oil reservoirs and require great expertise and expense to bring the oil ashore.

The Jurassic System has not lost the classical interest first discerned in England by William Smith, but surely it has also put on a new look.

REFERENCES

Cope *et al.* 1980*a* (*Special Report* No. 14, Jurassic, part 1).
Cope *et al.* 1980*b* (*Special Report* No. 15, Jurassic, part 2).
General and regional: Ager, 1976; Anderson and Dunham, 1966; Arkell, 1933, 1947*a*, *b*; Bristow and Bazley, 1972; Callomon and Cope, 1971; Hallam, 1958, 1969, 1971, 1975; Hallam and Sellwood, 1976; Hemingway, 1974*a*; Neves and Selley, 1976; Sylvester-Bradley and Ford, 1968, Chs. 11–16; Woodhall and Knox, 1979; Woodland, 1971; Worssam and Ivimey-Cook, 1971.
Lower and Middle Jurassic: Allen and Kaye, 1973; Dean *et al.* 1961; Duff, 1975; Green and Donovan, 1969; Hallam, 1960, 1971; Hudson, 1964; McKerrow *et al.* 1969; Martin, 1967; Morton, 1965; Mudge, 1978; Sellwood, 1972; Sellwood and McKerrow, 1974; Sykes, 1975.
Upper Jurassic: Anderson and Bazley, 1971; Casey, 1971; Cope, 1967, 1978; Gallois, 1976; Gallois and Cox, 1976, 1977; Howitt, 1964; Sykes, 1975; Talbot, 1973; Townson, 1975; Townson and Wimbledon, 1979; West, 1975; Wimbledon and Cope, 1978.
Sea areas: Ager, 1976; Brooks and Chesher, 1976; Finstad and Selley, 1976; Harrison *et al.* 1979; Howitt *et al.* 1975; Kent 1975*a*, *b*; Penn and Evans, 1976; Woodland, 1975.
British Regional Geology: Bristol and Gloucester District, East Yorkshire and Lincolnshire, The Hampshire Basin, The Tertiary Volcanic Districts, The Wealden District.

10

THE CRETACEOUS SYSTEM

During Cretaceous times the British Isles continued to move northwards, though at this time very slowly so that there was not much change in latitude from early in the period to early in the Tertiary Era – southern England being around 38° to 40° N. The more active Atlantic rift continued to expand so that an irregular oceanic strait developed to the south-west, and probably the Bay of Biscay also opened. The climate remained warm and equable and in Lower Cretaceous times at least there is evidence of abundant plant and animal life on land and in shallow waters.

The sedimentary history follows on from the Jurassic with records of shallow seas, some major areas of variable salinity, sands and mud-flats, and locally freshwater influxes. Some of the facies are also alike, especially the marine clays which here include the Gault, with lesser marine sands. But the detritus-free seas of late Cretaceous times did not give rise to oolites, shelly or coralliferous limestones; instead in a phase of unprecedented marine transgression there was deposited over 500 metres of that most distinctive and unusual calcareous facies, the Chalk.

Ammonites continue to be the most important zone fossils where they are available, but in addition to their absence in the variable Wealden facies of the south they are not plentiful at all levels in the marine rocks of the north-east. In the Chalk seas they gradually declined, family after family becoming extinct until all were gone shortly before the close of the period. In the non-marine rocks ostracods, spores and pollen are valuable guides.

The Cretaceous stages (Appendix, pp. 420–1, and Table 46) are named chiefly from France; the English traditional terms, lithological in origin, can be adapted to them fairly well. The relation between the Wealden and the marine sequence is only broadly determinable and, as noted in the last chapter, there are various problems concerning the Jurassic–Cretaceous boundary in this country. The boreal ammonites of the north-east are akin

to those of Russia and here the problems are largely those of minor
unconformities, poor exposure and rather sparse faunas surviving. In
southern England the boundary is taken at the marine transgression of the
Cinder Bed (Table 43, p. 305). In addition to its practical convenience the
episode can be linked with other transgressions (for instance in north-west
Germany) and the basal Cretaceous age of the Durlston Beds above has

Table 46. *Cretaceous stages with principal British lithological grouping*

	Stages		Lithologies
UPPER CRETACEOUS	MAASTRICHTIAN CAMPANIAN SANTONIAN CONIACIAN TURONIAN CENOMANIAN	SENONIAN	Upper, Middle and Lower Chalk
LOWER CRETACEOUS	ALBIAN APTIAN BARREMIAN HAUTERIVIAN VALANGINIAN RYAZANIAN		Upper Greensand, Gault Clay and Lower Greensand Marine rocks Wealden of the in south-east eastern with England upper Purbeck

some support from the microfossils. In Dorset this upper part of the
Purbeck facies consists of clays, thin limestones and evaporites, chiefly
gypsum. Ostracods and molluscs are fairly common, including some
freshwater types, and the small gastropod *Viviparus* is conspicuous in the
mottled limestones of the Purbeck Marble.

As in late Jurassic times the main foci of Lower Cretaceous sedimentation in the south were the linked basins of Wessex and the Weald; but
there was also a gradual and irregular expansion to the west and north
during the Aptian and Albian stages. After the late Jurassic recession of
the northern sea there was also minor irregularity in the Cretaceous
invasion, earliest in Norfolk and a little later in Lincolnshire and
Yorkshire.

No such history can be made out in the western and northern parts of
the British Isles, but the few remaining outcrops also suggest an increasing
marine expansion. There are some traces of littoral or shoreward facies in
the Chalk but no extensive palaeogeography is discernible. The extraor-
dinary absence of detritus from the bulk of the Chalk is a testimony,
negative yet eloquent, to the lack of transport from such low-lying lands as
may have survived.

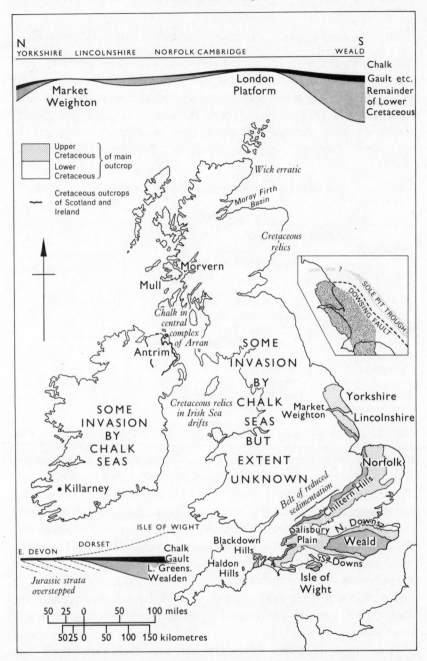

Fig. 68. Cretaceous outcrops. The top section shows generalized thickness variations, the bottom the decrease in thickness, and Albian overstep, from the Weald to east Devon. Inset on right: structures in seas off eastern England related to land and sea outcrops (cf. p. 349).

THE WEALDEN BASIN: MUD-FLATS AND SANDY INCURSIONS

After the basal marine incursion there was a short-lived phase of clays, limestones and evaporites, forming the Durlston Beds. Then for a long period, probably spanning stages up to the Barremian, the Wealden basin became a very shallow water region, dominantly of mud-flats. Salinity conditions were variable and more than one type of cycle can be recognized (Table 47). In the earlier history these were emphasized by

Table 47. *Wealden stages, lithology and cycles*

Stages	Formations	
BARREMIAN	Weald Clay: Upper division Lower division: clays	Several cycles on various scales
HAUTERIVIAN	{ Horsham stone clays	
VALANGINIAN	Hastings Beds: Upper Tunbridge Wells Sand	Possibly cyclic
and	Grinstead Clay Lower Tunbridge Wells Sand	Major cycle
UPPER	Wadhurst Clay	
RYAZANIAN	Ashdown Beds – much sand, local clays at base	Major cycle
LOWER RYAZANIAN	Durlston Beds (upper part of Purbeck facies), marine beds at base	

substantial sandy incursions, which basically distinguished the Hastings Beds; in later times the sands were much reduced to give the Weald Clay. The extreme shallowness is shown by the numerous layers of stems, roots and rhizomes of plants capable of colonizing the subaqueous muds; several are attributed to the 'horsetail' *Equisetites*. Trees were very uncommon in these conditions, but dinosaurs (especially the typical Wealden *Iguanodon*) could roam across the marshes and mud-flats, leaving the occasional footprint and skeletal remains as evidence. This picture is a revision of an earlier version of sandy deltas entering a lake or lagoon, partly because the clays did not, on further investigation, show evidence of being the deeper water facies.

The arenaceous influxes of the Hastings Beds came very largely from the north. The base of the sands is abrupt and the mud-flats changed to coalescent plains of braided rivers (Fig. 69); some of these can be distinguished by their detrital contents which include Upper Jurassic,

Carboniferous and Devonian rocks, with quartzites and possibly Lower Palaeozoic components to the east. This sudden change in regime has been attributed to an equally sudden rejuvenation in the source-land of the London Platform; since there is also indirect structural and geophysical evidence of a faulted margin and re-activated basement faults within the

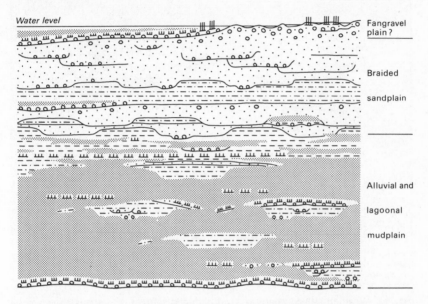

Fig. 69. Facies of a major cycle in the Hastings Beds Group. Clays with occasional siltstones and remains of *Equisetites* at the base pass upwards into more sandy beds with siltstones and pebble seams and then perhaps to an emergent gravel plain at the top, also colonized by plants; successive environments on the right. (Redrawn from Allen, 1976, p. 408.)

Weald itself, the 'platform' has also been interpreted as a horst block (or blocks) with periodic activity of the boundary faults. At the same time it acted as a barrier between the Wealden basin and the boreal sea to the north, but possibly only a partial or intermittent one. The more saline or brackish influences seen from time to time in the clay formations may well have come either across the lower parts of the London block, or round its western margin, rather than from southern France and the distant Tethys. However palaeontological proof, in the shape of boreal species in the Weald, is not forthcoming.

In their late stages the major sandy incursions developed into gravelly alluvial fans, particularly along the northern margin, and there was then a period of emergence and erosion, so that the 'Top Pebble Beds' reflect

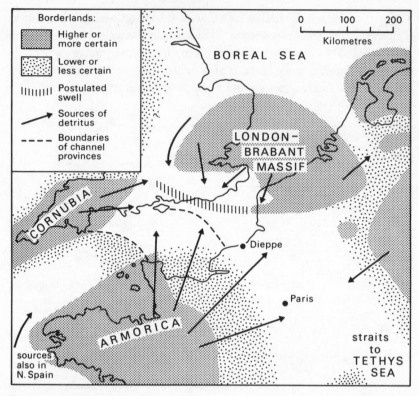

Fig. 70. Wealden palaeogeography and deduced sources of detritus entering the basins, including Wessex and the Weald, partly separated by a low swell or ridge. Boundaries between the eastern, central and western divisions of the English Channel also shown. (Adapted from Allen, P., 1965, p. 324, and 1976.)

residual or lag deposits with a falling water-level and subaerial erosion. The mud-flat regime was then re-established, its ostracod and molluscan faunas reflecting the varying salinities.

The change to the Weald Clay facies is somewhat obscure, partly because formations at the boundary are not well exposed. It seems that conditions were similar to those of the earlier clays but that sand influxes (and thus presumably the horst uplifts) were much reduced. Some detritus with East Anglian affinities drifted in from the north. Cyclic sequences can be widely recognized and are reinforced by salinity fluctuations; towards the top the marine effects are pronounced and include beds with oysters, echinoids and cirripedes.

The most important change in the later Wealden regime is seen in the western outcrops where the cycles incorporate detritus from a wholly new source. The sandy beds fade out or split up eastwards and the detrital minerals can be matched in Hercynian rocks – certainly from the Cornubian and Armorican (Brittany) massifs (Fig. 70) and perhaps from northern Spain as well; there is also some older 'basement' material, possibly Caledonian and perhaps Precambrian. Some detritus was derived from New Red Sandstone, such as that now found in Devon and the western Channel. Earlier, this western Hercynian source did not reach into the Weald but is known in the Wessex basin. The margin of the Weald basin on the south-west and south was probably a low ridge, or swell, from Portsdown (north of Portsmouth) to the dip-slope of the South Downs and thence across to the French coast south of Boulogne. This was a line of partial facies change, thinning and local unconformity during much of Lower Cretaceous sedimentation.

The Wealden in Wessex

A borehole on the Portsdown ridge recorded only 240 metres of Wealden but there is a marked expansion into the Wessex basin where the facies is rather different. There is no clear distinction between the Hastings Beds and Weald Clay and only the upper part of the latter is exposed, though a borehole section in the Isle of Wight extended down into the Purbeck. From here westwards there is general increase in the sandy components, and, in Dorset, a marked decrease in total thickness, the upper shales disappearing first. West of Lulworth the residue of 70 metres of sandy beds includes nearly 10 metres of conglomerates. In west Dorset no Wealden remains, having been overstepped by the Albian. In general this coarser detritus found in the west, probably not far from its source, reinforces the concept of Hercynian massifs, with some New Red Sandstone cover, as the main origin. For at least part of the later Wealden there was a good deal in common between Wessex and the western Weald in sedimentary regime.

APTIAN AND ALBIAN SEDIMENTS IN THE SOUTH

The Lower Greensand, Gault Clay and Upper Greensand have an irregular curved outcrop inside the Chalk downs from Folkestone westwards to Surrey and Hampshire and back to Eastbourne (Fig. 71). They also form much of the southern Isle of Wight and extend into Dorset, but only the higher Albian beds reach beyond the Jurassic and Wealden outcrops into Devonshire. Table 48 shows these groups and the standard stages; some of

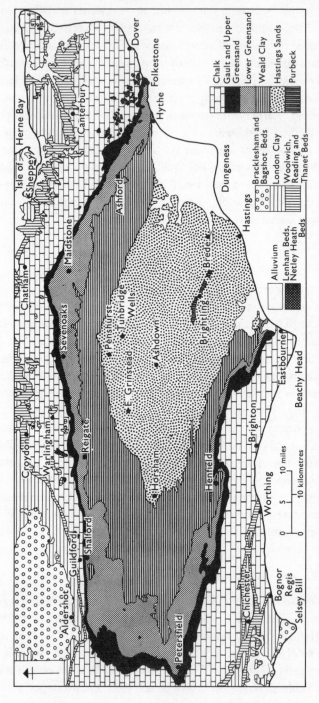

Fig. 71. Geological sketch map of the Weald and neighbouring areas. (Adapted from Gallois, 1965, pl. i.)

the lithological boundaries are diachronous, the most notable being the replacement westwards of the Gault Clay by the Upper Greensand. Furthermore, although the influx of lower Aptian shallow seas affected Wessex and the Weald nearly simultaneously there is no simple lithological equivalence between the two.

Table 48. *Aptian and Albian in the south, lithological divisions*

Weald		Stages	Isle of Wight
Upper Greensand and Upper Gault Clay		UPPER ALBIAN	Upper Greensand and
Lower Gault Clay		MIDDLE ALBIAN	Gault Clay
Folkestone Beds ⎫		LOWER ALBIAN	Carstone Sandrock
Sandgate Beds ⎬ Lower Greensand		UPPER APTIAN	Ferruginous
Hythe Beds ⎪		LOWER APTIAN	sands
Atherfield Clay ⎭			Atherfield 'Clay'

Lower Greensand

This name arose from a series of misconceptions, principally a confusion between these sands below the Gault Clay with others above it in the western outcrops. Surface exposures of the Lower Greensand are yellow or rusty brown from the oxidation of the iron compounds but in cliff sections or underground the glauconite is fresh enough to give a sporadic greenish tinge.

Fossils are usually rare in the sandy facies, probably because these have been leached and the shells dissolved. They may be fairly abundant in the harder or calcareous bands and from such rocks enough ammonites have been collected to establish the zonal sequence. Together with sporadic brachiopods, sponges, corals and echinoids they make up a representative marine Mesozoic assemblage, adapted to shallow and locally current-swept waters. The London Platform persists as the clearest source of detritus, with suggestions of periodic rejuvenation through boundary fault movements.

The relatively thin Atherfield Clay (20 m or less) is red, silty and fairly fossiliferous, the 'Perna Bed' at the base being a hard carbonate-cemented band characterized by large bivalves. A marked reduction takes place in south-east Sussex where only a remanié bed of nodules and rolled fossils remains. The sagging basin of the western Weald is again shown in the three succeeding sandy formations, which expand from east to west to

reach nearly 200 metres on the Surrey–Hampshire borders. The sands are
normally poorly cemented, often mixed with clay or silt and sometimes
cross-bedded or with seams of pebbles or phosphatic nodules. The harder
and better cemented beds are usually calcareous, or more rarely siliceous,
and form local impure limestones or cherts. Both types are exceptional in
their abundant fossils which in calcareous rocks include corals. The
pebbles of the Hythe and Sandgate beds continue to reflect sources in
rejuvenated London uplands. In the north-west local sandbanks contain
numerous derived Jurassic fossils, such as Oxfordian ammonites, and it is
thought that these probably came from an uplifted area to the north where
the later Jurassic of the upland had been stripped off. Fuller's earth is a
local component of both the Sandgate and the Folkestone beds. As in the
Jurassic examples the clay mineral montmorillonite is considered to be
formed from volcanic ash, and in the Aptian context this was probably
carried into the littoral seas by rivers draining the London Platform.

Both palaeontological and lithological evidence point to non-deposition
and erosion at several levels in the Lower Greensand. Local non-sequences
are found for instance at the base of the Atherfield Clay in east Kent,
below the Sandgate Beds in Surrey and where they overlap onto the Kent
Coalfield, and – most widespread – in the middle of the Folkestone Beds.
Such breaks are commonly accompanied by seams of phosphatic nodules
containing fossils derived from the beds below. They mark periods of little
or no sedimentation, rather than actual emergence, when the sand and
silts were winnowed away by currents.

The upper Sandgate Beds of Surrey show a re-advance across the
margin of the London Platform and this tendency continues in the latest,
Folkestone, regime. Although boreholes over the centre of this platform
show no Lower Greensand it is possible that it was a low swell,
intermittently submerged, where no permanent sediments remained. In
the Thames estuary region boreholes demonstrate a thin skin of Greensand above Palaeozoic and faulted Jurassic rocks and below the Gault
Clay.

Over the Portsdown ridge the Lower Greensand is reduced to 30
metres, but again the eastern part of the Wessex basin was a subsiding
region, and the fine cliff sections of the Isle of Wight expose some 245
metres. For the most part sedimentation seems to have been more
continuous than in the Weald and the only marked break is below the
Carstone. The latter is a term applied to certain Cretaceous rocks, a dark
brown coarse ferruginous sandstone with quartz pebbles and phosphatic
nodules.

In Dorset the Lower Greensand shows a more marked westerly reduction than the Wealden below or the Gault above. It amounts to only 60 metres in Swanage Bay, where ammonites are rare, some of the bivalves are apparently brackish types and the sands contain lignites. These marginal characters are enhanced farther west and at Lulworth Cove there is only a thin ironstone band with a brackish Wealden-like fauna. Nevertheless this bed is probably equivalent to the upper part of the Atherfield Clay in the Isle of Wight so that the overstep of the Gault Clay is combined with the absence of all the higher Aptian beds. West of Lulworth the Gault lies on Wealden.

The Gault Clay and Upper Greensand

Where both these facies are present, for instance in the western Weald, the greensand lies above the clay; but over southern England as a whole there is also a lateral transition and the sand replaces an increasing proportion of the clay westwards. Typical Gault Clay is dark grey, stiff and tenacious; phosphatic nodules are the main lithological variant and a particularly persistent band marks the junction between Upper and Lower Gault. As in other Cretaceous formations they commonly represent a break and winnowing of the sediments. The term Upper Greensand covers a wide lithological range from glauconitic sandstones and siltstones to calcareous marls and cherts; the facies reflects the same westerly influx of detritus that is a persistent, though locally variable, feature of Cretaceous sedimentation in southern England.

The clay fauna is a rich one with ammonites and other molluscs, echinoids and crustaceans; in the clearer waters sandy sea floors supported brachiopods, bivalves, sponges and bryozoans, but fewer ammonites. The type sequence of the ammonite-bearing Gault is at Folkestone where almost all the 40 metres of middle and upper Albian consist of clay with only a small addition of glauconitic sand at the top. As the beds are traced westwards along the foot of the North Downs there is an increase in the total thickness and also in the sand component. Incomplete evidence in the western Weald suggests a total of over 100 metres, including an expanded Upper Greensand.

In a wider context there seems to have been a marked extension of the muddy Gault seas fairly early in their history, with minor irregularities of the sea floor affecting deposition and thickness. At the base of the Upper Gault a period of more pronounced erosion planed off the beds below, with renewed faulting in the region of the Thames estuary. Over the

London Platform there is clear continuity in Upper Gault times, with clay thicknesses around 60 metres, but also much thinner amounts of the Lower Gault and some basal sandy detritus. The full connection between north and south therefore may well have been achieved within the lower Albian period but in the initial shallow seas little permanent sediment survived.

The Gault overstep to the west

In the Isle of Wight both clay and greensand facies are present but their difference is accentuated; in the south of the island some 35 metres of 'greensand', including massive sandstones with seams of chert, overlie 30 metres of clay. This trend continues into Dorset, and the overstepping base of the Gault from Swanage to Devon (Fig. 68, inset) is one of the clearest transgressions in British stratigraphy. In east Dorset the clays overlie Lower Greensand, but probably with unconformity; thence westwards there is a downward transgression onto remnants of the Wealden, Purbeck, Portland and Kimmeridge beds; along Lyme Regis bay the Albian outliers overlie Middle and Lower Jurassic. At Lyme Regis itself a few metres of the clay facies persists (on Lias) but westwards into Devon only the greensand facies remains.

In Devonshire the principal Upper Greensand outliers form the Blackdown Hills (Fig. 57) where the glauconitic sands include *Exogyra* and chert masses near the top; shells beautifully preserved in chalcedony were once collected here. Farther west the last Cretaceous outpost in South-West England is the Haldon Hills, east of Dartmoor, at 240 metres O.D. The Albian components consist of 80 metres of glauconitic and pebbly sands with *Exogyra* and a few corals; the uppermost beds include a small amount of Cenomanian sands. A different littoral facies of foraminiferal limestones borders the nearby Tertiary basin of Bovey Tracey. At present these variable inshore deposits lie well below the summit of Dartmoor (580 m O.D.) so that at this time, late Albian and early Cenomanian, the seas are most unlikely to have submerged that massif. In the latest Cretaceous transgression the position may well have been different.

Lower Cretaceous earth movements

The main tectonic effects of late Jurassic and early Cretaceous times are shown as faulting in the sea areas. Those of southern England are later, late Aptian or early Albian, and are clearest in south Dorset where they complicate the larger pattern of overstep. Small sharp east–west folds and major faults were planed off by the mid-Albian transgression so that north of Weymouth, for instance, the underlying beds vary in a short distance from Oxfordian to Wealden. Substantial uplift at this time is also suggested along the north-westerly continuation of the Portsdown ridge, because in a borehole in west Hampshire the Gault rests on Kimmeridge Clay. Similarly north of the marginal fault of the Vale of Wardour (south Wiltshire, Fig. 24) the underlying formation is Oxford Clay. It is thus probable that faulting and uplift were fairly widespread in South-West England at this time and perhaps farther afield. There is the block-faulting in the Thames estuary and possibly some of the displacements or erosional uplifts in the Midlands include a Lower Cretaceous component. In the south-west warping continued into the basal Cenomanian, affecting deposition and facies.

THE NORTHERN SEA

North of the London Platform, in Norfolk, Lincolnshire and Yorkshire, all the Cretaceous rocks are marine, deposited near the western margin of the large boreal sea that extended with irregularities and fluctuations across the North Sea to north Germany, Denmark, Poland and Russia. Exposures in Norfolk and Yorkshire are decidedly poor, those of Lincolnshire a little better. They have also been crucial in determining the Jurassic–Cretaceous boundary in this boreal marine facies; the very small amounts of uppermost Jurassic were mentioned in the last chapter.

The sequence set out in Table 49 crops out at the foot of the Lincolnshire chalk wolds and represents a typical nearshore marginal facies; deposition was slow, with interruptions and pauses when phosphorite formed nodules on the sea floor and minor erosion phases reworked the earlier sediments. The initial Cretaceous marine invasion of southern England had no exact equivalent in Lincolnshire, and after an erosional gap sedimentation set in a little later with the Upper Spilsby Sandstone. Thereafter the principal facies were sandy, often with pebbles, muds and lesser oolitic ironstones; the last includes the Roach as well as the Claxby Ironstone. The Tealby Limestone is no more than a thin impure sandy bed, sandwiched between thicker clays.

A broad cyclic pattern can be detected, from sandstones to ironstones, clays, thin limestone and then the same in reverse order, with non-sequences in the Aptian Stage (commonly poorly represented in this province), especially below the Sutterby Marl where more than one Aptian zone is represented by layers of phosphatic nodules. The Carstone, a pebbly ferruginous sandstone, is a thin but widespread rock type at this level and passes up gradually into the Red Chalk. The latter is a curious facies confined to the northern province and principally of middle and upper Albian age, though lesser red or pinkish beds are known in places higher up. Typically the brick-red rock contains scattered quartz pebbles,

Table 49. *Lower Cretaceous succession in central and south Lincolnshire*

Lithological formations	Stages	Thickness (metres)
Red Chalk Carstone	ALBIAN	11
Sutterby Marl and clays with non-sequence	APTIAN	4

Fulletby Beds,
  Roach, ironstone
Tealby: upper clay
     limestone
	BARREMIAN	31

lower clay Claxby Ironstone and clays	HAUTERIVIAN and VALANGINIAN	18
Upper Spilsby Sandstone with nodule bed	RYAZANIAN	11

———————————— Minor unconformity ————————————
(Lower Spilsby Sandstone, Portlandian or Volgian)

small belemnites, bivalves and terebratulids; the colour is due to hematite but its origin is obscure. An erosional source from an earlier red facies has been suggested (e.g. Trias, exposed in the North Sea Basin), but the colour boundaries are transgressive and a diagenetic cause is perhaps more probable.

At the southern end of the Lincolnshire outcrops the Cretaceous rocks disappear under the Fens and the Wash to reappear on the Norfolk coast at Hunstanton (Fig. 77). Here the Carstone and thin Red Chalk are similar, but the underlying beds are different, though still shallow water and variable. They consist dominantly of sands below and more clayey types above (Table 50). Near Hunstanton the former rest on an eroded surface of Kimmeridge Clay and include cross-bedded quartzose sands that have been worked for glass-making. Farther south outcrops are negligible but fortunately temporary exposures have shown a fuller succession. Here the greyer more glauconitic sands, like the Spilsby Sandstone (which they somewhat resemble), span the Jurassic–Cretaceous boundary. There are similarly gaps and signs of contemporaneous erosion but in these marginal waters, on the fringes of the London Platform, local variation is common and there is no precise similarity between Norfolk and Lincolnshire. Above the succeeding thin clays the Carstone is strongly transgressive, cutting out all the underlying beds southwards. Conversely the thin Red Chalk of Hunstanton passes into thicker grey beds in south-west Norfolk and thence to normal Gault Clay in Cambridgeshire.

Turning now northwards, as the Lincolnshire beds approach the

Table 50. *Lower Cretaceous succession in Norfolk*

Lithological formations	Stages	
Red Chalk (passing up into Gault Clay) ⎫		⎫
Carstone ⎬	ALBIAN	⎪
————————non-sequence————————		Maximum thickness 113 m
Fulletby Beds ⎫		⎬
Dersingham Beds ⎬ including	BARREMIAN and HAUTERIVIAN	⎪
Snettisham Clay ⎭		⎪
Sandringham Sands (upper part) ⎫ with nodule bed ⎬	VALANGINIAN and RYAZANIAN	⎭

Humber nearly all become thinner and then disappear, but the Red Chalk remains with a very small amount of Carstone below. Near Market Weighton there is a small section where these remnants can be seen lying on Lower Lias – the only formations to survive across this formidable swell since early Jurassic times. In the Yorkshire basin on the northern side all the rocks between the Kimmeridge Clay and the Red Chalk are surprisingly different; all are comprised in the Speeton Clay Formation, chiefly exposed in badly slipped sections on the coast north of the Chalk at Flamborough Head (Fig. 62). The bulk consists of grey clays with seams of cementstone, ferruginous and phosphatic nodules. Glauconite is found at many levels and pyrite chiefly in the lower clays. The stratigraphical sequence is based on belemnites and crushed ammonites, with some help from the foraminifera. Other fossils are ostracods, bivalves and brachiopods (including *Lingula*); one bed is rich in crustaceans and another in *Pentacrinus* ossicles. This is another version of a Mesozoic clay, comparable to the Gault but earlier and thicker.

Derived Kimmeridgian fossils are again found in the basal nodule layer, but although Yorkshire continued to be a strongly subsiding basin, the erosional gap is larger than in Lincolnshire and the earliest clays are slightly later than the inshore sandstones there. Four beds of bentonitic clays have been found in the lowest stage (Ryazanian), representing primary ashfalls in this case with little or no redistribution. The Speeton divisions above include all the succeeding stages; towards the top of the Aptian a layer with glauconitic sands and nodules probably corresponds to the Carstone and above that grey and pink marls are succeeded by a Red Chalk facies which here continues up into the Cenomanian Stage.

The clays are difficult to measure at Speeton itself but over 100 metres have been estimated. In a borehole a few kilometres inland, however, over 400 metres were recorded. Thus the clays in the Yorkshire basin were substantially thicker than the more variable inshore facies on the East Midlands Shelf south of the Market Weighton swell. Neither the basins nor the swells of the Mesozoic seas were established on a single uniform pattern and the same variation extended to neighbouring areas of the North Sea.

OUTLIERS AND REMNANTS BETWEEN NORTH AND SOUTH

We now return to the long intervening stretch where thin, widely separate patches are found chiefly as outliers beyond the chalk escarpment from Wiltshire to Bedfordshire. In stratigraphical position they lie between the two pervasive clay formations, Kimmeridge below and Gault above. In sedimentary terms they were deposited in shallow waters near the western end of the London Platform during two principal transgressions. The facies are rarely those of the type areas but are more littoral and sandy; moreover erosional phases have removed much of the thin spreads, during and after Cretaceous times.

The basal Durlston transgression is traceable sporadically, largely as thin sands, from the vales of Wardour and Pewsey as far as north Buckinghamshire. The sands lie either on the Lulworth Beds (lower Purbeck) or overstep onto the Portland. The outlier of the Whitchurch Sands, eight kilometres north of Aylesbury, is an example. Wealden sands have also been tentatively identified in this county and near Oxford, for instance the 20 metres of the Shotover Ironsand, but their freshwater molluscs have no correlative value.

The upper part of the Lower Greensand (upper Aptian and lower Albian) is more widely and securely recognizable in Wiltshire and Berkshire. The Seend Ironsand near Devizes was once worked for its iron content and produced a large fauna, but the best known outlier consists of

50 metres of sands and gravels at Farringdon in Berkshire. Within a small area these beds lie on Corallian and Kimmeridge Clay and are famous for their sponge gravels at the base; other fossils include bryozoans, brachiopods and echinoids. The cross-bedded sands and pebbles accumulated in channels on the floor of a shallow clear sea.

More Aptian faunas are known from sandy beds near Oxford and Aylesbury, and to the north-east there come in the thicker and more continuous outcrops of Bedfordshire. They include the Woburn Sands with an unusual assemblage – a land-derived flora of conifers, cycadophytes and early angiosperms. Above lie the lenses of the Shenley Limestone, with sand grains and pebbles scattered in a calcareous rock; the brachiopods and other shells grew on the uneven surface of the sands below.

The Lower Greensand continues through Cambridgeshire but gradually becomes thinner; the underlying fragmentary groups are no longer found so that there is a larger erosional gap above the Kimmeridge Clay. In the Fens a small capping of greensand accounts for the upstanding Isle of Ely and similar beds are banked up against the reef patch of Corallian at Upware. In Norfolk, as noted earlier, Aptian sediments are very sparse or absent.

The sporadic sediment and faunas of this intermediate belt have wider implications than provisional identification or age. The marine Durlston components of the Whitchurch Sands, together with some derived uppermost Jurassic ammonites in the basal Woburn Sands, show that the seas extended round this end of the London Platform, at least for a time, around the Jurassic–Cretaceous boundary. And since they have been traced so far there arises at least a strong possibility that the basal Durlston (Cinder Bed) invasion of the southern basins entered by this northern route, rather than by the long passage from the Tethyan seas.

On a country-wide scale the western limits of Lower Cretaceous sedimentation are also problematic. In South-West England a line is commonly drawn not very far beyond the present Wealden and Purbeck outcrops, partly on account of a presumed source of detritus in Devon and Cornwall and lack of Lower Cretaceous rocks in the central English Channel, and partly because of the marked thinning in these formations. Such thinning also affects the south Wiltshire inliers in the vales of Wardour and Pewsey.

However, it becomes much more difficult to continue this westerly limit northwards, or to deduce at all satisfactorily the extent of the restricted seaway (a 'Bedford Straits'?) between the shelving edge of the London

Platform and a land to the north-west. This is principally because the
latter concept was based on some kind of enlarged Welsh Massif, inherited
from Jurassic geography. As we saw in the last chapter that massif is now
much more doubtful. It would probably be best to acknowledge that while
there is evidence of fringing intermittent Lower Cretaceous sediments
along this belt we know virtually nothing of any western margin or of the
western extent of the northern sea itself.

JURASSIC AND CRETACEOUS ASHFALLS

Traces of Mesozoic vulcanism are widespread in north-western Europe,
apart from the major Jurassic lava field in the North Sea. Probably they
are ultimately related to the earlier stages of North Atlantic rifting, similar
to the more pronounced Tertiary effects of the later stages. In Britain a
number of ashfalls have been noted (Fig. 72) as fuller's earths or
vulcanogenic clay, for instance of Bathonian age in the Fuller's Earth near
Bath (and possibly clays in the East Midlands), Upper Jurassic in Skye,
Ryazanian in the Speeton Clay of Yorkshire and Aptian in the Weald. It is
quite likely that detailed examination will reveal more, though not all clay
minerals of the smectite group, which includes montmorillonite, are of
volcanic origin. Independent evidence is desirable, such as the inclusion of
volcanic glass fragments or high temperature minerals of igneous rocks.

The stratigraphical age of the Jurassic and Cretaceous examples in
Britain is fairly secure, at least in terms of the stages. Any volcanic centre
as a source, however, is much more difficult to identify. None is known
from land outcrops and there are problems in establishing comparable
ages and magmatic affinity. On balance, although igneous rocks are
known from several sea areas none has so far proved a certain source. In
particular age determinations from submerged or buried rock masses tend
to be few or provisional.

THE CHALK

In its thickness, extent and purity the Chalk is unique in the British Isles
and also in north-western Europe. It occurs in the Anglo-Parisian Basin
and the English Channel, and stretches across the North Sea to the
Netherlands, Denmark, north Germany and Poland. Other similar but
not identical 'chalks' on other continents reflect the same great marine
transgression over the continental edges in Upper Cretaceous times. Some
kind of marginal facies, commonly sands, is found near the bordering

landmasses such as Cornubia, Scotland, Brittany and the Ardennes. But most of these belong to the Cenomanian and Turonian stages, and how far the Campanian (late Senonian) seas may have submerged portions of the continents is largely conjectural. During the latest Cretaceous times the

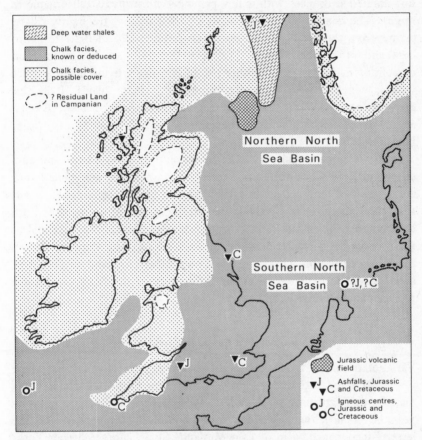

Fig. 72. Map suggesting the maximum extent of the Chalk seas in the late Cretaceous transgression; however the remaining land areas are very tentative. Some centres and products of Jurassic and Cretaceous vulcanism also shown. (Data taken from Hancock 1975, Fig. 3, and Harrison *et al.* 1979.)

seas shrank again and upper Maastrichtian and early Tertiary sediments are not widespread. The fullest European sequence is in the North Sea.

In England the Chalk usually forms gently rolling downs and uplands with a scarp edge overlooking Lower Cretaceous or Jurassic country. It underlies the Quaternary and Tertiary of East Anglia and the London Basin, where it was earlier a source of artesian water. Rocks of Chalk age,

though not all of that lithology, are the most important of the few
Cretaceous outcrops in Ireland and Scotland.

Our familiarity with the Chalk, in quarries and cliffs from Dorset to
Yorkshire, is apt to obscure its true singularity. Pure white chalk is a very
fine-grained carbonate with a few principal constituents, all organic in
origin. The coarser fraction consists of shell debris, frequently calcite
prisms from the bivalve *Inoceramus*; this fraction also includes foraminifera and other microfossils. The very fine fraction is derived from
coccoliths (nannofossils) – minute planktonic algae, living in the surface
waters, which secrete rings and groups of calcite plates. It is this
dominance of coccoliths that particularly distinguishes the chalk, places it
among the pelagic sediments and separates it from the more familiar major
limestones.

Chalk is not normally rich in fossils and the very large museum
collections, often beautifully preserved, are rather misleading since they
are a result of long collection rather than an original abundance. Apart
from the microfossils bivalves are the most plentiful, together with
echinoderms, belemnites and brachiopods. Ammonites are rarer and as
zone fossils do not extend far up into the Chalk. The standard British
zones (Appendix, pp. 420–1) are based on southern England; in Yorkshire
and Lincolnshire there are minor differences in the faunas and some
replacements among the indices.

Regional variations and lithology

Variation in thickness over the English outcrops reflect two factors. There
were still some differences between subsiding basins and intervening
massifs, distinguishing principally a 'Wessex–Paris Basin' (including by
now the Weald), the East Anglian or London and Brabant Massif, the
lesser Lincolnshire basin or East Midlands Shelf, and the deeper Yorkshire basin; the last had the greatest affinity with the much larger complex
North Sea basins. These structures and their relative subsidence governed
the original thicknesses but the amount now present is much more
affected by post-Cretaceous erosion, so that the age of the highest
surviving Chalk varies conspicuously from one region to another.

Soft white chalk, typical of southern England, was deposited as a
low-magnesian calcite, unlike most limestones. Since this mineral is stable
at surface temperature and pressure there was little early diagenesis and
the soft friable character largely survives. In the harder chalks, found in
Yorkshire and especially north-east Ireland, the coccolith plates are

welded together; there was originally a small amount of aragonite and the pores are infilled with secondary calcite. The most obvious non-calcareous component is flint, which was formed as biogenic amorphous silica, later transformed into chalcedonic quartz. It may infill pore-spaces and also replace calcite and so include some fossil remains or trace-fossils.

Deductions on the depth of the Chalk sea are derived principally from analogies between Cretaceous and modern floras and faunas. The abundance and extent of the coccoliths, settling on a bottom undisturbed by currents, and the distribution of the sponge faunas suggest a range between 100 and 600 metres depth, but there will have been substantial variations, including shallower belts at the margins. Marginal sediments are rarely preserved in this country, however, or not in a chalk facies.

The distribution of the successive stages in north-western Europe shows that although the Upper Cretaceous seas spilled widely over the continental margins the process was not wholly regular. Thus after the marked early and late Albian transgressions, there was a pronounced but rather irregular overspill in the Cenomanian and then a regression in the early to middle Turonian. The final transgression reached an acme in the Campanian Stage, succeeded by an equally marked Maastrichtian regression. Clearly these oscillations, with probably some continuing differential subsidence and residual submerged topography, will have affected the seas of the Chalk, their depth and conditions.

During deposition the sea floor was commonly occupied by a thin layer of soft calcareous ooze but an important feature is the later development, slightly below the sediment surface, of hardgrounds. These harder layers, which fade out downwards, include nodular chalk or chalk 'pebbles' with coatings of glauconite or phosphate. Some also provide biostratigraphical evidence of condensed deposits, such as the 'Green Beds' (glauconite-coated nodules) in the Upper Chalk of the Isle of Wight. Hardgrounds and signs of submarine erosion are more common than was once recognized; multiple examples give rise to the various nodular or conglomeratic 'rocks' of the Chalk. Such conditions were more prevalent in the shallower waters and over the massifs where subsidence was less, especially the submerged London and East Anglian Platform.

The best known of these hard lithological bands are:

Top Rock	top of *planus* Zone
Chalk Rock	base of *planus* Zone
Melbourn Rock	low in *labiatus* Zone
Plenus Marls	*gracile* Zone
Totternhoe Stone	low in *rhotomagense* Zone

The traditional British grouping of Lower, Middle and Upper Chalk is related to these lithological horizons, in that the base of the Middle Chalk was drawn at the Melbourn Rock and that of the Upper Chalk at the Chalk Rock. Naturally these divisions fit only approximately into the zonal system, itself heterogeneous in origin. Even within the English outcrops the Chalk Rock cannot be recognized in Lincolnshire and Yorkshire.

Southern England

In Upper Cretaceous times this province, defined principally on faunas, overlapped the now submerged London and East Anglian Platform. In many outcrops the lowest Cenomanian is separated from the Albian below by a minor break, but the detrital components tend to resemble the earlier facies, whether Gault Clay or Upper Greensand. Chalk Marl refers to a soft rock with a substantial addition of silt or clay; glauconitic chalk may contain quartz sand as well. Even where there is no striking addition the Lower Chalk tends to be slightly muddy and grey rather than white. The rhythmic deposition that is widespread at many levels is more conspicuous here than in the purer chalks because of the colour variation between the bands with less or more marl. Phosphatic nodules appear locally at the base and concentrations of sponges or sponge spicules but flint is not common.

Variations in thickness of the Lower Chalk show that there were still some differences in subsidence. In south-east England 60 to 80 metres is common, but the accentuation of the Wessex basin is shown by the 107 metres in the Portsdown borehole. There is a general reduction northwards into East Anglia, accentuated at Hunstanton on the Wash coast where only 17 metres occur; this thinning chiefly affects the lowest bed (the Chalk Marl) which rests unconformably on a comparably reduced amount of Red Chalk. The two distinctive beds of the division are widespread – the Totternhoe Stone, gritty with shelly fragments and glauconite grains, and the Plenus Marls, a grey layer with the belemnite *Actinocamax plenus*.

The Cambridge Greensand is a local but unusual form of basement bed, where calcareous clays lie on an irregular eroded surface of Gault Clay. The phosphatic nodules in the hollows were once worked for agriculture and included remanié fossils from at least two Gault zones below. In addition there is a rather mysterious assemblage of 'erratics' – far-travelled boulders and pebbles. Some may have come from the Midlands or Wales but others have no known source; it is possible that they are secondarily derived from a formation such as the New Red Sandstone, but how they got into the early Cenomanian sea remains obscure.

The most important lateral facies change among the Cenomanian beds is towards the south-west and is thus parallel to, but less extensive than, the similar trend in the Albian Stage below. Moreover the present Chalk outcrops have been more depleted by Tertiary erosion than those of the underlying beds. In Dorset there are two interacting factors. Latest Albian and Cenomanian deposition was affected by a low ridge, or swell, in mid-Dorset which trended north-west and was thus aligned with post-Cretaceous structures in the same region. Over the swell the Upper Greensand is reduced and the lower Cenomanian absent except for some remanié fossils. The second influence is shown by the diachronism of the Chalk Basement Bed – a thin phosphatized conglomerate with numerous ammonites. This rises westwards through three ammonite zones; beneath are some further (Cenomanian) greensands and remanié beds, above the normal greyish Lower Chalk. The relationship is in fact a local demonstration of the major westward transgression. That this reached as far as the Haldon Hills is shown by the uppermost greensand there with an early Cenomanian fauna.

As would be expected there is more uniformity in the Middle and Upper Chalk than in the lower division with the proviso, already noted, that the thickness of the Upper Chalk is strongly affected by later erosion. The Middle Chalk reaches 80 to 90 metres in the south and is largely soft white chalk, with flints appearing near the top. The latter feature continues throughout the Upper Chalk, and the two 'rocks' are confined to the *planus* Zone at the base. Where fairly complete the Upper Chalk is by far the thickest of the three divisions. In East Anglia 320 metres are preserved, and more in the Isle of Wight (400 m) despite the greater erosion there, a comparison that again demonstrates the greater subsidence in Wessex.

In the south-west erosion has reduced the Upper Chalk to small outliers in east Devon, neither rising above the basal Coniacian zone. However that the seas did extend much farther, and for a long time, is shown by the derived fossils in more than one group of later gravels. The most extensive are those that cap the greensands of the Haldon Hills and from which casts have been collected of fossils up to the *testudinarius* Zone (upper Santonian).

Similar deposits at Orleigh Court, five kilometres south-west of Bideford in North Devon, extend the record up to the upper Campanian (*mucronata* Zone). If the thickness of Chalk represented by these zones was similar to that of East Anglia, then Devon and Cornwall were very largely submerged in the last great transgression. Finally this aspect is reinforced by the derived fossils, up to Maastrichtian in age, in Tertiary gravels near Bournemouth. It is the kind of indirect but valuable evidence only available where very durable rocks enclose the fossil remains.

East and North-East England

North of the Wash the Chalk of Lincolnshire and Yorkshire belongs to a separate province though the distinction is less in this more uniform facies than in Lower Cretaceous strata. Some of the southern zonal indices can be used, especially in the middle levels, but other species are too rare and have to be replaced by northern ones. Not all the lithological markers extend to the north-east either, so that although the terms Lower, Middle and Upper Chalk have commonly been employed they may not have precisely their southern boundaries.

As in the south the Lower Chalk is the most varied, with changes conspicuous in the region of the Wash, at the northern edge of the Market Weighton structure and, rather less clearly, at its southern margin. Thicknesses thus vary in Yorkshire from 18 to 42 metres, the coastal outcrops being substantially the thicker. Among the grey partly nodular chalk, reddish rocks form a conspicuous variant. They are sufficiently pronounced at the base to be called the Cenomanian Red Chalk, separate from the obvious Albian bed of that name below. One or more pink bands also occur farther up and the cross-cutting relationships of some of these reddish units reinforce the view that the pigmentation is diagenetic.

In addition to the marl bands there are beds rich in detrital shell fragments, like the Totternhoe Stone of East Anglia, and two with abundant *Inoceramus*; flints are largely lacking. Minor erosion levels and bored surfaces are found and a number of cyclic sequences can be detected, with repetitive lithological characters, stromatolites and faunas. At about the junction of Lower and Middle Chalk the Plenus Marls are overlain by the more conspicuous Black Band – dark grey laminated shales resting on a pronounced erosion surface and thought to represent a temporary muddy sea floor in stagnant conditions.

In Yorkshire the Middle Chalk amounts to about 112 metres and the Upper up to 300 metres; in Lincolnshire there is much less because later erosion has cut down far into the upper division. Above the Black Band an obvious tripartite division consists of flintless chalk, relatively thin, a thick succession of flinty chalk spanning the junction of Middle and Upper

Chalk, and a further thick development of flintless chalk, which reaches probably up into the *pillula* Zone. Upper Chalk faunas are not normally prolific but there are two rich sponge beds. Where the tabular flint seams and marl bands are common they are found to provide reliable lithological markers over both Yorkshire and Lincolnshire.

In its thickness, hardness and faunas the chalk of the Yorkshire basin shows affinities with that of the North Sea. Some resemblances have even been traced across into north Germany, but these are principally in the Cenomanian and Turonian stages. Some northern characters, in all three chalk divisions, are found in the northern outcrops of Norfolk and this county is in some respects in an intermediate position between north and south. Maastrichtian chalk, as distinct from remanié fossils, is found in England only as glacially contorted masses on the north Norfolk coast at Trimingham (Fig. 77).

TRANSGRESSIONS IN IRELAND AND SCOTLAND

Only Upper Cretaceous beds are found in Northern Ireland and the Hebridean region and in distribution they follow the earlier Mesozoic rocks; the relative importance of the two countries is reversed, however, and the Irish outcrops are the more extensive and informative. They underlie the Antrim Basalts as a nearly continuous strip on the south-east, east and north, and are seen at several places in the west. Even so the outcrops are varied and incomplete, which adds to their interest rather than detracts from it.

There is a fairly simple lithological distinction into greensands below and chalk above, but both vary in thickness, and to some extent in age, over the long strip-like outcrops. The name Hebridean Greensand has been applied to the lower facies but this is a particularly heterogeneous assemblage; the Chalk is also known as the White Limestone, very hard and well cemented. A composite succession is given in Table 51. The lowest, Cenomanian greensands are now seen only in the south-east, for example, from Larne to the Belfast region, but may once have been more widespread; however the ridge of Dalradian schists in north-east Antrim and some of the more upland country to the west were probably not submerged. A minor non-sequence interrupted this greensand complex and then followed a major regression and erosion period. Turonian sediments are unknown, but there is a trace of them in a derived phosphatized ammonite in the basal Senonian detritus in the north. This introduces the second major transgression, followed by a gradual extension of the seas. At the base there is commonly an early Senonian greensand facies and some overlying glauconitic chalk; the process was irregular however and some parts in the west were not covered until later. Much of the white chalk belongs to the *mucronata* Zone, which as elsewhere was a period of marked expansion. The small pocket of chalk far to the south, near Killarney, belongs to this age and the whole of Ireland

Table 51. *Upper Cretaceous of north-east Ireland*

Stages	Lithology	Outcrops
LOWER MAASTRICHTIAN }	White Limestone	} Composite from North and East Antrim
Includes CAMPANIAN and SANTONIAN zones	White Limestone with hardgrounds and flints Glauconitic Chalk (? with non-sequence) pebble beds Senonian Greensand	
(TURONIAN absent)		
CENOMANIAN	Glauconitic Sandstone ——— non-sequence ——— Yellow sandstones, etc. Glauconitic sands and basal conglomerate	} East Coast and Belfast
	(Lias and older rocks)	

may well have been submerged. In Antrim the chalk contains flints at intervals throughout, sponge beds and a number of hardgrounds. The lower Maastrichtian follows conformably, though the stage is restricted to some small downfolded and faulted outcrops on the north coast.

A particular feature of these Irish outcrops is their demonstration of late Cretaceous transgressions and regressions; in England the earlier ones, Aptian and Albian, were the most extensive, with the Cenomanian advance seen in the south-west. In Antrim there is the basal Cenomanian influx, the conspicuous Turonian unconformity and erosion, and the early to late Senonian expansion – each transgressed surface carrying its sandy or conglomeratic basal beds. Rather unexpectedly these Irish chalks seem to have more faunal affinities with southern England than with Yorkshire.

The west side of Scotland probably had a similar history for although the outcrops are small and scattered there is again a lower sandy group, a thin remnant of white chalk and Turonian beds are absent:

Silicified Chalk   probably *mucronata* Zone, several metres
Cenomanian   thin brown clays with brachiopods and white quartzose
                     sandstone (12 m)
                     glauconitic sandstone with shelly calcareous bands (9 m)

This sequence is found in Morvern, opposite the north coast of Mull, and there are further representatives on that island, on Eigg and Skye. The strongly transgressive Cenomanian overlies Jurassic and Trias in its several outcrops and, on the Morvern uplands, the Moine Schists. The white sandstone is an unusual facies; it is virtually unfossiliferous and in its well rounded, well sorted grains and lack of accessory minerals probably represents the prolonged working on the shoals of a shallow sea.

The extent to which the uplands of Wales, northern England and Scotland may have been submerged at the height of the transgression is a subject of much speculation but virtually no decisive evidence. It is mainly a geomorphological problem; even a thin Mesozoic spread could have provided the cover on which early Tertiary rivers were superimposed, to give the drainage pattern that is still noticeably discordant on the underlying rocks. Nevertheless there is no firm support for this spread – often termed the 'Chalk cover' – nothing found such as the chalk residues, flints and remanié fossils which have afforded valuable clues elsewhere. Also there are other possible versions of the denudational history, invoking later Tertiary erosion surfaces and gradual uplift, especially in Neogene times. Accordingly while the question remains interesting it is not primarily a stratigraphical one.

CRETACEOUS ROCKS OF THE SEA AREAS

Cretaceous stratigraphy inherited the two main provinces established in Jurassic times but as the chalk seas gradually spread their distinction became less. In southern England the Lower Cretaceous outcrops are only a small sector of the Anglo-Parisian (or Wessex–Paris) Basin. In the centre and south-east of that large downwarp the facies is both more marine and more argillaceous than, for instance, the Lower Greensand or the Hastings Beds. The latter however have a counterpart in the Boulonnais and a similar detrital belt overlaps, irregularly, the south-west edge of the Ardennes. At this time the marine connection of the main basin was far to the south-east, with the Tethyan seas.

In this situation the Cretaceous sequence of the eastern Channel is simple, being essentially similar to that of Sussex and Kent. However in

the centre, between the Cotentin Peninsula and Cornubia, Lower Cretaceous rocks are largely missing, so the residual Hercynian masses
formed upland ridges or a barrier that not only supplied detritus into
Wessex and the Weald but also bounded the Wessex–Paris Basin in this
direction. Nor is there much information farther west, though Lower
Cretaceous inliers reappear north and west of Finistère. They are reported
to be thick and mainly in a Wealden facies, one with spores, but glauconite
in the more westerly outcrops suggests some marine incursions. Presumably here the Western Approaches led towards the growing Atlantic and
its marginal epicontinental seas. An important part of this history is the
opening of the Bay of Biscay (i.e. the rotation of Spain to somewhere near
its present position). But although much of this movement is likely to have
been within the period it is not very precisely dated and perhaps was late
rather than early Cretaceous.

Chalk is present in all the Channel sectors but although the complement
is fuller than on land it is very rarely complete. Some 450 metres is found
south of the Isle of Wight and similarly south of Plymouth; the Maastrichtian Stage is present in both places and in the second is conformable with
the Danian (lowest Tertiary) above. North and west of the Cotentin the
higher basement region is shown by the overstep of the upper stages, and
though white chalk is the usual facies such transgressions are accompanied
by glauconitic sands. Turonian beds are not recorded in parts of the
western Channel, probably reflecting the widespread regression during
that stage. One might expect the fullest sequence to come from the
Western Approaches, but in fact the sub-Eocene unconformity increases
in this direction; the Channel did not subside into a simple downwarp at
the end of the Mesozoic and some irregularity and erosion is found here in
Tertiary stratigraphy also.

Less information is available in the Celtic Sea though the Chalk is
obviously widespread. Lower Cretaceous in a non-marine facies is reputed
to be thick in the northern part; it forms a gas reservoir (p. 405) and may
well be present farther afield. An isolated volcanic episode early in the
Cretaceous produced the phonolite of the Wolf Rock, off Land's End.
From the Celtic Sea northwards the records become very sparse, though
Upper Cretaceous is presumed to extend into St George's Channel and
Cardigan Bay. It has not been found in the Bristol Channel, Tremadoc
Bay or the northern Irish Sea. Moreover in spite of the well scattered
remnants of transgressive chalk and greensand in northern Ireland, Arran
and the Hebrides, none is recognized for certain in the offshore basins
west of Scotland. In a different setting there is a large Mesozoic outcrop

north of Cape Wrath and west of the Orkney platform and this may include Cretaceous sediments; they have also been inferred in the West Shetland Basin. Nevertheless the remnants on the Scottish and Irish mainlands do make these absences rather curious. It is conceivable that the thin transgressive spreads (possibly of the Cenomanian or Campanian transgressions) were almost all removed in the uplift and erosion that intervened before the early Tertiary eruptions.

The southern North Sea is a major part of the Boreal Province in Europe, which just reaches into eastern England, includes part of the Netherlands, Denmark and a complex medley of basins in north Germany and then extends across Poland (very largely concealed) to wide outcrops in Russia. The Lower Cretaceous ammonites have been particularly studied but other groups show regional distinctions, some surviving into the Upper Cretaceous. To the south lay the reduced uplands of the Hercynian belt: the Ardennes and Rhineland, the Harz Mountains and Bohemian Massif. Detritus from these is recognizable along the southern shores of the basin, and in Germany, for instance, even the Upper Cretaceous includes varied clastic components. A volcanic centre is known subsurface in the Waddensee region of the north-west Netherlands (Fig. 72) but the age (late Jurassic or early Cretaceous) is not very clear.

On the English side there are two main elements. The restricted inshore shelf of Lincolnshire (or of the East Midlands) was bounded on the north by the Market Weighton block – or more precisely its northern hinge. This structural trend is continued offshore as the Dowsing fault line (Fig. 68, inset), with the common south-easterly trend, ultimately to pass into Dutch waters. In parallel fashion, the Yorkshire basin to the north passes into the south-easterly elongate Sole Pit Trough, in which the Lower Cretaceous is very similar to that of Speeton.

Over the North Sea as a whole the Lower Cretaceous is commonly thin and has not received very much attention. The earth movements initiated in middle and late Jurassic times continued into the Aptian and Albian stages so that Upper Jurassic and Lower Cretaceous are linked in being locally thin and incomplete, commonly faulted (with halokinetic effects) and succeeded by a phase of erosion and transgression. This occurs especially on raised fault blocks; in some of the depressed areas sedimentation was continuous. Where found in the southern basin the facies resembles that of Speeton, clays with red chalk above. A succession of deep water shales has been recorded from the northern basin, and along the Scottish coast erratics have been brought onshore from more than one Lower Cretaceous stage. In the Moray Firth black shales overlie the

Kimmeridgian clays, and some wedges of unfossiliferous sands may have been derived locally at this time.

Upper Cretaceous history in the North Sea partakes of the general increasing uniformity. Hard white chalk is typical and is up to 900 metres thick in the Central Graben; the feather edge approaches closely to the Scottish coast. The Mid North Sea High and the Ringkøbing-Fyn High were covered, though more thinly. Some fault-block effects persisted even into the Campanian Stage but this was unusual. Inversion structures are part of Upper Cretaceous history in the southern basin; they are seen onshore in the south and also include the Sole Pit Trough. The Cleveland Dome in north-east Yorkshire, from which the chalk has now been removed, may be a similar structure.

The later chalk seas deposited white chalk very widely, including parts of Denmark, Poland and south Sweden, but as elsewhere the last stage (upper Maastrichtian) was one of pronounced regression, though continuity with the Danian chalk is found in some of the central rifts. In the far north there is an interesting transition in the Viking Graben from chalk to marls and then to shales (Fig. 72), presumably a deeper water facies, which for some obscure reason lay beyond the accumulation of the coccolith sediments. In the last stages the North Sea Basin became simpler, the faulting and facies variation reduced, foreshadowing the large elongate depression of the Tertiary.

A satellite view at this time would be very foreign to our eyes – a great waste of waters, occasionally whitening in the shallows. Perhaps to the north and north-west there was a low archipelago, remnants of Precambrian schists and quartzites. Then the seas gently retreated; the last chapters are imminent as the outlines of land and sea become more and more familiar. British stratigraphy is approaching its latest phase – less than 70 million years to go.

REFERENCES

Rawson *et al.* 1978 (*Special Report* No. 9, Cretaceous).
General: Allen, P. 1965, 1969, 1972, 1976; Arkell, 1947*b*; Bristow and Bazley, 1972; Casey, 1961, 1971; Casey and Rawson, 1973; Hancock, 1961, 1969, 1975, 1976; Hancock and Kauffman, 1979; Harrison *et al.* 1979; Howitt, 1964; Kennedy and Garrison, 1975; Neale, 1974; Woodhall and Knox, 1979; Worssam and Ivimey-Cook, 1971.
British Regional Geology: East Anglia, Hampshire Basin, Northern Ireland, The Wealden District.

11

TERTIARY ROCKS AND HISTORY

The Mesozoic Era ended some 65 million years ago and in formal terms the succeeding Cainozoic Era is now divided into Tertiary and Quaternary Suberas. In the former the divisions are thus:

Neogene System
$\begin{cases} \text{Pliocene Series} \\ \text{Miocene Series} \end{cases}$

Palaeogene System
$\begin{cases} \text{Oligocene Series} \\ \text{Eocene Series} \\ \text{Palaeocene Series} \end{cases}$

The names in the right-hand column were long ago designated as systems, when Lyell based Tertiary nomenclature on the increasing proportion of living molluscan genera; it became clear, however, that they were far smaller units than the earlier systems and hence in part the reorganization. Within the series the stage names are a problem; most were derived from France, Belgium and Germany and have been so variously interpreted that they are sometimes omitted altogether. Table 52 only gives a lithological succession but the stages previously used are mentioned in the Appendix (p. 419).

Part of these problems are faunal, in that after the disappearance of the ammonites there was no single outstanding group of stratigraphical fossils. For a long while molluscan assemblages were employed, as in the classic early work in the Paris Basin, later some of the large foraminifera such as *Nummulites* in the Eocene were used as well. The value of microfossils in oil geology and borehole sections has more recently made these the primary biostratigraphic indices, especially the planktonic foraminifera and the even smaller nannoplankton. The mammalian faunas, so important in North America, are less useful in Europe where continental facies are less extensive; however they do occur in some outcrops, especially in France, with pollen and spores. The European geochronology of the Tertiary Subera is largely based on isotopic data from glauconite in the sediments, with some additions from igneous rocks.

As in other parts of the world the British Isles were affected by some marked changes in plate movements and continental relations during Tertiary times. In particular Greenland drew away from Europe early on in a major episode of rifting and vulcanism, now recorded in Scotland, Ireland and along other continental margins of the North Atlantic. The gradual northward progression also continued and accelerated, so that from an Eocene latitude of about 40° N, southern Britain arrived at 51° N in modern times.

PALAEOGENE SEDIMENTS AND FLUCTUATIONS

It was seen in the last chapter that the chalk seas had already contracted in the latest Cretaceous and in all land outcrops of southern England there is a marked unconformity, with local tilting, below the Tertiary beds, the lowest of which are incomplete. The same applies to much of the English Channel but in the deeper parts of the North Sea there is continuity from the Maastrichtian to the Palaeocene. On resumption sedimentation was very different from the preceding carbonates, with shallow water clastics dominant; it is possible that this change was related to the new phase of North Atlantic rifting, the uplift of the rifted margins and hence increased erosion.

There was a marine gulf in the western Channel, widening into the adjacent Atlantic Ocean, already established at that latitude. It connected with the southern North Sea along the central and eastern Channel, but from time to time this connection was broken by a low ridge or swell from South Devon to the Cotentin Peninsula, a feature still emergent in the Palaeocene and early Eocene. In these phases the Anglo-Parisian-Belgian Basin was a partly land-locked sea centred on the eastern Channel, from which marine incursions extended into northern France and southern England.

In this situation the main outcrops on the English side are in the Hampshire and London basins and the Channel (Fig. 73). The first two are separated by the Weald anticlinorium; this is deduced to be mainly a mid-Tertiary structure, though perhaps with some gentle upwarping and planation preceding early Palaeocene sedimentation.

The Hampshire Basin

The principal sections here are on the west and east coasts of the Isle of Wight (Fig. 73, inset), along the mainland cliffs of Bournemouth and into

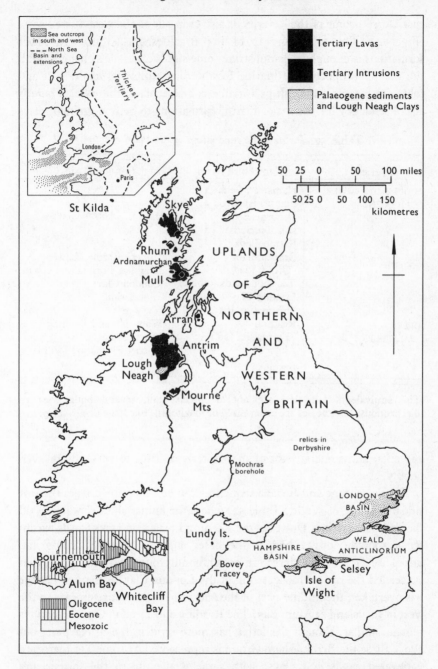

Fig. 73. Tertiary outcrops. Inset above: relation of Anglo-Parisian-Belgian Basin to that of the North Sea, sea outcrops shown in south-west only. Inset below: the Isle of Wight and neighbouring mainland.

east Dorset; some of the lower beds also crop out on the west Sussex coast
(Fig. 74). The total outcrop of the Bracklesham Group and lower
formations is enough to demonstrate some significant facies changes, with
the greater proportion of marine fossiliferous sediments in the east and
more freshwater, brackish or variable components in the west. There is
also a tendency for the major marine formations to be fine-grained, clays

Table 52. *Palaeogene successions in southern England*

Series	Hampshire Basin	London Basin
MIDDLE and LOWER OLIGOCENE	Hamstead Beds Bembridge Beds Osborne Beds (312 m) Headon Beds Barton Beds	(Absent)

EOCENE	Bracklesham
	  Group (237 m)
	 'Bagshot Sands' (10 m)
	 London Clay
	  Formation (81 m)
		Upper, Middle, Lower
		  Bagshot Formation (115 m)
		 London Clay
		 Formation
		  (155 m)

UPPER PALAEOCENE	Reading
	  Formation (26 m)
		Woolwich and Reading
		  Formation (20 m)
		 Thanet Formation (30 m)

	(Chalk)	(Chalk)

The equivalents indicated are not all precise and several boundaries are
diachronous; thicknesses are generalized or maximum. For stage names see p. 419.

and silts, and some at least of the westerly detritus to be sandy or even
pebbly.

The Palaeocene and Eocene sequences can be exemplified from the east
side of the Isle of Wight (Table 52) where the faunas show that nearly all
the beds are marine. However the Reading Formation lacks fossils except
for a sporadic oyster bed at the base, and the 'Bagshot Sands' (an
imprecise facies term here) are an unfossiliferous sandy incursion that
succeeded the major transgression of the London Clay. A marked change
has overtaken the middle part of the succession where it reappears in the
west of the island at Alum Bay. The Reading and London Clay formations
remain similar, though the latter has more sand in the upper part, but
above these the Bracklesham Group is represented by some 240 metres of
variegated sands and clays, with some glauconite in the former and
occasional marine microfloras but also some non-marine components. The
clays contain abundant vegetable remains and some distinct plant beds.

Westwards, along the Bournemouth cliffs and into east Dorset, there is a further accentuation of the facies changes. Gravels in the Reading Beds contain flints and pebbles of Cretaceous cherts; only 30 metres of rather sandy London Clay remain, giving place above to 85 metres of bright red sands, pipe-clays and coarse cross-bedded sandstones; these, sometimes retaining the term Bagshot, are probably equivalent to the upper London Clay and lower Bracklesham beds to the east. The Bournemouth Formation above includes some marine facies, some indeterminate, and also leaves of palms, conifers, ferns and deciduous trees.

The marine incursion of the Barton Formation follows in the central

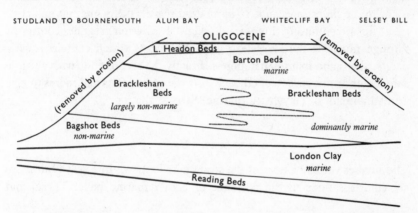

Fig. 74. Diagram to show the facies changes in the Eocene beds of the Hampshire Basin; relative thicknesses are suggested but the diagram is not to scale.

outcrops, dominantly as clays with abundant molluscs and some nummulites. Towards the top, however, there is a progressive replacement by freshwater forms, and a major regressive or continental phase set in over the uppermost Eocene and lower Oligocene.

The Headon Beds crop out on the Hampshire mainland and in the New Forest but the higher Oligocene strata are largely confined to the northern half of the Isle of Wight. The facies is also commonly finer than in the earlier formations, with much clay, calcareous marl and beds of impure limestone; fluviatile sandy incursions are rare. The whole reflects quiet coastal lagoonal conditions and lake precipitates, but even so with some more saline or marine phases. Mammals, reptiles and plants have been collected from the Lower Headon Beds on the Hampshire cliffs but the Middle Headon sandy clays contain brackish fossils; moreover a thin bed with a purely marine fauna, including corals, is found in the east of the Isle of Wight and also in the New Forest, as the Brockenhurst Beds. A more persistent regression is indicated in the Upper Headon Beds, which contain marls and limestones rich in the freshwater shells, *Lymnaea* and *Planorbis*.

Similar facies continue through the Osborne and Bembridge beds, though with a minor marine incursion, thick with oysters and other shells, in the Bembridge marls in the east of the Isle of Wight. Both 'semitropical' and temperate types characterize the Bembridge floras, with insects, land shells, mammals and reptiles. The thin freshwater Bembridge Limestone is remarkably persistent – a widespread phase of still shallow waters. The small oogonia (reproductive bodies) of the alga *Chara* occur in the limestones and, as in modern meres, the genus was an active agent in precipitating lime muds. The freshwater regime continued through much of the Hamstead clays and sands which contain aquatic plants, shells and mammalian bones. Finally the record ends with a gentle submergence; the seas flooded back over the coastal flats and meres to give a small amount of clays with marine shells at the top.

The London Basin

The smaller vertical extent of the strata here is shown in Fig. 75 but there are compensations in the presence of earlier marine beds. These, and

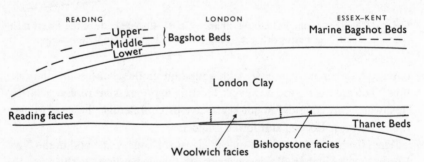

Fig. 75. Diagram, similar to Fig. 74, to show the principal Palaeogene formations and certain facies of the London Basin; relative thicknesses are suggested but the diagram is not to scale.

typical London Clay, are best seen in the easterly outcrops around the Thames estuary; in the west, at Reading and beyond, the facies more resemble parts of the Hampshire Basin. After the major late Cretaceous and early Tertiary uplift the seas first invaded the London Basin for only a short distance and the Thanet Formation does not extend beyond south Essex and east Surrey. In most places the unfossiliferous glauconite sands amount to about 14 metres but double this is recorded at Herne Bay in Kent where shell beds also occur. Over much of the outcrop the sands are underlain by a basal layer of green-coated flints, the Bullhead Bed. Unlike those in later conglomerates they are largely unworn and probably represent a clay-with-flints residue, such as occurs on existing chalk surfaces, incorporated during the Thanet invasion without much transport.

The succeeding incursion of the Woolwich and Reading Formation was much more extensive and the double name covers three rather than two variants. The marine Bishopstone facies of glauconitic shelly sands in north Kent is again restricted to the east. The non-marine Reading facies – sands and mottled clays, locally with plant remains – is found extensively in the west. In the London region, and forming the Woolwich Beds of the type outcrop, there is an intermediate facies incorporating brackish and estuarine shells. All three are underlain by the marine Bottom Bed, green-coated flints, glauconitic sands and clays.

The London Clay is by far the thickest formation and has the largest outcrop, a triangle from east Kent to Berkshire and northwards (under Quaternary) well into East Anglia. Typically it is a brown, grey or bluish deposit with occasional silty and cementstone layers. As such it is best displayed in its thickest development in the east; 140 metres are known in south Essex. The westerly thinning and increase in the silty layers probably represent partly a genuine reduction in sedimentation and subsidence and partly a diachronous junction with the Bagshot Sands above, so that sands were being deposited, possibly inshore, while clays persisted in the deeper sea to the east. There is more than one type of fossiliferous, pebbly or sandy basement bed beneath the clay. The best known is the Blackheath Beds of south London, the pebbles being quantities of well rounded flints; the Oldhaven Beds are a more sandy variant, with a more marine fauna, in east Kent. Both transgress southwards over the Woolwich and Thanet formations onto the chalk of the North Downs and show strong channelling at the base.

Although fossils are not plentiful throughout the London Clay a large varied assemblage has been collected from certain parts, notably towards the top on the south side of the Thames estuary. Molluscs are dominant but there are also bryozoans, brachiopods, annelids, fishes, foraminifera and dinoflagellates (nannoplankton); the rarer invertebrates include cirripedes and crustaceans. This is a fairly representative marine Tertiary fauna, though not a shallow one, the sea being perhaps 180 to 200 metres deep. The plants however are outstanding. It seems that the fragments of a coastal flora were carried out to sea and, becoming water-logged, sank onto the bottom muds. Apart from drifted wood the specimens consist principally of the fruits and seeds of dicotyledons, palms and conifers, the most abundant being the coconut-sized fruits of the stemless palm, *Nipa*, a modern genus of Indian deltas.

The climatic implications of this flora have long been outstanding because a large proportion of the genera (not species) are inhabitants of modern lowland tropics and a comparable climate was deduced. That the temperature was much warmer than that of present times is accepted and is borne out also by the faunal affinities, but an Eocene hinterland of a true tropical rain forest presents difficulties. The latitude (about 40° N) was well outside the modern tropical belt and nearer that of the Mediterranean and there is the significant proportion of non-tropical trees. It has also been emphasized that the present climatic belts, in what is more like an interglacial period, are not a good guide to the less pronounced range of early Tertiary times when there were no ice-caps. We should thus perhaps deduce a markedly warm and wet climate on the Eocene lowlands but not tropical in a modern sense.

The Bagshot Beds, or Bagshot Formation, overlie the London Clay over wide stretches west of London where they reach 120 metres in thickness. Farther east there are only outliers capping hills in Essex and the north London heights such as Hampstead and Highgate. Fossils are very few (and chiefly as moulds) throughout but the sands are probably all marine, with some intertidal levels; *Nummulites* is found in the middle Bagshot division and the whole corresponds to much of the Bracklesham Group of Hampshire.

CONTINENTAL DEPOSITS IN THE WEST

Beyond the main outcrops and farther from the seas there are a few isolated items. Outliers of unfossiliferous sands and gravels, sometimes called 'Bagshot Beds', continue much farther west than the remainder, rather in the same way that the earlier Cretaceous sandy facies extended beyond the clays. In central Dorset coarse gravels overlap the local Reading and London Clay formations and fill in solution hollows in the chalk. A larger version is the Haldon Gravels which cap the greensand of the Haldon Hills (p. 332). Here some 20 metres of rolled flints, cherts and Palaeozoic rocks are interspersed with seams of sand and clay, reflecting erosion on Dartmoor nearby. The only fossils are those derived from the now vanished Chalk (p. 343) so that the age, probably lower Eocene, is only inferred.

The most important outcrop in this region consists of the Bovey Tracey Beds or Bovey Formation, filling in a depression on the south-east slopes of Dartmoor (Fig. 57). Although the outcrop is only 16 kilometres long the basin was deep; geophysical evidence indicates a floor at 1200 metres below the surface and 300 metres of the beds have been penetrated in a borehole. The depression was gradually silted up by a stream or streams bringing detritus from the Dartmoor granite and hills. Sands were dumped with masses of wood and gravels near the margin; clays formed from the decomposed feldspars and have been used for pottery; there also are seams of lignite and a variety of plants and pollen. The dominant genus is the redwood *Sequoia*, with thick growths of the palm *Calamus* and water plants such as *Stratiotes*.

At Petrockstow in central Devon there is an analogous small elongate basin, again deep, through which boreholes have reached a Carboniferous floor at 660 metres depth. Both this basin and that at Bovey Tracey lie along the major Sticklepath–Lustleigh Fault, one of several north-westerly fractures, and it is likely that sedimentation was accompanied by downsagging on the fault line. Still farther in the same direction an offshore Tertiary basin, north-east of Lundy, is known from seismic evidence; a borehole here encountered seams of lignite. The age of these various deposits is not very clear as the plants are long-range forms; however the pollen suggests an early Oligocene age and the deeper parts of the Petrockstow and Bovey basins might range down into the Eocene.

Another interesting aspect is the rarity of such deposits. During periods of emergence the hilly country of any age might have contained many such silted-up basins, as do modern uplands. But such features tend to be geologically transient and all too often are swept away by later phases of erosion. The preservation of these aligned basins and their infilling probably reflects their exceptional downwarping, with further faulting in mid-Tertiary times.

North Wales

The Jurassic sequence in the Mochras borehole, on the coast of Tremadoc Bay, has already been mentioned (p. 310) and it is overlain by 525 metres of Tertiary beds that form a link between those of the south-west and the Lough Neagh Clays of Antrim. They overlie the eroded and weathered surface of the Lias and all are non-marine with silts and clays dominant. Conglomeratic bands and some red beds occur in the lower part, thin lignites and plant beds in the upper, the whole representing a varied accumulation on a flood plain, with marshes and shallow lakes. The conglomerates include some Lower Palaeozoic rocks and cobbles up to 10 centimetres in diameter, reflecting a major uplift of the Welsh Massif along the intervening boundary fault.

The age of the Tertiary beds is not known very precisely, but from the pollen late Oligocene has been tentatively suggested, ranging perhaps into early Miocene at the top. The exceptionally thick sequence also has some tectonic implications. The total throw of the fault against the Cambrian rocks of the Harlech Dome must be at least 4500 metres. A displacement of this order was presumably effected in stages. Some downfaulting may have preceded the Upper Cretaceous, comparable to that inferred below the Chalk farther south; but it is also likely that Oligocene deposition was itself accompanied by movement along the fault so that the base of the continental sediments now lies some 600 metres below sea-level. The faulting, however, was local – a deeply depressed basin-sag – because no fault of such dimensions is recognizable to the north, where it would enter the Cambrian outcrops on the north side of Tremadoc Bay.

THE TERTIARY VOLCANIC PROVINCE

While sedimentation was being re-established in the Anglo-Parisian-Belgian Basin more dramatic events took place in the north-west. Here the early Tertiary vulcanicity is part of the greater North Atlantic Province, caused essentially by changes in the process of North Atlantic rifting. Up to very late Cretaceous times or the beginning of the Palaeocene the rift had continued northwards as a narrow opening between south-west Greenland and Labrador; then there was a major change in direction and a more active split developed between east Greenland and Europe. The concurrent vulcanism, typically basaltic, is now seen on the parted continental edges, with the Rockall plateau and the Faeroe Islands as outlying remnants on the European side. The modern vulcanicity in the central rift of Iceland marks the latest phase and at the same time the position of the present mid-Atlantic ridge.

The exposed rocks on the European side, in Britain principally in western Scotland and northern Ireland, suggest a relatively short vigorous episode from perhaps 12 to 10 million years – ages based on isotopic determinations and reinforced by the predominantly reversed magnetic polarity. Crustal stresses were probably building up in this part of Laurasia during the late Mesozoic and a broad continental stretch became slightly uplifted in very late Cretaceous times. When this rifted the fracture broke, irregularly, across a heterogeneous basement already carrying scars or lines of weakness from Caledonian and Precambrian orogenies with some Hercynian and Mesozoic faulting. The major centres of eruption were thus particularly focussed along previous lines of weakness, major faults, or at the intersections of these. The dyke swarms, both parallel to the rift margins as in Greenland, or normal to it, the dominant north-westerly trend in Britain, testify to the varied extensive stress patterns. One major line has been traced from Lundy Island northwards to lie parallel to the Hebridean centres and out to St Kilda and the Faeroes; several others are postulated, less extensive.

In the British Isles there are three main aspects of this igneous activity: the basic lavas, the varied central intrusions and extending beyond these a range of minor intrusions, especially dykes. In some areas these three, as they are now exposed, fall into a broad time-sequence; the lavas are intruded by the plutonic rocks and both are cut by the dykes – or at least by a late phase of dyke injection. However in many other areas this succession does not hold and intrusion and extrusion were probably simultaneous for much of the time.

The lavas of the North Atlantic Province are dominantly tholeiitic, as for instance in the Faeroes, but there are important and distinctive variations. Thus the Scottish types are almost exclusively alkali basalts or their associates, for example, hawaiite, mugearite, benmoreite and trachyte. In Ireland also alkali basalts are dominant but some tholeiites occur in the Middle Basalt Formation. The lavas cover much the largest areas, especially in Antrim (Fig. 56), and once they must have been far more extensive, welling up gently from shield volcanoes of Hawaiian type. Some 2000 metres have been estimated on Mull and as much as 8000 metres survive in eastern Iceland.

The eroded roots of the volcanoes are seen in the central intrusions. In Scotland they are found on Skye, Rhum, Ardnamurchan and Mull (Fig. 9). To the south there is Arran, and the Blackstones Bank is a submerged centre 45 kilometres south-west of Mull. Those of Ireland include the Mourne Mountains, Slieve Gullion and Carlingford, while Lundy is a lone outpost much farther south. Out in the North Atlantic the islands of Rockall and St Kilda are comparable centres. Among these many have had a complex history, with both acid and basic intrusions, often in the form of ring structures; areally the acid rocks (granites and granophyres) slightly predominate.

The minor intrusions are chiefly dolerites; they include some massive sills, such as those approaching 100 metres in thickness that penetrate the lava piles of northern Skye, but the dyke swarms are much more extensive. In Ireland they reach far to the west in the Dalradian country of Donegal and Mayo. South-west Scotland contains numerous examples while the swarm centred on Mull includes the Acklington Dyke of Northumberland and the Cleveland Dyke of Yorkshire. Similarly the swarm from the Mourne Mountains extends into North Wales and includes some small intrusions in the Isle of Man and the West Midlands.

The North Atlantic Province, in Britain and elsewhere, is thus character-ized by its overwhelmingly basic magma and the origin of the acid rocks comes into question. In contradistinction to the Caledonian and Hercy-nian granites – which are obviously in a different tectonic setting – certain of the central intrusions are accompanied by a marked positive gravity anomaly, even where basic intrusions are few or absent. Accordingly the lighter rocks at the surface must be compensated by denser, basic or ultrabasic, rocks at depth. It is probable that the granitic magma originated from the fusion of material in the upper levels of the crust; isotope ratios and melting relations suggest that such a source could include Lewisian gneiss or Torridonian arkose. The heat needed for the fusion would have been supplied by the basic magma itself.

Associated sediments and age of the vulcanicity

After the retreat of the chalk seas in northern Britain there followed the widespread uplift already mentioned, here accompanied by some local deformation. The early lavas then flowed out over a mature undulating landscape where fertile wooded country was interspersed with lakes and rivers. In Antrim the intervening period was relatively short and Chalk as high as lower Maastrichtian has been found beneath the basalts. In Scotland there was more erosion and the lavas transgress from the thin Cretaceous remnants onto all three Jurassic divisions and thence onto Cambrian and Precambrian; in Skye there was also major faulting in this interval.

A variety of sediments is found either beneath the first lavas or interbedded with them in the Inner Hebrides. Sands, lignites and tuffs underlie the basalts on Skye, though as a rule tuffs are rare. On Mull there was a large crater lake in which some layers of pillow lavas formed. The cliffs of Canna expose over 120 metres of lavas, agglomerates and fluviatile conglomerates, suggesting a basaltic plateau traversed by intermittent vigorous streams.

The plant beds of Mull and Ardnamurchan give a glimpse of the contemporary forests – oak, plane, hazel, ginkgo and magnolia – noticeably distinct in composition and ecological type from the strandline 'tropical' genera of the slightly later London Clay. In Antrim the main terrestrial interlude is shown by the Interbasaltic Bed which separates the lower and middle basalts. It is a substantial zone of weathered lava and decomposition products and the surface was colonized by a rather similar flora of hazel, alder, maple and conifers, monocotyledons and ferns.

The biostratigraphical evidence from these floras is not very satisfactory. In Scotland no agreed age has been derived either from the leaf beds or their contained pollen. The Antrim lavas are the best provided for in this respect, with the Maastrichtian chalk below, probable Palaeocene pollen from the Interbasaltic Bed, and the Lough Neagh Clays above. Fortunately isotopic data have considerably improved the situation, though as in older rocks there may be some misleading late dates caused by slow cooling.

In our present context, however, it seems that the majority of results show that the pronounced igneous activity was largely within the Palaeocene. Dates obtained from the major plutonic centres range from about 65 to 57 million years and though those from the lavas are less well substantiated they cover a similar range. The late dykes extend down to about 52 million years, and the later intrusions of Skye, Rockall and Lundy were probably not much earlier. Antrim is exceptional in that the earliest lavas (possibly 66 m.y.) may have appeared just within the latest Cretaceous. There are also some provisional data from farther west to suggest igneous activity began there in late Cretaceous times, for instance from rocks recovered from the Blackstones Bank igneous centre, Rockall Bank and Rockall Trough; certain small intrusions in western Ireland might be in a similar situation. However the full age pattern confirms previous views that the great bulk of this important volcanic activity was part of early Tertiary history, later to be followed by the deep erosion that now exposes the plutonic masses.

The Lough Neagh Clays

No Tertiary sediments survive above the lavas of Scotland but those of Antrim are partly overlain by the Lough Neagh Clays (Fig. 56), lake clays which formed in a depression after a further period of erosion and warping. To the east of the present lough they lie on the upper basalts but westwards overstep onto lower levels and then onto the Mesozoic rim of the lavas. At present their outcrop covers some 500 square kilometres and has been penetrated by boreholes to 350 metres, so that the formation is a relatively substantial item in the Tertiary of the British Isles.

Lithologically the sediments are chiefly pale grey silty clays with beds of sands and pebbles in the top and bottom parts; there are also many lignite seams, some accompanied by ironstone nodules. In the lower levels plants are abundant and there are some freshwater shells. The plant remains, especially spores and pollen, have much in common with those in parts of the Mochras borehole (p. 359) and elsewhere in Europe. The comparable floras show a fair range but on balance a late Oligocene age has been attributed to the Lough Neagh sediments.

MID-TERTIARY MOVEMENTS AND NEOGENE RELICS

Some type of Tertiary earth movement – faulting, folding or simply uplift – affected much, or probably all, of the British Isles. Only where Oligocene strata are present can the effects be dated more precisely; however since the mid-Tertiary (? middle to late Oligocene) structures are known to be vigorous in the south it is common to correlate others with this period. The term Alpine was once used for these structures and though the major plate movements along the Alpine and Pyrenean chains may well have had repercussions in the more stable regions to the north it is difficult to draw precise parallels.

Although in southern England there are some compressional effects the dominant influence seems to have been vertical movements between adjacent blocks, along old lines of weakness. The surface expressions are seen as the Weald anticlinorium, the London and Hampshire synclinoria, the Isle of Wight monocline, and the Purbeck anticline with thrusting; several have counterparts in the central and eastern sections of the English Channel. The Weald is a large-scale 'inversion' structure like those of the southern North Sea (p. 271).

The pronounced Hercynian thrusts and folds of South-West England are cut by major dextral transcurrent faults of the same period; one of the largest, the Sticklepath–Lustleigh Fault, was mentioned in relation to the Bovey Tracey basin. Parallel faults, with throws of up to 150 metres, are known in the Lias and Trias of the north Somerset coast and the Tertiary dykes of Lundy are also cut by north-westerly fractures. Transcurrent faults of similar trend offset the Hercynian structures in south-west Wales. Across the Channel in Brittany and the Paris Basin there are further north-westerly faults and as the Wealden structures cross over into the Boulonnais they curve into a parallel trend.

In the Tertiary Volcanic Province the mid-Tertiary effects are best defined in Antrim where the Lough Neagh Clays provide an age reference; the faults cutting both the clays and the underlying basalts have throws of up to 300 metres and several follow a Caledonoid trend. A similar situation is found in the Hebrides, for instance in the basalt plateau country of Skye and Raasay, though here some faults are larger; that separating the two halves of Raasay (Mesozoic and Tertiary on the south-east, Precambrian on the north-west) is a Caledonoid structure of at least 750 metres' displacement.

In between southern England and western Scotland structures affecting Jurassic and Cretaceous outcrops have also been attributed to Tertiary movements. Most are relatively small but an east–west belt of fractures in south-east Yorkshire (bounding the Market Weighton block, or East Midlands Shelf, on the north) has throws of up to 300 metres accompanied by marked disturbance in the Chalk. The several major faults in the Triassic plain of Cheshire and the Midlands lack a clear age reference and were probably caused by repeated movements. The Caledonoid Wem Fault is one of the largest, with a post-Liassic throw of 1500 metres – possibly part-Mesozoic and part-Tertiary.

At all events it is likely that many of the older faults that still have a topographic expression in the uplands of Britain were re-activated in Tertiary times, so that the present form and structure of the country carries a Tertiary inheritance, with Neogene uplift and erosion as a significant component.

Neogene relics

The extreme sparseness of Neogene deposits in the British Isles follows logically from this uplift and erosion; even in the English Channel the sequence is reduced and incomplete. From this time, until the arrival of the Pleistocene ice-sheets, the major centre of sedimentation lay in the North Sea. The few land outcrops are either coastal or, if inland, chiefly in solution hollows.

The Lenham Beds of Kent are the best known example in the second category, occurring mainly in solution hollows in the Chalk, but also as a small patch on Beachy Head on the Sussex coast. The marine faunas of these ferruginous sands, which occur largely as molluscan moulds, have been compared to the upper Miocene of Belgium. Their altitude also has some implications, for at present they lie at 190 metres O.D. on the North Downs; thus they seem to entail a late Miocene invasion of the London

Basin of which no other indication survives. Then a gradual uplift followed, which in eastern Britain was part of a regional tilting, complementary to the major subsidence in the North Sea.

The Coralline Crag of Essex belongs to the margin of that sea; it is a small outcrop at the south end of the East Anglian crags (Fig. 77), the remainder of which are taken to be Quaternary. The white and yellow cross-bedded sands include great numbers of molluscs, at some levels whole, at others reduced to shell debris; lesser components are echinoids, brachiopods and barnacles. The whole, which is under 20 metres thick, was laid down as banks of shelly sand in a shallow current-swept sea in late Pliocene times. That the climate was still relatively warm is shown by the Mediterranean affinities of the fauna and the absence of the boreal species that appear in the later crag assemblages. At the base of this and some other crags a residual conglomeratic deposit, the Suffolk Bone Bed or 'Nodule Bed', includes some Mesozoic fossils, flints and brown sandstones, or 'Boxstones', containing casts of Tertiary molluscs; these presumably were derived from some earlier marine inshore deposit (possibly Oligocene or Miocene) that no longer survives.

An exceptional Neogene remnant has been found in Derbyshire. Among the many solution hollows in the Carboniferous Limestone, a very few have been found where plant-bearing clays have collapsed into the cavity and have been covered with glacial debris. The pollen from the clays is that of a heath and woodland vegetation and a rather tentative age is deduced around the Miocene–Pliocene boundary. In Ireland similar solution hollows are known in Tipperary. In themselves such Neogene relics are not perhaps momentous, but they do suggest that there may be other evidence, scattered over the limestone outcrops of Britain, where local conditions were favourable and the Pleistocene glaciation was protective, rather than destructive as it was more commonly.

The Neogene and later uplifts of the British Isles have some geomorphological implications, in that successive shifts of base-level must have affected denudational history. Two types of landforms could have resulted – marine planation platforms and the terraces or erosion surfaces that flank the inland hills. Many of the erosion levels have been recognized as Pleistocene and some of the earliest and highest might be Pliocene, or even earlier; but since deposits of this age are so very sparse, age data are lacking and the landforms cannot add much to Tertiary geological history. The most conspicuous examples include the coastal platforms in South-West England, especially north Cornwall, and more doubtfully some in South Wales.

TERTIARY ROCKS OF THE SEA AREAS

Tertiary stratigraphy in the eastern and central sections of the English Channel (Fig. 1C) has clear links with the land outcrops on either side and the distribution is similarly influenced by mid-Tertiary structures; the Bembridge to Valéry-en-Caux line (p. 268) is in fact a continuation of the Isle of Wight monocline, first trending eastwards and then swinging round to south-east. The major outcrop here is the large Dieppe Basin, almost reaching the French coast and forming a continuation of the Hampshire Basin. The sequence from the Thanet Formation to the Bracklesham Group (and possibly higher) is essentially similar to that in the Isle of Wight but is probably all marine. Minor outcrops of early Tertiary beds crop out in a faulted syncline in the central section but again there is little above. No Neogene at all is known and the major Oligocene regression is suggested by very limited outcrops of freshwater limestones north-west of Le Havre.

In the western Channel, after the comparative uniformity of the late chalk seas, a different regime developed. Nearly all the Tertiary beds sampled are marine and include glauconitic and calcareous sands, lying above basal Palaeocene calcarenites. Dominantly this was a region of clear shallow seas widening into the Atlantic; the foraminifera (including *Globigerina* and *Nummulites*) have affinities with those of Aquitaine. Even so there are some breaks in the sequence and towards the Hercynian fringe off Brittany Eocene algal limestones suggest very shallow waters. Again the Oligocene is represented only by freshwater limestones and the later Miocene marine influx follows unconformably; the 'Globigerina Silts' and calcareous facies form an overlapping outcrop in the west that widens and thickens towards the edge of the continental shelf. Seismic reflections and rare dredgings suggest that Tertiary rocks form part of the upper continental slope where they show some down-slope faulting.

Less is published on the Celtic Sea but patches of both Palaeogene and Neogene overlie the Chalk and probably become much more extensive towards the south-west. A tongue of Tertiary also reaches up into the southern Irish Sea, parallel to the thick Mochras sequence on the shore of Tremadoc Bay. The margins of the continental shelf off Ireland and west Scotland have some similarity to those of the Western Approaches, in that Tertiary strata drape over the edge and down the slope and the lower parts are faulted. Farther out the Porcupine Sea, between the main continental edge and the Porcupine Bank, is probably in part a Tertiary basin. Around the Hebridean islands there are stretches of lavas on the sea floor, for instance a wide outcrop south-west of Skye and another west of Mull. A deep small basin of Tertiary sediments overlies the former and has been compared to the Lough Neagh Clays.

The North Sea

As might be expected the Tertiary sequence here is much thicker than on the British mainland, though in the south with resemblances to that in the northern Netherlands, and nearly all is marine. All the main divisions are known at one place or another (Fig. 76) but in varying degree according to

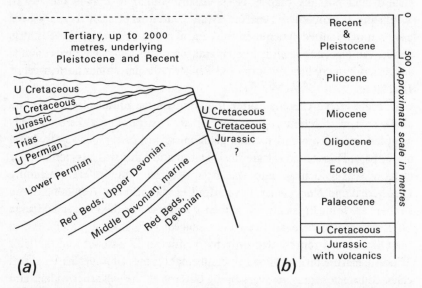

(a)   (b)

Fig. 76. Generalized cross-sections through two North Sea oilfields. (a) The Argyll oilfield, with the edge of the Central Graben on the right (redrawn from Kent, 1975b, p. 464). (b) The Forties oilfield, showing the thick Tertiary sequence and the position of the Middle Jurassic lavas and tuffs. (Adapted from Walmsley, 1975, p. 479.)

their economic importance. Thus there is not much information on the Neogene, whereas the Danian limestones at the base and some overlying Palaeocene sands have proved to be valuable hydrocarbon reservoirs. The sands include a coastal–deltaic sequence in the central belt north-east of Aberdeen and deeper water turbidite sand-fans; the detrital source was apparently to the north-west in the region of the northern Scottish Highlands or Orkney Islands. For the most part however the remaining Tertiary is a mudstone sequence with only thin sandy and calcareous beds.

Several ash layers are known, especially in the Palaeocene. Those in the centre and north were presumably derived from the Hebridean volcanoes, carried by the dominant westerly winds. In south-east England there is mineralogical evidence for thin ashfalls in the Thanet Beds and at the base of the London Clay. Another Lower Eocene source for a widespread horizon in Denmark and north Germany has been deduced in the Skaggerak.

Maximum thicknesses of Tertiary strata are recorded in and near the Central and Viking grabens, with thicknesses around 2400 metres and possibly more since Tertiary and Quaternary have not always been distinguished. A thinning over some salt domes suggests that halokinetic movements persisted into the Tertiary but otherwise by then the whole North Sea had become a single subsiding basin. This subsidence and thick infilling reflects its position as the main sedimentary catchment for northern Europe in a regime where the earlier tectonic stresses had disappeared, perhaps with the final rifting between Europe and Greenland. By Neogene time Europe became blocked out into its present landmasses and the North Sea continued in its sedimentary role.

REFERENCES

Curry *et al.* 1978 (*Special Report* No. 12, Tertiary).
General: Curry, 1965; Daley, 1972, 1973; Kent, 1975*b*; Knox and Ellison, 1979; Knox and Harland, 1979; Woodland, 1971, 1975.
British Regional Geology: Hampshire Basin, Northern Ireland, South-West England, Tertiary Volcanic District.

QUATERNARY HISTORY:
FROM PRE-GLACIAL TO GLACIAL STAGES

For very many years the outstanding feature of Quaternary history was considered to be widespread glaciation, so that it was sometimes called the Great Ice Age and is still the best known of the various ice-ages that have left their traces on the earth's surface. In terms of duration, however, even in British stratigraphy this was misleading: first because it was not one long freeze-up but a series of alternating warm and cold periods; and second because in Britain the cold periods only developed extensive ice-sheets in the last three (cf. Table 53). Consequently there was a long pre-glacial period, much more than half of Quaternary time, before British till was formed. For the most part the land area was slowly rising while the North Sea Basin continued to sink. Marine shelly sands and clays were laid down on the fringes of the basin and appear in East Anglia as the 'crags', all of which are thought to be Quaternary except for Coralline Crag of the preceding chapter. Their floras and faunas bear witness to the chilling climate and the same is true elsewhere in Europe. However this does not mark the first development of all ice-sheets in the northern hemisphere; there were probably major glaciers in North America well before and the growth of the Antarctic ice-cap has been traced back much farther, perhaps to the late Miocene.

The international approach to Quaternary classification has shifted radically in the last decade from the continents to the ocean floors, related especially to deep-sea cores associated with magnetic reversals. Deep-sea cores have already begun to revolutionize Quaternary studies, since the planktonic foraminifera found in them can give direct indications of past surface temperatures, while the ratios of the oxygen isotope ^{16}O and ^{18}O in the tests of the benthonic foraminifera are now known to reflect the growth and melting of the polar ice-caps. An ocean-based stratigraphy is thus emerging, compared with which that of the lands may well turn out to be seriously deficient.

Quaternary and Pleistocene are almost equivalent but the former includes the small Holocene Series at the top.

LOWER QUATERNARY MARINE BEDS

In East Anglia (Fig. 77) the lowest Quaternary formation is generally taken to be the Red Crag. Its fauna shows a marked increase in cold or boreal species compared with that of the Coralline Crag and there are also

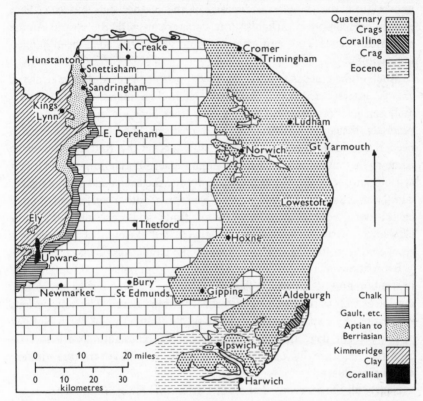

Fig. 77. Geological sketch map of East Anglia, showing outcrops of Jurassic and Cretaceous rocks, Coralline Crag and Quaternary crags; glacial and interglacial deposits not shown but some important quaternary localities are given.

the earliest remains, in this country, of true horses, elephants and oxen. The lower boundary in East Anglia is also convenient in that a minor phase of erosion and disconformity elapsed between the Coralline and Red crags.

The classic descriptions in East Anglia date from the turn of the century and were largely derived from cliff sections and lesser inland exposures. More recent revision has been based on borehole sequences which show that, above the Red Crag, the various other named crags (Norwich,

Weybourne, etc.) are not satisfactory units, often relating to specific facies. The current stratigraphy depends primarily on evidence from pollen and foraminifera; in turn the fossils were governed by climatic fluctuations so that the stages also represent alternating cold and temperate phases. A somewhat similar early to middle Pleistocene history is known from the pre-glacial deposits of the Netherlands but even here precise correlation is difficult. Any comparison with the classic Alpine sequence – always an unfortunate prototype – has long been judged impossible.

Much of the crags is sandy, with shells, pebble seams and large banks of gravel, interspersed with clays and silts. The marine facies contain molluscs and foraminifera; some of the finer beds are called, rather less precisely, estuarine while freshwater deposits and beach gravels have also been recognized. Where exposed at the surface the Red Crag owes its name to the oxidation of the iron compounds; in boreholes, of which that at Ludham (Norfolk) is the most important, the sands and silts are grey. The later crags extend beyond the Red Crag onto a surface of eroded chalk, with a basal conglomerate of flints and occasional bones of elephants, mastodon and deer. An articulated skeleton of a whale was recovered from the quiet water clays of the Chillesford Beds in Suffolk.

Estimations of age and climate are derived from pollen and foraminifera in the borehole sequence. Pollen may drift a long way out to sea so that marine as well as freshwater beds reflect the land floras. The cold periods are marked by a sparse heath vegetation, not unlike that of north Norway at the present day, and the warmer ones by temperate spruce forests. Similarly there are fluctuations in the climatic affinities of the foraminifera and the molluscs.

Nearly all the outcrops that survive from this lower Pleistocene sea are confined to East Anglia, but that it invaded the London Basin is shown by remnants in Surrey and Hertfordshire, both belonging to the Red Crag. The Netley Heath Beds contain blocks of ferruginous sandstone with a Red Crag fauna and lie at about 180 metres O.D. on the North Downs. They do not show fluviatile or marine characters at present and are associated with flint pebbles, clays and sands, so they may have been disturbed or redistributed by periglacial action. Similarly disturbed sandstone at Rothamsted (Hertfordshire) is at a lower level (under 120 m O.D.) but reinforces the evidence for this marine invasion.

In Cornwall the marine shelly clays of the St Erth Beds have long been something of a puzzle. They lie at about 44 metres O.D. on the neck of land between St Ives Bay and Mount's Bay and have not been exposed for very many years. In the last century over 250 molluscan genera were collected, with Mediterranean affinities and lacking littoral species. Quiet offshore waters and a late Pliocene age were deduced. More recent examination of the foraminifera and pollen, in the clay matrix, makes lower to middle Pleistocene more probable.

EROSION SURFACES

During the long period from the culmination of the mid-Tertiary earth movements to the arrival of the late Quaternary ice-sheets, most of the mainland of the British Isles was gradually rising – a slow tectonic effect complementary to the better known continued subsidence of the North Sea Basin and to a lesser extent of other areas, especially the Irish Sea. Such movements must have gradually lowered base-levels around the country and if, as has been widely suggested, the movements were somewhat episodic, then their traces may be looked for in elevated erosion of planation surfaces. These may flank the uplands or border the hills near the coast, especially where they cut across relatively hard rocks; they may also be sought as elevated terraces along the major rivers, chiefly outside the glaciated regions.

Many such geomorphological features have been described and discussed over the years, but in the context of this book – stratigraphy and geological history – the whole subject of erosion surfaces is fraught with difficulties. In the last decade these difficulties have been accentuated, partly because the effectiveness of various erosion agents to produce well defined surfaces has been questioned, and partly because the classic examples of denudation history in the south-east, including the relationship of the Thames terraces to the glacial periods farther north, has been challenged.

That certain types of surface do exist in certain areas is self-evident – they may be visible to casual observation, reinforced by altimetric surveys or substantiated by statistical analyses. But these do not dispose of several other problems. There are questions of identification and correlation of surfaces, especially where fragmentary, and also of later warping or the lack of it. If late Neogene and early Pleistocene tectonic effects included even mild warping, as some authorities believe, then correlation by altitude is hazardous at best and over long distances useless. Other debates centre on the main erosive agency, subaerial or marine, and many platforms lack diagnostic evidence of either. But most serious in our present context is the fundamental problem of dating. The number of adequately defined surfaces that are clearly related to diagnostic fossiliferous deposits are all too few and radiocarbon techniques can only be applied to very late examples.

Some of the clearest platforms are those that plane off the complex Hercynian structures around the south-western coasts, especially of Cornwall. Above the raised beaches carrying glacial debris they occur at about 130, 230 and perhaps 300 metres O.D., with other more debatable higher levels. A marine planation for the lower levels is commonly, but not universally, accepted and evidence of age is slender. The St Erth Beds (p. 372) may belong to the same stage as the 130-metre planation (the 430-foot platform), and thus the latter *may* be early to middle Pleistocene but the deductive links are weak. Similar platforms in South Wales, at about 60, 120 and 185 metres O.D. also lack age data. However in recent years there has been a tendency to ascribe more of the coastal plateaux to Pleistocene erosion rather than to Neogene.

Probably the most satisfactory evidence in this field comes from the middle and lower terraces of certain southern rivers, such as the Thames, Avon and Severn, together with more isolated examples from the Cam and the Trent. Some of the more reliable items appear in Table 53 and are seen to range from late Wolstonian, after the recession of that ice from the Midlands, to late Devensian and early Flandrian. In the middle and lower Thames valley some 10 to 12 terraces have been recognized, forming a descending sequence – the most extensive falling eastwards at a low gradient – which ranges from about 170 metres O.D. in the earliest and highest, western levels near Reading to only a few metres above the sea-level in the lowest and latest examples in London and the margins of the Thames estuary. Nevertheless even in these much studied examples there are complexities in the unravelling and correlating of the terrace deposits, which may show much lateral and vertical variation. Palaeontological support of age is still confined to the lower terraces, of Hoxnian and later stages. Even those long-respected datum levels, the Boyn Hill (100-foot) and Taplow (50-foot) have lost some of their authority and it has been suggested that these names are unsatisfactory when extended far from their type localities.

In such a fluid situation what of the higher earlier surfaces commonly observed in many parts of the country? It must be concluded that the farther these lie above, or away from, fossiliferous sediments the more uncertain their age becomes, and they can add little to our knowledge of stratigraphy. In future years it is more likely that the authoritative stratigraphical history of the Neogene and pre-glacial Pleistocene periods will be based on the sequences penetrated in the seas around the British Isles and the erosional history of the land areas become only a marginal facet of this larger frame.

GLACIAL AND INTERGLACIAL GEOLOGY: DEPOSITS AND METHODS

In the early days of Quaternary investigation emphasis was laid on the glacial deposits because they were so unusual and the cold fauna – 'woolly' rhinoceras, mammoth, reindeer, etc. – caught the imagination. Accordingly the deposits and periods were named by glaciations, including the unfortunate Gunz, Mindel, Riss and Würm of the Alps. With the exception of the last (the Würm Glaciation, Weichsel or Devensian) these were never a good standard, if only because in the mountains each glaciation tends to destroy the evidence for its predecessors. There is real difficulty in comparing and correlating glacial deposits in their own right, and now that we know far more about the interglacial beds these have become the more significant, and in effect it is their names that form the guide lines in correlation.

Interglacial deposits are naturally varied. They include coarse to fine fluviatile sediments, lake silts and muds, peats and littoral to estuarine types near the coasts. They contain the sporadic remains of many animal and plant groups and the organic-rich muds and peats have provided the most valuable radiocarbon dates. The fullest sequence in Europe comes from the great northern plain, where during the refrigerations the Scandinavian ice-sheets advanced from the north into the Netherlands, Denmark, north Germany and Poland. East Anglia is in some sense a westerly continuation of this plain, and with the Midlands provides some of the best British sequences. In between the Quaternary of southern North Sea basin is thicker than on land, though not much detail is published on its history and composition.

If we take British glacial and interglacial stratigraphy together, by late Quaternary times a new regime had developed. In place of the restricted marine crag facies (and very little indeed elsewhere) we have to hold in view the very diverse processes of deposition and erosion through which this complex series of terrestrial deposits was formed, and often removed. For such deposits a greater range of stratigraphical methods have been employed than in the more orthodox systems. Nevertheless the aim is the same: the establishment of a standard dated sequence, with which less perfect sections may be compared. The difference between 'orthodox' and this glacial–interglacial stratigraphy is chiefly that in the latter the sections showing uncontrovertible evidence of conformable dated beds are few and short.

Tills, fluvio-glacial and periglacial deposits

At no time during the glacial periods was the whole of the British Isles covered with ice-sheets; even at their greatest extent southern England, south of a line from the Severn to the Thames, was free from permanent ice and periglacial conditions obtained. Within the glaciated areas it has long been observed that the Quaternary history is less easy to follow on the uplands, where erosion rather than deposition was dominant. In some areas different suites of erratics provide evidence of different directions of ice-movement, but superposed tills, or tills and interglacial deposits, are very rare and only the last glaciation has left substantial records. In places it has been relatively easy to distinguish the last glaciation with its fresher deposits and landforms from one or more older ones. Hence arose the terms Older and Newer Drift – the latter being, in a more modern terminology, Devensian.

Although till was formed within the area of the ice-sheets its exact nature and origin are not very clear. Little of it appears to correspond to a true ground moraine, which in modern conditions is a much thinner deposit. To a large extent till was probably carried in the lower layers of the ice, and so was chiefly englacial rather than subglacial. Thick lowland till was probably deposited near the ice-margin where there was little actual movement. Where the tills spread out on the lowlands some attempts have been made to recognize and correlate them by their lithology and erratic content, and hence such terms as Chalky Boulder Clay or Scandinavian Drift. The system has its drawbacks, however, if only because more than one ice-sheet in eastern England may have passed over chalk lands and incorporated debris from them. More recently it has been shown that, in certain circumstances, the erratic pebbles tend to be aligned with their long axes parallel to the direction of ice-movement. In a circumscribed area the stone-orientation of successive tills may be significantly different and help to distinguish them.

Till and other glacial deposits are thus unusual, in the same way that major glaciations are unusual geological events. But they are much more, and more interesting, than just a thin veneer – a minor nuisance to the geologist who is in search of solid rocks. Over 80 metres of glacial and interglacial deposits have been recorded at Crewe and a similar thickness

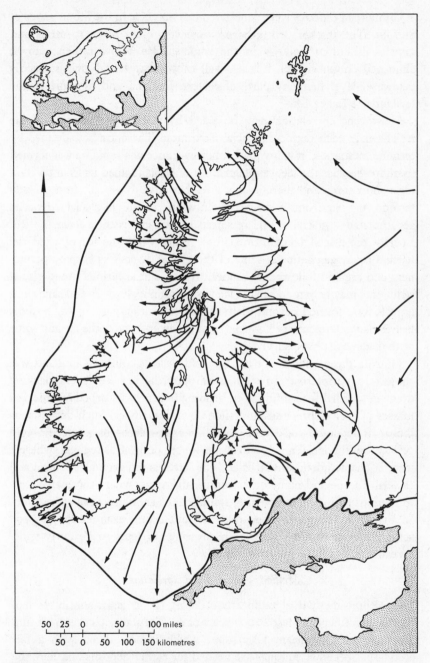

Fig. 78. The extent of maximum glaciation and generalized directions of ice-flow. Inset: the ice-sheets of Europe. (Adapted from several sources.)

at Cromer; 143 metres were penetrated in a borehole in the Stour Valley of Suffolk. The thicker drift records, however, often relate to sands and gravels in buried hollows in the rock surface below rather than to abnormal amounts of till. It is also well known that the sub-drift surface is below sea-level in many parts of eastern England and in some of the lowlands of Lancashire.

Pleistocene correlation may be aided by tracing the margins of the ice-sheets at each period of glacial maximum. These are best defined by terminal moraines; in many places, however, no such moraine was formed (perhaps because the ice was stagnant or only maintained its final position for a short time) and there is only a feather edge to the till. Immediately beyond the ice-front a range of outwash, or fluvio-glacial, deposits accumulated – glacial debris re-sorted by streams issuing from the ice. Locally pro-glacial lakes formed for a time, when the ice blocked the normal drainage channels; some of the glacial diversions became permanent and are still followed by modern rivers. These various fluvio-glacial sediments may help to locate the ice-front, but not all water-laid sands and gravels were formed beyond it. In the lower melting parts of an ice-sheet heavily laden streams are known to run on, in and under the ice and some of their deposits have survived.

Outside the ice-sheets during the cold periods southern England was subject to permafrost. In such conditions the surface layers, lacking drainage, are liable to solifluxion; bedding becomes disturbed or unsorted masses of rock debris and mud slide down hill slopes. Such deposits, as 'head' or 'coombe rock', can sometimes be related to a specific cold period. Solifluxion tends to affect the wetter periglacial regions, whereas loess is characteristic of the drier. The latter is important in central and eastern Europe and many other continental areas, where successive loess formations and their intervening soil horizons are a means of correlation. In Britain it has rarely been recorded but recent examination, by soil scientists in particular, show thin coverings of loess to be fairly widespread.

Methods of dating and correlation

The isotopic method of radiocarbon dating (p. 4) is an admirable tool for the deposits of the last 40 000 years or so. This takes us well back into the Last, or Devensian, Glaciation, which is now dated with a fair precision. Special techniques are needed for earlier deposits and here the results are fewer and more uncertain. The use of oxygen isotopes as the most universal method has already been mentioned (p. 370), though there will be major problems in relating this sequence or the geochronological time-scale to deposits formed on the continents or on their margins.

Changing levels due to eustatic causes appear at first sight to be a very satisfactory method of correlation because the results should be world-wide. The volume of water locked up in ice-caps during glacial maxima must have lowered sea-level by several scores of metres. The amount is much debated because of the difficulty of estimating the thickness of the ice, but some 100 metres has been given as a lower estimate. In practice any such correlation is complicated by other factors. Many parts of the world were affected during Tertiary times by tectonic depression and elevation, notably along the Alpine chain, and the Alps themselves and the Apennines have risen tectonically in Quaternary times. Then the weight of the ice-caps depressed, isostatically, the landmasses that were thickly covered and after the ice melted these lands gradually rose, as Scandinavia is still doing; even in Scotland lesser analogous effects are shown in the late glacial and Flandrian raised beaches. Isostatic effects are not obvious in England but it would be rash to rule out mild tectonic influences entirely. Within a limited region something has been done by tracing various late Quaternary river terraces, provided these occur in a regular order. In fortunate circumstances they may be related to a known till, but the actual dating will normally depend on radiocarbon results from organic silts, etc., or finding reliable fossils in the terrace deposits.

At this point we may hark back to the principles and terms of stratigraphical correlation as set out in Chapter 1, and summarize Quaternary methods along these lines. The British stages are chronostratigraphic units (p. 12) and are ultimately governed by climatic fluctuations, the cold periods and the warm periods. However in order to establish a succession and effect correlations it is necessary to distinguish between one cold, or warm, period and the next. The earlier methods were mainly lithostratigraphic and included attempts to correlate named tills; but as in other systems these have gradually given way to biostratigraphic methods, based primarily on the interglacial floras and faunas with some assistance from radiocarbon dating in the latest stages. Not that all is neatly documented now – far from it, as the 'Gipping problem' of a later section illustrates. Nevertheless over the years biostratigraphy has made much progress in Quaternary problems.

Palaeontology

The fossils most commonly employed have been land mammals,

man and his implements, and pollen; while to these have been added valuable results from molluscs and insects, especially beetles. For most groups Quaternary time was too short for the evolutionary replacement of one species by another and many are still living. Climate is again the main factor and ages are mostly deduced from the absence, presence or abundance of certain species, found by experience to characterize one interglacial or another.

The most important mammals that roamed over northern Europe during interglacial periods, or outside the ice-sheets, were elephants, hippopotamus, rhinoceros, deer, horses, oxen and rodents; they include southern forms such as the hippopotamus and northern such as the reindeer, but some groups now restricted to warmer climates had a wider range in Pleistocene times. The elephants are particularly useful and one of the few groups in which the species were sufficiently short-lived to characterize parts of the period:

Elephas (Mammuthus) primigenius	late Pleistocene
Elephas (Palaeoloxodon) antiquus	middle Pleistocene
Elephas (Arkiodiskodon) meridionalis	early Pleistocene

These stratigraphical ranges are only approximate and the first two, for instance, overlapped in the last interglacial. Mammalian remains are found in terrace gravels, interglacial beds and some types, abundantly, in caves; in the last situation, however, they are often difficult to place in the stratigraphical sequence. Molluscan shells occur in a variety of fluviatile, littoral and terrestrial deposits and the delicate skeletal remains of beetles in lake silts and peats.

The hominids as a group are rare as fossils and in spite of all the searches of anthropologists only relatively few examples of Pleistocene man are thoroughly known. Of these only one is British – Swanscombe 'Man' (in fact probably a female) – an incomplete skull from the 'Boyn Hill' or '100-foot' terrace, or later part of the Hoxnian interglacial period. Despite their great evolutionary interest, skeletal remains are usually dated by the beds in which they are found rather than vice versa. Artefacts, or implements, on the other hand are locally abundant. In Britain they were commonly fashioned out of flint, admirable material for flaking and easily to hand in the south and east of England. Today an expert flint-knapper can produce a hand axe in about five minutes, and Pleistocene man probably made such tools, discarded them when blunted and made some more. It is thus not surprising that though skeletal remains are so rare, artefacts on some sites are plentiful.

The analysis of pollen from interglacial deposits still remains the most far-reaching biostratigraphical method. Pollen grains are very resistant to decay and tree pollen in particular can be carried long distances by the wind; thus the analysis of pollen in peat or lake clays reflects the main constituents of the vegetation up to some 100 kilometres around. The flora of each interglacial or warm period passed from a cold sparse tundra type of vegetation to temperate forests and back again, tree pollen being dominant in the central warmer section and 'non-tree pollen' before and after (Fig. 79). If this were all it would not serve to distinguish one

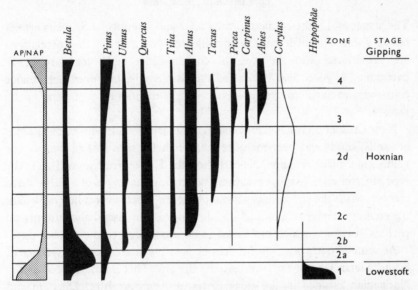

Fig. 79. Composite pollen diagram through the Hoxnian interglacial period. The thicknesses of the black lines indicate percentages of total tree pollen. The panel on the left shows the relative proportions of tree pollen and non-tree pollen (AP/NAP). The currently-used stage names for Gipping and Lowestoft are Wolstonian and Anglian respectively. (Adapted from West, 1963, p. 175.)

interglacial period from another; but fortunately the detailed proportions of the dominant forest trees (chiefly birch, pine, spruce, fir, elm, oak, hornbeam, lime, alder and yew) show patterns of change that do seem to be somewhat different in each. The late Devensian and Flandrian floras can be traced in the same way.

From detailed pollen analysis it appears that the vegetational changes in northern Europe were sufficiently widespread for a broad correlation of pollen zones to be possible from East Anglia to northern Germany. Even so, purely stratigraphical restrictions attend this method as elsewhere and the critical sections in which pollen-bearing beds are clearly related to an identifiable glacial deposit are not many. The piecing together of short or isolated sequences, which sometimes has to be done in other stratigraphical successions, is a particularly necessary, delicate and sometimes controversial exercise in Quaternary studies.

GLACIAL AND INTERGLACIAL HISTORY: THE STAGES

Cromerian to Ipswichian

The stages and some of the typical glacial and interglacial deposits are set out in Table 53, but it must be emphasized that many of the correlations are provisional and the ages of certain older beds are particularly debatable. A good deal is centred on East Anglia but even here major controversies have emerged, one of which is mentioned at the end of the section.

Near Cromer on the Norfolk coast an estuarine and freshwater sequence of sands, muds and peat includes marine shells, mammalian bones, tree stumps and plant remains. Near the top the Upper Freshwater Bed is the type of Cromerian Stage, containing pollen and mammals of a fairly warm climate. With the succeeding arrival of the ice-sheets over much of Britain the cold and temperate oscillations developed into glacial and interglacial periods, a regime which has lasted in all probability to the present day.

At some period, possibly Anglian or possibly later, there is evidence of Scandinavian ice not far from eastern Britain. This is chiefly seen in the Norwegian erratics in the early drifts of Aberdeenshire, Durham and Yorkshire. They may have been lodged there by the Scandinavian ice-sheet itself abutting on our coasts or, perhaps more likely, picked up by British ice travelling along the coast over earlier outwash deposits from a Scandinavian sheet. Some such sheet was presumably responsible for the general deflection of the British ice-streams, mostly southward parallel to the coast as far as the Wash and then advancing over East Anglia.

At its maximum extent (Fig. 78) the Anglian glaciation with its till – the Chalky Boulder Clay – overrode the Chiltern scarp; lobes extended into the northern side of the Thames basin but without leaving end moraines. The River Thames was permanently displaced to the south. Near Lowestoft the Corton Sands, marine with shells, probably represent a minor recession or interstadial within the main glaciation. Farther west there are few deposits that can be certainly attributed to this stage but the lowest, the Buddenhall Clay of the Midlands, has some pollen-dated support.

The succeeding Hoxnian Stage is sometimes called the Great Interglacial by analogy with the Holstein on the continent. The type section at Hoxne in Suffolk exposes carbonaceous clays and silts in a small lake basin on the surface of the Anglian till. The pollen sequence demonstrates a change from cold to temperate conditions and back again and the section is topped by periglacial deposits. A temperate climate is also suggested by other interglacial beds, correlated with those of Hoxne with varying degrees of certainty – such as the pollen-bearing Nechells lake silts and clays (Birmingham) and the sands and silts at Clacton (Essex). The latter include the warm water shell *Corbicula* together with *Elephas antiquus*, *Hippopotamus* and Clactonian artefacts. There is widespread evidence for a high sea-level in this temperate stage; at Clacton itself, in the Nar Valley of west Norfolk, where the land was submerged to a depth of 16 metres with marine molluscs, and at Kirmington in Lincolnshire where there are estuarine and marine clays at 25 metres O.D. The Boyn Hill terrace of the Thames is also attributed to this stage.

The Wolstonian glaciation takes its name from the West Midlands, though here is seen as outwash deposits rather than substantial till. Much of the Chalky Boulder Clay of the East Midlands has been associated with this stage, together with the Older Drift of northern England and one or more tills in Wales and elsewhere around the Irish Sea. In South-West England there are small patches of till at Barnstaple in North Devon, on the Cornish coast and in the Scilly Isles. These also have been tentatively ascribed to this glaciation and presumably reflect a temporary extension of the ice across the Celtic Sea or Bristol Channel.

The West Midlands was a meeting place of ice-lobes from several sources. The lower Severn drainage was blocked by western ice, while that from the east and north-east blocked drainage towards the Wash and the Humber. There was thus produced Lake Harrison, one of the largest English pro-glacial lakes, which left its traces in bedded sediments. After the retreat of the Wolstonian ice the Severn drainage was restored and the present Warwickshire Avon came into existence as a tributary.

By the Ipswichian Stage most of the modern river valleys were established and the interglacial sediments are commonly related to them. In the type section, at Bobbitshole near Ipswich, pollen-bearing clays lie in a channel cut into an underlying till. Comparable deposits are known from Stutton (Suffolk) and in the third terrace of the Cam, the former being rich in southern molluscs. Pollen from the Upper Flood Plain terrace of the Thames (Trafalgar Square) is Ipswichian in composition and here also belong certain lower terraces of the Severn and the Avon. The sea-level

Table 53. *Representative Quaternary successions in England and Scotland, based on Mitchell et al. (1973) with minor additions*

Stage	East Anglia	Southern England	Midlands	Lincolnshire and East Yorkshire	Northern England	Scotland
FLANDRIAN	Peat, alluvium, estuarine clays, raised beaches and submerged forests					
DEVENSIAN	Hunstanton Till Loess and periglacial features	Coombe Rock Loess, peat	Wolverhampton (Irish Sea) Till Four Ashes Gravel Chelford sand and peat	Holderness Tills Dimlington Silts Loess and head	Main Dales Glaciation Lower and Upper Boulder Clays of Durham	Loch Lomond, Perth and Aberdeen readvances Strathmore–Moray Firth Glaciation
IPSWICHIAN	Interglacial deposits at Ipswich and Cambridge March Gravels	Interglacial deposits at Ilford, Trafalgar Square Raised beaches Cave deposits	Gravels of Avon No. 3 Terrace, etc.	Buried cliff and beach deposits at Sewerby	Interglacial deposit at Austerfield (Yorks) Kirkdale cave deposit	
WOLSTONIAN	Outwash gravels in north Norfolk Solifluxion deposit at Hoxne	Fremington Till (Devon) Coombe Rock 'Taplow' terrace	Glacial deposits, Wolston Deposits of Second Welsh Glaciation	Basement Till of Holderness Chalky Till of Wolds	'Older Drift' of the Pennines Till with Scandinavian erratics (Durham)	Indigo Till, Banffshire and Aberdeenshire

Stage	Interglacial deposits at Hoxne, Marks Tey, Nar Valley, Clacton	Interglacial deposits at Swanscombe	Interglacial deposits at Quinton and Nechells (Birmingham)	Interglacial deposit at Kirmington (Lincs)	Peat, Shetland
HOXNIAN				Interglacial deposit at Kirmington (Lincs)	Peat, Shetland
ANGLIAN	Lowestoft Till Corton Beds Cromer Tills	Chalky Boulder Clay of Finchley and Hornchurch Winter Hill Terrace	Bubbenhall Clay ? First Welsh Glaciation		
CROMERIAN	Upper Freshwater Bed (upper part of Cromer Forest Bed)	? Cave deposits of Westbury sub-Mendip ? Slindon (Sussex), raised beach			
PRE-CROMERIAN	Remainder of Cromer Forest Bed	High-level gravels of uncertain age			
PRE-CROMERIAN	Marine 'Crags', sands, silts, clays of East Anglia	Deposits at Netley Heath and Rothamsted			

Several of the suggested correlations are tentative or provisional, especially in the earlier stages; there is some doubt about the validity of the Wolstonian Stage, particularly as a glacial episode. Pre-Cromerian stages have been named (in upward order) Waltonian, Ludhamian, Thurnian, Antian, Baventian, Pastonian, Beestonian; others may be defined in the future.

Deposits within a box are not necessarily in stratigraphical order.

does not seem to have been generally so high as in the Hoxnian interglacial and the East Anglian localities are not far above modern sea-level. However a submergence of eight metres is suggested by interglacial beds on the Sussex coast.

Having traced glacial and interglacial history so far, though with some simplification, a note must be appended on the 'Gipping problem'. This is an alternative interpretation of recent years and since it concerns the East Anglian sequence it is cardinal in British Quaternary stratigraphy. Briefly the Gipping Till (once the type of the Gipping Stage, now renamed Wolstonian) was defined in the valley of that name in Suffolk and was also the till considered to underlie the Ipswichian interglacial beds a few kilometres to the south. Nevertheless it had never been found in satisfactory superposition above either Hoxnian interglacial or Anglian glacial deposits. The revised version maintains, therefore, that there is no Gipping Till and in East Anglia only one major Chalky Boulder Clay resulting from one glaciation – the Anglian; further, that this till underlies both the Hoxnian and the Ipswichian temperate deposits so that these belong to the same interglacial stage, the last or pre-Devensian, although including a minor cold period.

Such a view, based on detailed mapping in East Anglia, demands serious consideration, but when compared with the established scheme of Table 53 it is seen to throw out of gear not only the earlier East Anglian sequence, but correlation with the Midlands and the Wolstonian glaciation. The position of the latter depends substantially on the Hoxnian age of the underlying Nechells lake deposits, but these also have been investigated in detail.

This book is no place to pursue these intricacies far, nor is it likely that the Gipping problem will remain, at least in its present form, for a long time; it will be solved one way or the other or given a new slant as new evidence is presented. But the reasons behind such a divergence of opinion are worth noting. They concern problems in the identification and mapping of tills and in the lateral variation native to glacial and inter-glacial deposits; they raise doubts about the reliability of palaeontological data (here the pollen of Nechells and Hoxne) as a primary tool in correlation; and the alternative possibility has to be considered, or refuted, that a major glaciation in the Midlands, Wales and the north could be represented in Suffolk by only a minor cold phase. Similar questions have enlivened Quaternary stratigraphy in the past and are unlikely to be wholly absent in the future.

During Quaternary times Ireland has been somewhat isolated from the rest of the British Isles – far more so than England from the continent. At present the central belt of the Irish Sea lies below the 100-metre contour except for a narrow isthmus between northern Antrim and the south-west Scottish Highlands and in Pleistocene times there would have been little easy access across it either in periods of low or high sea-level. Marine molluscs of Red Crag affinities (basal Pleistocene) have been found derived in late glacial outwash deposits in north Wicklow. The interglacial faunas and floras are depleted or carry their own native Irish species; Palaeolithic artefacts are unknown. Much therefore depends on the richer pollen sequence of the Hoxnian Stage, or Gortian in Irish nomenclature. Several localities are known in the southern part of the country, together with the type locality in south Galway. It is possible that an earlier glaciation is represented by erratics stranded on the southern coasts but very little is known about it.

The Wolstonian Stage is represented by a major glaciation – the Munsterian. Nearly all the country was covered by ice, and multiple tills (perhaps three glacial phases) have been recognized locally in the south-east. The main ice-centre was probably the Dalradian uplands of the north-west, but here as elsewhere the history of the upland regions is difficult to interpret, as are many of the upland tills; the best sections tend to be on the coast. The Wicklow mountains, a source of granitic erratics, were a minor dispersal centre and the west Cork and Kerry mountains supported an independent ice-cap.

Presumably the Munsterian glaciation was separated from the last, the Midlandian, by an interglacial phase equivalent to the English Ipswichian, but remarkably little is recognizable from it – so far only some fragmentary estuarine sands on the Wexford coast. They include some shells and pollen and have been disturbed by later ice.

The Devensian, or Last, Glaciation and Flandrian Stage

During the last major glaciation the ice-sheets did not reach so far south as in the earlier ones; the glacial topography is clearer and the deposits are less weathered. They are also found in the valleys, in contradistinction to the Older Drift, which, in the Pennines for instance, forms weathered and dissected patches on the interfluves. End moraines are relatively common, such as the Escrick and York moraines of the Vale of York and that which takes a sinuous line across southern Ireland, for instance in Limerick and Tipperary. The irregularity of the Devensian ice-front (Fig. 80) is caused especially by the lobes that extended southwards down the lowlands and

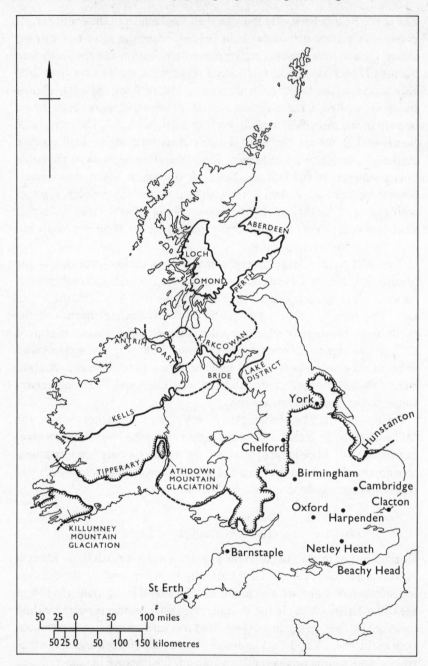

Fig. 80. Map showing the limits of the Devensian (Last) Glaciation and, tentatively, three successive re-advances or standstill lines, including an alternative interpretation between Antrim and south-west Scotland. (Adapted from several sources.)

the failure of the ice to override some of the uplands, such as the North Yorkshire Moors, southern Pennines and Leinster mountains. In spite of this greater topographic clarity, however, the line is not wholly agreed (South Wales especially has its variants) and it may be that different parts represent different Devensian phases rather than the ice-front position at any one time.

Part of the difficulty is the rarity of fossiliferous deposits under the Newer Drift, all the major examples quoted in the last section lying outside its limit. However radiocarbon dating has now clarified much of later Devensian history and three divisions can be defined broadly in years:

late Devensian 26 000–10 000 ⎫
middle Devensian 50 000–26 000 ⎬ years ago*
early Devensian before 50 000 ⎭

Equally it has become clear that although the stage as a whole was a cold period, with such indications as solifluxion, cryoturbation and ventifacts, it only became truly glacial in Britain towards the end. In east Yorkshire peats below and above the Devensian tills give dates of about 18 000 and 13 000 years ago, so that the ice-sheets here only lasted some 5000 years and perhaps less. This is very near the southern margin, but the general picture of a late vigorous Devensian advance, followed soon by deglaciation, has support from dated material under the tills of Cumbria and North Wales.

Ireland is famous for its wealth of glacial deposits and landforms but a stratigraphical approach is made difficult by the extreme rarity of interglacial deposits or reliable fossiliferous horizons. Provisionally most of the major drifts are ascribed to the Last Glaciation, or the Midlandian Stage in Irish terminology. There seem to have been various ice-advances and multiple tills but much of the country was covered by ice at the maximum (Fig. 78) with the main dispersal centres in the highlands of the north and west, recognized by their erratics. As before smaller ice-caps centred on the mountains of Kerry and Wicklow, each partly ringed by local moraines. At the height of glacial advance Scottish ice, with erratics from the Clyde area, spread far into Antrim and Londonderry and with Irish Sea ice invaded the eastern coast. A late Scottish re-advance impinged on the north Antrim coast. Numerous drainage diversions can be traced to glacial origins and the midland plain, especially in the north and west, exhibits one of the world's largest drumlin fields; eskers abound in the south and east.

* These dates, and all others in this chapter, are in terms of radiocarbon years which, for geological purposes, may be regarded as roughly equivalent to true time.

Outside the ice-sheets in southern England part of this cold stage is represented by the lowest and latest terraces of the Severn and the Avon, and in parts of the Thames valley. A late glacial, low sea-level resulted in the cutting of the steep narrow buried channels below the present Thames and Severn estuaries. During the recession of the ice-front water was again held up between the ice and higher ground. Pro-glacial lakes were formed around the Leinster mountains and in north Yorkshire. The eastward drainage from the Vale of Pickering to the North Sea was permanently blocked by drift and the Yorkshire Derwent now flows south-westwards into the Ouse. One of the most famous glacial diversions belongs to this stage when the previous Severn drainage, northwards to the Irish Sea, was blocked by ice from that direction. The waters of the upper Severn cut the Ironbridge gorge and still take this devious way into the lower Severn valley.

During the latest Devensian phases subsidiary oscillations in climate can be recognized, the more temperate being called interstadials. They were separated by standstill periods or by more active re-advances of the ice-sheets; these are best seen in Scotland but also occur in north-east Ireland. Correlation of these advances has proved difficult, partly because of problems in mapping the local tills and their margins and partly because pollen and radiocarbon results are few. It is generally accepted that in Scotland the Loch Lomond re-advance was the latest and that there were earlier ones, with margins farther to the east and south. Thus an Aberdeen re-advance and a Perth re-advance have been identified, and two more in parts of the Southern Uplands, including the Lammermuir Hills, but their precise definition and correlation (is there for instance a Perth–Aberdeen line?) is debatable. The principle picture, however, stands. The ice-held regions gradually and irregularly became smaller, retreating in latest Devensian time to the western Scottish Highlands, which then as now received the greatest amount of precipitation.

Quaternary deposits with much glacial debris are known from the seas around the British Isles; in the North Sea they are remarkably thick and partly marine. Up to 1000 metres has been recorded in the central depression of the southern North Sea, with some resemblance to the Dutch sequence. Most of the published determinations refer to the Devensian Stage.

The Flandrian Stage is presumed to begin when the last major ice-sheets melted from Scotland, some 10 000 years ago. Early in the stage there was relative emergence over much of Scotland and peats have been found representing pollen zones IV and later (zones I, II and III lie within the late Devensian). About this time the open park-like or tundra vegetation over lowland Britain gave place to denser forests, dominated at first by pine and birch and later by successions of various trees. The pollen zones also show that a climatic optimum warmer than the present day took place about 5000 years ago.

Some 7000 years ago peat, or 'moor log', was formed on the site of the Dogger Bank, at that time a lowland bordering a reduced North Sea. Since then this region has been resubmerged in the main Flandrian eustatic rise in sea-level, which culminated 5000–5500 years ago. It is reflected in the higher and oldest post-glacial shorelines of Scotland, where subsequently through a combination of isostatic uplift and minor eustatic oscillations lower benches were formed, conspicuous now as the raised beaches of many coasts, both in Scotland and north-east Ireland. Flandrian submergence is shown farther south in the peats or 'submerged forests' and there were other geographical effects. During the glacial periods when ice stretched from Britain to Scandinavia the northern drainage of Europe, from the Elbe to the Rhine, was blocked; the waters were probably deflected into a westerly course down a broad valley that lay on the site of the upper English Channel. At the height of the Flandrian transgression, in the Neolithic period, this valley, already deepened perhaps by glacial overspill, was drowned and the Straits of Dover were formed.

Quaternary geology, in all its diversity, does not lend itself to any simple or generalized conclusions but one aspect may be taken as an end-piece. There is no significant division between the geological past and the present. It is not so much that, in a uniformitarian manner, we may interpret the past by reference to the present – a generalization that even in Phanerozoic stratigraphy is not universally supportable, as witness the difficulty of interpreting till or the absence of modern oolitic iron ores; it is rather that Quaternary processes, sediments, faunas and floras pass smoothly into those of modern times.

Palaeolithic cultures give place to Mesolithic, Neolithic, Bronze Age and Romano-British, and the historian takes over from the archaeologist. The distribution of modern animals, and still more plants, has been strongly influenced by the fluctuations of ice-sheets, climates and sea-level, opening some routes of migration and closing others. The present land surface in both glacial and periglacial regions has been affected by the same factors, and weathering and denudation are already altering the latest drifts in the same way that they altered the older ones.

Flandrian sediments are separated only arbitrarily and for convenience from late glacial on one hand and modern on the other, and the three combine to form a record of continuous sedimentation in the Fenlands and the Wash. Possibly the term 'modern' in itself suggests an unreal boundary. As soon as a sediment comes to rest it is subject to physical and chemical changes – such as the expulsion of water from the pore spaces, the disturbance of bedding structures by organisms and the change in chemical conditions (especially towards reduction) as soon as one or two millimetres of covering layers have accumulated. A few centimetres down in the bottom sediments of the Humber, the Wash or the Firth of Clyde, the muds and silts are in the first stages of diagenesis, which in time will transform them into lithified rocks, comparable to the shales and siltstones of the Jurassic or the Coal Measures. Given yet deeper burial, enhanced temperature and pressure, they may become slates and schists such as those among the Dalradian groups.

REFERENCES

Mitchell *et al.* 1973 (*Special Report* No. 4, Quaternary).
Bowen, 1974, 1978; Hays *et al.* 1976; Hudson, 1977; Penny, 1974; Shotton, 1977; West, 1963, 1968.
British Regional Geology: Quaternary or Glacial Geology in several handbooks.

13

EPILOGUE

This chapter reviews some more general topics, theoretical and practical, and allows a few more personal reflections.

THE DATA OF PALAEOGEOGRAPHY: QUESTIONS OF SCALE

From one aspect the stratigraphy set out in the preceding chapters can be seen as a series of changing geographies; some sources are world-wide, others depend on the collection and analysis of precise local detail.

The microcosm

First, to balance the earlier generalizations, we will look at a few examples of detailed studies. Outstanding here are those on the Wealden sediments and environments, which could only be summarized in Chapter 10 – see P. Allen (1976) and references therein. For nearly 40 years Allen has concentrated on the sedimentology of the Wealden in the Weald basin, also drawing on tectonic and palaeontological evidence. One recent facet only is quoted here as a manageable example – the sources of the detrital tourmaline (Allen, 1972). This is a common accessory mineral in arenaceous rocks, being extremely indestructible; its particular virtue is that the mineral grains and tourmalinized rocks can be dated by an isotopic method (^{40}Ar–^{39}Ar) and thus provide information on the upland source areas and their phases of tourmalinization, a post-orogenic effect.

Tourmaline is common in the Wealden of Wessex but not till later in the Weald itself, in the sandstones of the Weald Clay. In both areas the source was westerly, but Cornubia (Fig. 70) cannot supply the older pre-Hercynian grains, nor those of high grade metamorphism, such as kyanite and staurolite. Armorica must be added and probably northern Spain as an additional source. These old massifs were partly mantled by New Red Sandstone and it may be that there were others, now submerged in the

Western Approaches or even in North America, if the North Atlantic rift was not yet a full barrier. Tourmaline is thus one brick in the edifice of Wealden environments and the basin itself is favourably placed. Global

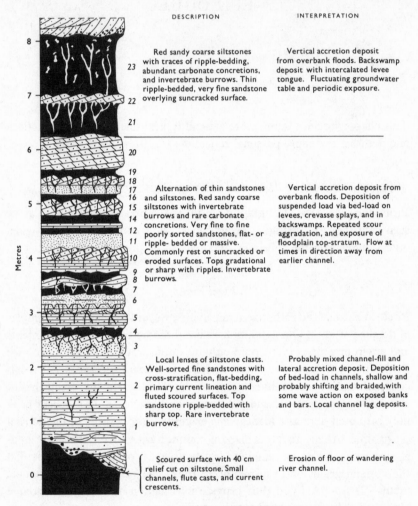

Fig. 81. Generalized succession and interpretation of a Breconian cyclothem at Mitcheldean, Gloucestershire. (Redrawn from Allen, 1964, p. 188.)

tectonic influences by now were reduced to the opening of the Atlantic and the Bay of Biscay, and although there is debate on the relief or influence of the neighbouring landmasses the positions of most are clear.

Palaeogeography during the Upper Palaeozoic was exceptionally varied and several examples could be taken where local and detailed examinations lead to wider estimates of environment. Six cycles in the Lower Old Red Sandstone were analysed by J. R. L. Allen as part of a systematic study of the Anglo-Welsh Province. Fig. 81 shows a characteristic section and its interpretation. The age is known to be Breconian (p. 426) since this is one of the few quarries that have yielded vertebrates of that stage (*Rhinopteraspis cornubica*). The sediments, their structures and environments are examined in detail and compared with modern regimes. Allen concludes (1964, p. 190) that 'This cyclothem is probably fluviatile in origin. In general succession it closely resembles a modern alluvial sequence, in that uniform coarse beds underlie coarse and fine deposits, followed by uniform fine sediments.'

Carboniferous stratigraphy abounds in environmental interpretations and the stratigraphic framework is well supported by the zonal schemes. In the Dinantian modern assessments have been made of the limestones and their faunas (e.g. Mitchell, 1972; Ramsbottom, 1973), while precise field descriptions are necessary to define the stage boundaries (George *et al*. 1976, p. 6). The summarized R_I sequence (p. 202) draws on recent analyses of Millstone Grit sedimentology and environment by Allen (1960), Walker (1966) and Collinson (1969, 1970). In 1965 De Raaf *et al*. dealt similarly with the Westphalian of North Devon, a pioneer work in that province.

The economic value of the Coal Measures has produced an exceptional range of data. Among them the problem and definition of cyclothems have provoked much discussion (e.g. Duff *et al*. 1967, p. 132; Schwarzacher, 1975, p. 253). Some type of cyclic sequence is common in the Westphalian measures, but so also is variation and rarely if ever is the complete 'ideal cyclothem' actually encountered.

However here we consider the clearest repetitive item in such a sequence, the marine band and its faunas, taken from hundreds of detailed borehole sections. Fig. 82 shows the advance, acme and retreat faunas characteristic of a major marine band, such as might be found in almost all British coalfields and in France, Belgium and Germany; the lesser bands tend to lack some or all of the 'acme' genera. In the example shown, the Vanderbeckei or Clay Cross Marine Band, the concentration of the most fully marine species coincides with the thickest development, some four to five metres, in North Staffordshire and the East Midlands. Towards the southern margin and to the north *Lingula* (? intertidal as at present) is more common. A similar transition is found from south-west Wales to Bristol.

Some extensive deductions follow from this combined lateral and vertical distribution. It supports the reliability of marine bands as correlative markers; it reinforces the general view of Westphalian palaeogeography – the vast coastal flats and few residual 'islands' of slightly higher ground; it presents some evidence on salinity. In this last context emphasis may be placed on the very varied faunas, drawn from eight phyla, which broadly appear and disappear in the same order. Even

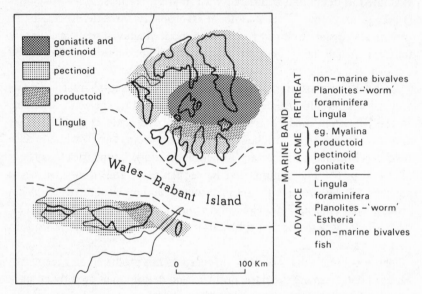

Fig. 82. Map showing the faunal assemblages at the acme of the Vanderbeckei Marine Band from the Pennines to South Wales, the coalfields outlined. On the right, faunal sequence in an idealized marine incursion. (Data from Calver, 1968, pp. 10, 38; 1969, p. 248.)

so it is instructive to find that in the Vanderbeckei band during the regressive phase there are thin intercalations of non-marine ostracods and bivalves between the pectinoid–goniatite faunas. This may mean that both groups could tolerate brackish conditions, or that there were minor fluctuations, local advances and local retreats. Salinity problems are rarely simple.

On a different scale the interrelations of sediments and plant assemblages have produced some closely argued deductions on the non-marine facies – a valuable integration of botanical and geological research. The environments include swamps, lakes, flood plains and deltas, all on a small scale and compiled from quarry sections (Scott, 1978, 1979). It is also deduced that certain genera, or assemblages dominated by them, grew in different situations and in drier or wetter conditions. Thus the sphenopsid (horsetail) *Calamites* probably grew in the wettest conditions on the stream or lake banks, the lycopods dominated the swamp forest and the pteridosperms the better drained plains. There are also changes in the spore content of the coals themselves, reflecting changes of successive plant communities. They are analogous to the pollen sequences of the interglacial peats and indeed the latter could be taken as an excellent example of palaeontological minutiae which nevertheless have substantial environmental and stratigraphical value.

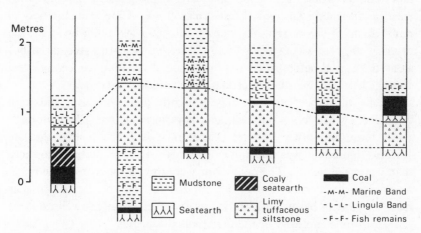

Fig. 83. Six borehole sections in the Scottish Namurian (Fife), showing the relation of a tuffaceous siltstone to a coal seam and associated cyclic sediments. (Redrawn from Francis, 1961, p. 205.)

Borehole sections have also demonstrated the complex interrelationships between Carboniferous volcanic vents and their tuffs and the surrounding sediments. In east Scotland there are many levels of tuffs and tuffaceous siltstones derived from small impersistent volcanic cones, and lesser examples in the English Midlands. Some ashfalls extended 26 kilometres from their source and after the eruption the volcanoes tended to collapse forming small caldera structures. The details of one tuffaceous siltstone in Fife show that this bed interrupts the normal Coal Measures cyclothem at different levels in different boreholes (Francis, 1961, p. 205); in particular the coal and seatearth are found above and below (Fig. 83). Ashfalls are normally rapid and taken to represent fairly reliable time-markers, which would suggest that in this case the cyclic sequence, or at least the coal formation, was diachronous (cf. Duff *et al.* 1967, p. 151).

Another result concerns the dominant wind direction, deduced from the elongate pattern of fall-out, and this agrees with that forming the dunes in the Lower Permian aeolian sandstones (p. 246); both were easterly in present orientation, or the north-east trade winds of their period.

The mineralogy of carbonate–evaporite sequences tends to be complex in its original form and becomes more so on diagenesis. The outstanding Zechstein deposits were outlined in Chapter 8, especially from the work of D. B. Smith (e.g. 1974). The late Jurassic (lower Purbeck, p. 304) of the Dorset coast is at first sight very different. Nearly all the beds are limestone but again microscopic structures demonstrate a complex sequence of diagenetic changes (West, 1964, 1975). Chief among these is replacement of secondary anhydrite by calcite or less often by silica; the primary sulphate was gypsum. Four successive facies are recognized but along the coastal outcrops there is much lateral variation, reflecting small changes between subtidal, intertidal and supratidal conditions.

Locally the tidal flats were raised slightly above sea-level and were colonized by conifers – the remnants now forming the 'Fossil Forest'. From these and other climatic indicators it is probable that the late Jurassic hinterland was not more than partly arid, or perhaps no more than warm temperate. The conditions were thus distinct from the much more extensive Permian examples, but all illustrate the delicate balance between shore or basin configuration, water-level, salinity and climate that governed the complexity of the precipitated minerals.

Detailed mineralogy is also required for the recognition of tuff beds, ashfalls or bentonites, in the identification of the clay minerals or relics of volcanic glass. Those of Mesozoic age were mentioned earlier (p. 338) and several are now known in the Lower Tertiary both from land outcrops and in borehole cores from the North Sea (Knox and Ellison, 1979; Knox and Harland, 1979). In the latter the tuffs provide useful marker beds which can assist the correlations based on the microfossils; the subject is thus linked with offshore exploration and economic geology (p. 416).

Although these examples of environment, palaeogeography and supporting evidence are varied they have some features in common, in that they relate largely to shallow marine, marginal or intertidal, and fluviatile conditions. The modern analogues thus have the advantage of being reasonably accessible and in recent years have been much investigated. For various reasons the earlier systems have not been productive along quite the same lines. There are some detailed interpretations, such as that of the erosion of the Ballantrae massif (Williams, 1962), the history of the Ordovician volcanics and their vents in south-east Leinster (Stillman *et al.*

1974), or the distribution of Silurian shelf faunas (e.g. Ziegler *et al.* 1968).

Overall, however, it is the turbidite facies, with its structures, internal divisions, proximal and distal versions, that is outstanding among Lower Palaeozoic sediments, for instance in Wales or southern Scotland. While these can provide much detail on the actual infilling of a basin or trough and the accumulation at plate margins, the configuration, outline or palaeogeography of these elements often remains obscure. In Britain the problems are accentuated because of the debated plate arrangements preceding the last closure stages of the Caledonian orogeny. Put in another way, once the Caledonian chain is formed, stretching across Laurasia from Newfoundland to Norway, a major part of the stratigraphical jigsaw puzzle falls into place. The smaller-scale evidence outlined above can be fitted into the larger frame. What would be most valuable among the earlier detrital or turbidite facies would be unequivocal evidence of the source regions (is it Dalradian or Lewisian, from Anglesey, Rosslare or Brittany ?). Although some logical inferences can be made (e.g. p. 114), proof is not easily achieved.

It might be thought that Precambrian rocks were under still more severe handicaps, viewed from this particular angle, and it is true that the lack of fossils precludes much correlation and hence the difficulty of deciding whether certain sediments and conditions were (or were not) contemporaneous from one part of an outcrop to another. Nevertheless some rock groups have been treated along the same lines as their Phanerozoic analogues. The Torridonian is a conspicuous example, with regional analyses of environment, of detrital sources and more detailed individual studies (e.g. Stewart, 1969; Williams, 1969; Lawson, 1976).

Dalradian sediments also provide examples, such as the Jura Quartzite (Islay Quartzite of Table 7; Anderton, 1976). The sands and finer clastics are now quartzites and slates; the former exhibit various types of cross-bedding and the whole many characters of tidal shelf sediments; they are deduced to have been swept by intermittent storms and were probably in connection with the open sea or a gulf of the ocean. There is no direct indication in Dalradian palaeogeography of where such a sea might be (cf. p. 56) though structures in the quartzites themselves suggest on balance a south-easterly direction.

In its type region, on and near Islay (Fig. 13) the Portaskaig tillite is not much deformed and its glacial characters are convincing. Fig. 84 shows a glacial erratic fallen into bottom sediments and periglacial structures. The large varied boulders in the tillites and interspersed bedded sediments (including varves) support the deduction of successive major ice-sheets, grounding in shallow waters.

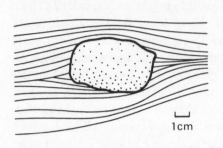 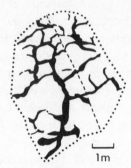

Fig. 84. Detail from the Portaskaig tillite, Dalradian. On the left a granitic boulder dropped into a varved siltstone; on the right, the pattern of sandstone wedges, in plan. (Redrawn from Spencer, 1971, pp. 39, 41.)

The macrocosm

At the other end of the scale there are the global effects in which processes in one part of the world have repercussions in many other regions. Within the history of mankind these may include large earthquakes or exceptional volcanic eruptions, though in the geological record these only appear as a tuff layer or a few metres' displacement on a fault.

In the stratigraphical record marine transgressions and regressions occur on various scales and the evidence may be only local, continent-wide or world-wide. Those on a small or medium scale have for instance been deduced in Jurassic and Carboniferous sequences in this country, and a eustatic cause has been invoked, as distinct from tectonic warping or simply changes in sedimentary regime. The causes remain matters of discussion, one problem being that it is easier to identify transgressive phases than regressive ones. The eustatic argument also needs to be reinforced by analogous contemporaneous events found on other continents, since changes in ocean-level should be world-wide, and a detailed zonal scheme is desirable in order to demonstrate this. The Namurian example (Fig. 44) and those from the Lias have the advantage of ammonoid zones.

A few much larger transgressions are generally accepted, the most extensive and celebrated being that in the late Cretaceous, in Europe culminating in the Campanian Stage (p. 420) and in North America a little earlier (Hancock and Kauffman, 1979); it is also recorded where marine sediments have been carried far into the interior of several continents. The Permo-Triassic is generally regarded as a regressive period, though perhaps broken by a lesser transgression causing an abrupt incursion in northern Europe and the formation of the Zechstein Sea (Smith, 1979).

Early Cambrian was another time of transgression though the age data are not nearly so precise, and conversely there is evidence for regression in the late Ordovician. In British stratigraphy the former can be seen in the basal Cambrian facies of north-west Scotland and the Midlands, but not in North Wales or the Dalradian trough; the latter is reflected in the absence of Ashgill beds in parts of the Welsh Borders and the extensive Llandovery transgression that succeeded it (McKerrow, 1979). A good deal may depend on the local topography of the shelf seas and hence the extent to which such changes in sea-level left a permanent record. Nor should we expect analogous effects in the more unstable areas; there is the marked unconformity that preceded the upper Llandovery submergence in south Mayo, but it was accompanied by major uplift and erosion and the causes, within the orogenic belt, were presumably tectonic.

There remains the problem of ultimate causes of eustatic oscillations and only two seem to be large enough in terms of raised or lowered sea-levels (Donovan and Jones, 1979). The withdrawal of oceanic water to form continental ice-caps, and their subsequent melting, is an acknowledged cause of Quaternary oscillations – a glacio-eustatic mechanism. The rate of change could have been relatively rapid, up to one centimetre per year. Parts of northern Africa were affected by a major glaciation in late Ordovician to early Silurian times, as were parts of all the southern continents (then assembled as Gondwanaland), in the late Carboniferous to Permian. It is therefore possible to adduce a glacio-eustatic cause for events in other parts of the world, including north-west Europe, during these periods. It is very doubtful, however, whether the early Cambrian transgression bore any relation to the melting of the Varangian ice-sheets, the time-gap being too large.

From the Trias to the early Tertiary it appears that there were no polar ice-caps and another cause must be sought in this time-span. The most likely, though difficult of proof, is an enhanced spreading rate at the mid-ocean ridges, so that as the volume of the ridge increased the seas spilled over onto the edges of the continents. Alternatively a simple length-wise extension of a ridge, or the total ridge system, might suffice. The changes in the ocean floor would operate much more slowly than those of the ice-caps and they seem to be the best explanation of the large slow, somewhat irregular, Cretaceous transgression. The post-Campanian regression persisted into Tertiary times and was accentuated as the ice-caps developed, first in Antarctica and later in North America.

It is reasonable to suppose that there were analogous sea-floor spreading changes in pre-Jurassic times, but since that floor has disappeared no evidence is available. However, combined at certain periods with glacio-eustatic oscillation, and possibly with mild tectonic warping or sedimentary effects, the cumulative results could well be complex and not attributable to a single cause. Among the macrocosmic or global phenomena transgressions and regressions are likely to remain a fruitful source of debate.

The concept and application of plate tectonics might well be taken as the most influential aspect of the macrocosm. It appears in simplified form in Chapter 1 and in attempted interpretation in Chapter 4, so not much further need be added here. However, in the years of rapid growth and adoption (the revolutionary child is almost come of age) some changes can be detected. Early on the hypothesis was applied very widely to a great variety of geological situations whereas more recently the significance of that geological variety has received more attention, and some modifications or additional tectonic patterns have been considered or emphasized (cf. Ziegler, 1977, p. 84).

The great progress in research has confirmed several components, notably of the ocean floor and median spreading ridges but also of processes at passive rifted margins. It is the convergent margins and their resulting orogenic mountain belts that have remained more difficult, some frankly controversial in their interpretations. The Mediterranean section of the Himalayan–Alpine belt is a classic instance, in its complexity of microcontinents, diverse basins, heterogeneous crustal foundations and in its multiple hypotheses of origins and history – the kinematic jigsaw puzzle of Laubscher and Bernouli (1977, p. 1). The Andes, once a much quoted prototype, have been judged on a basis of many years' geological research to be 'most complex and hardly suitable for generalizations' (Gansser, 1973, p. 126).

Although we need not press analogues between present and past too strongly, such experienced views may at least engender caution when interpreting older mountain belts, and in this I am fortified by similar reserves expressed by fellow writers on stratigraphy (Anderton *et al.* 1979, p. 276). Caution however is not lack of interest, far from it. Indeed I believe that it is especially the balancing of scrupulously recorded facts, for and against a given hypothesis (whether stratigraphical, tectonic or any other), that makes a synthesis both more fascinating and genuinely scientific.

Vertical movements such as block-faulting are a noticeable component within some mountain belts and also on aseismic continental margins. Kent (1977) has drawn attention to such movements from Permian to mid-Cretaceous times on the borders of several continents, including Europe and the North Sea. After the mid-Cretaceous these effects were replaced by slow simple subsidence, with transgression over the continental margins. Such changes are thought to reflect some genuine world-wide tectonic control, slower than plate movements and little understood.

In addition to such global questions the geological field has broadened in other ways. As research methods become more complex there is a tendency for fragmentation, and the earth sciences are no exception: the different ends of the multi-discipline corridor begin to develop different languages, often based on necessary terms of precision. Yet it becomes more important that the palaeontologist should understand the processes of the sedimentologist; that the stratigrapher should grasp not only the demands of his own nomenclatural code but the implications of petrochemistry; that the structural geologist be able to balance the sparse age data in his orogenic belt, whether these be isotopic or the remains of microfossils. There have been some attempts at overcoming the barriers between specialists, but more could be done. The next section comes down to some hard facts but these also have their questions of theoretical and practical understanding.

THE RELATION TO ECONOMIC RESOURCES

The occurrence and development of the earth's resources are such an important part of modern geology that some reference should be made to them here (cf. Knill, 1978). It is however in a limited context – to exemplify briefly those rocks, minerals and fluids that have particular relevance to British stratigraphy as outlined in the foregoing chapters. They can be classified as follows:

> energy resources, mainly coal, oil and natural gas
> building stones, etc., limestone and dolomite, refractories
> evaporitic minerals
> Mesozoic sedimentary ironstones
> epigenetic metalliferous ores
> superficial deposits and aggregates
> groundwater

From these I am selecting only a few major examples. Moreover what constitutes an 'economic resource' may change from decade to decade, or sometimes from year to year, according to price and demand on the world's markets. Accordingly output figures are not given here (some may be found in the references), but over the past 150 years the output of metallic ores from the British Isles has tended to decline, though with many fluctuations. British coal production reached its peak early in this century but has now fallen to less than half the maximum. Conversely much more of the industrial mineral resources are being actively pursued. Another advance of the last few decades is the greater interchange between theoretical and practical aspects, and the information stemming from the search for economic resources has added substantially to the understanding of the geology in and around these islands, as it has in other countries. Many references are general or relate to England; Duff (1965) deals with Scotland, Thomas (1972) with Wales, Williams and McArdle (1978) with Ireland.

Energy resources

In spite of its great economic importance coal can be dealt with briefly since the stratigraphy of the Coal Measures was covered broadly in Chapter 7. On a smaller scale, however, items such as seam thickness, lateral continuity and quality, the type of roof and floor, and the effects of faulting have much practical importance (Williamson, 1978). The clays, shales and coal itself are incompetent strata and liable to minor disturbance, both natural and induced during mining. Detailed studies of the associated sediments and structures may allow obstacles to be forecast and hazards lessened. Rather paradoxically, some problems are accentuated by technical advances because mechanized coal-cutting and self-advancing roof support systems are less flexible than the older methods, which could work through more faulted ground.

In modern conditions mining in tectonically complex regions, such as South Wales, or where seams are interrupted by igneous rocks, as in parts of Scotland, is at a disadvantage. New coalfields are therefore likely to be most productive in relatively stable areas where faulting is not intense and dips are fairly low. Concealed measures found recently in the Midlands were mentioned earlier (cf. Francis, 1979) and the great value of the Selby Coalfield lies in the fact that the Barnsley seam – traditionally the mainstay in south Yorkshire – is here from 2.0 to 3.25 metres thick and at an accessible depth. Opencast mining is profitable for shallow seams, or for exceptional coals such as the anthracites of South Wales.

Apart from its key contribution to energy demands coal mining occasionally adds to other resources. Thus in some areas mine drainage is a useful adjunct to water supplies, more than that derived from the Coal

Measures aquifers themselves, while modern exploration entails a detailed grid of boreholes, providing data on the overlying formations as well as the Coal Measures.

The hydrocarbons, oil and gas, are derived from the organic matter trapped in sediments and the source of petroleum is widely held to be oceanic plankton. Small amounts of hydrocarbons are not uncommon; it is the large economic fields that are rare. In regions of marked deformation the fluids will have been lost at the surface long since, a risk that increases with the age of the rocks; over 80% of the world's oil has been found in Jurassic and later strata which have been only gently deformed (Warman, 1978, p. 43).

The basic conditions for large accumulations are simple: the fluid is held in a reservoir rock, porous and permeable, and will tend to flow in the direction of least pressure, commonly up-dip; it must be confined by some impermeable stratum, the cap-rock, and there must be a structural or stratigraphical trap to prevent flow to outcrop. Among the world's oilfields sandstones are the commonest type of reservoir, while limestones, especially dolomitized and reef rocks, come second. Stratigraphy is thus of prime importance and a knowledge of lateral facies change. The petrography of the potential reservoir rocks is also crucial and much attention has been paid to diagenesis in sandstones and the way it affects porosity and permeability (cf. Taylor, 1978).

In Britain the mention of oil and natural gas promptly conjures up the North Sea but the full picture is a little more extensive. There has been much exploration in western seas and around Ireland, though not very much is published. The Kinsale gasfield in the northern Celtic Sea, 70 kilometres south-east of Cork, is based on a reservoir in Lower Cretaceous fluviatile and deltaic sands, Aptian or Albian. In the south-eastern part of the Irish Sea the important Morecambe gasfield is not yet in production, but its reservoir rock is in Permo-Triassic sandstones. Oil has been found in the West Shetland Basin but its extraction entails technical and economic problems.

On the British mainland small oil and subordinate gas accumulations have been known since 1919, chiefly in anticlines (cf. Falcon and Kent, 1960). The largest group is in the East Midlands where the reservoirs are principally in the sandstones of the upper Namurian and lower Westphalian. Small but increasing amounts have been derived in the south, especially near the Dorset coast, from Jurassic limestones and sandstones. Production in Dorset began in 1959 and Wytch Farm, south-east of Wareham, is the largest oilfield yet discovered on land in Britain.

Elsewhere the reservoirs are Magnesian Limestone (Zechstein) in north Yorkshire and Permo-Trias in west Lancashire. Gas was obtained at the end of the last century from a Carboniferous anticline (Cousland) east of the Midlothian Coalfield (p. 223) and again in the 1950s.

Most of these are very small, however, and it is the fields of the North Sea that have raised the United Kingdom to rank not far below the major world producers. The region conforms to the outline set out above – a gently subsiding basin of shallow water sediments, little altered and not much folded, here set at the edge of a continent; the main structures are belts of rifting (cf. Chapter 8 and Fig. 58). The individual structures governing the various fields are varied and often complex, but a number occur along rifted margins, for instance of the Central or Viking Graben, where a later cover is draped over the reservoir beds, which themselves may be bounded by faults or unconformities (e.g. Fig. 76(*a*)).

The main reservoir strata are as follows (some have appeared in earlier chapters):

1. Rotliegendes sands, Lower Permian – the main source of the gas wells of the southern North Sea, such as those off eastern England and the north-west Netherlands. The Argyll field, in Scottish waters, has some oil at this horizon.
2. 'Bunter' sandstones, or lower Trias – a lesser source exploited in the Hewett field, also for gas in the same region.
3. Sands of Rhaetic, Lias and ?Middle Jurassic age – several large northern oilfields in and near the Viking Graben, north-east of the Shetland Islands.
4. Oxfordian to lower Kimmeridgian sands – a few oilfields in British waters, east and north-east of Aberdeenshire (Piper and Montrose).
5. Over the Cretaceous–Tertiary boundary, in Maastrichtian and especially Danian limestones (p. 368) – a major group, including Ekofisk, on the Norwegian side of the Central Graben.
6. Palaeocene and lower Eocene sands, in the Forties oilfield east of Peterhead, Aberdeenshire, and in the Frigg gasfield (Norwegian) on the eastern edge of the Viking Graben.

In most cases the cap-rocks are impermeable shales and marls, but that over the Rotliegendes sands is mainly the thick Zechstein evaporites. The locations of many named fields appear on the maps mentioned at the end of the chapter and some details in Woodland (1975) and Kent (1975*b*).

In earlier years the Oil Shales in the Dinantian of Scotland (p. 189) were worked for oil, though the hydrocarbon was not liquid oil or petroleum but kerogen, which had to be distilled to produce oil. However the individual seams are thin and the method was expensive; the industry is now extinct in Britain. Smaller seams of oil shale are known in the Kimmeridge Clay but any direct exploitation is unlikely. Their chief role is as a possible source rock from which oil has migrated into Jurassic and later reservoirs.

The last organic resource is a little unexpected, but the very extensive peat deposits of central Ireland have been used to fuel power stations for several years and at present these provide 23% of the national electricity demands of Eire; peat here is thus a much more prolific source than coal.

Uranium ores cannot yet be said to constitute an economic resource in this country, but exploration in northern Scotland has detected small amounts that might be potentially economic in the future (Michie and Gallagher, 1979). The stratigraphical relation is also interesting. Thin carbonaceous, or more rarely phosphatic, beds occur in the Middle Old Red Sandstone of Caithness and occasionally uranium contents up to 2000 ppm have been detected in them, related to phosphate content. At the basin margin there are some enrichments in secondary uranium minerals where the Lower Old Red Sandstone overlies the Helmsdale Granite and similar minerals also occur in the granite itself, for instance in fractures or silicified zones. The primary source is taken to be magmatic and the secondary minerals tend to be precipitated in reducing conditions, which the undisturbed lake bottom waters would provide from time to time. There are similar occurrences in Orkney (Michie, 1972) near Stromness, where there are small outcrops of a pre-Devonian metamorphic basement. Relatively high uranium contents have been recorded from a fault breccia and also in the basal arenaceous facies overlying the schists.

Building stone, limestone and dolomite, refractories

Building stone is now much less in demand than concrete aggregate but in the past a great variety was employed locally for modest housing, the true quality stones being worth transporting (if possible by water) for ecclesiastical and state buildings or the houses of the wealthy. The Jurassic limestones were most in demand and in this context the outcrop has been called the limestone belt (Clifton-Taylor, 1972). Other more regional rocks include the lower Triassic sandstones, from the Midlands to north-west England, Scotland and Ireland, while the finer-grained Carboniferous sandstones have supplied many industrial areas. Certain dense, well bedded parts of the Magnesian Limestone had a wider repute and were used up and down eastern England and as far south as Westminster. Igneous rocks have been valued for their hardness or ornamental appearance but do not feature widely as a building stone except where they can be quarried nearby, for instance in Aberdeenshire or Wicklow.

Brick-clays are also widespread and include Pleistocene glacial clays and those from the Coal Measures. However in recent years this resource has narrowed to one chief producer – the lower part of the Oxford Clay, partly on technical and partly on economic grounds. The plastic clays have desirable qualities in their carbonate and organic contents, the latter being high enough to reduce the coal consumption for firing. Much of the stratigraphical and faunal research in the Oxford Clay has been based on the very large brick-pits.

Since fissile or cleaved rocks suitable for roofing are more restricted than building stones, slates have also been worth more extensive transport. Four examples may be given, from the past and the present. The uppermost Charnian rocks (the Swithland Slates) were much used in the Midlands and those of the Upper Devonian (e.g. from Delabole) in Cornwall. The green Ordovician tuffs of the Lake District are now employed as an ornamental stone as well as for roofing. In the nineteenth century, however, the country was flooded by the Lower Cambrian slates (p. 66) from North Wales; these are uniform and economical with the advantages of mass production.

The uses to which limestone may be put are enormous (cf. Hull and Thomas, 1974) – in the cement and chemical industries, for agricultural lime and as an aggregate. Among the thickest limestones the Lower Carboniferous maintains its pre-eminence as a source of lime and steel-making flux. It is quarried notably in the Derbyshire Block, adjacent to industrial regions and also to the limestone-lacking Midlands. The Great Scar Limestone and thicker Yoredale limestones have also been extracted, as have those of Wales, Bristol and the Mendips. The Yoredale facies also extends into Scotland, but despite useful properties the Cambro-Ordovician rocks of the North-West Highlands are too remote for major working. The softness of the Chalk precludes it from some uses, but much is quarried for cement, also for lime and in the production of whiting, etc.

Economically dolomite has much in common with limestone, but in addition it is used for making refractory bricks for furnace linings. The main source is the belt of Magnesian Limestone, parts of which are high grade rock, approaching pure dolomite in composition. There have also been smaller or older workings in the outcrops of Carboniferous Limestone, for instance in the South-West Province. In the field of refractories, ganister and fireclay occur as thin beds in most of the major coalfields, conveniently at hand to centres of industry.

Evaporitic minerals

Although these minerals often occur in the same sedimentary basins as carbonates and may be interbedded with them, they differ in being much more limited in distribution and, because of their solubilities, are for the most part extracted from depth. The sources of salt, gypsum, anhydrite and potash are almost all Permian and Trias (Chapter 8). Of these the great Cheshire salt basin (p. 256) is much the largest, though there are smaller workings or potential reserves in Staffordshire, Worcestershire, Somerset and north-east Ireland – all these being within the Keuper Marl facies. In north-east England the halite is a major part of the Zechstein cycles. The chemical industry is the major consumer and salt is normally brought to the surface as brine, but one mine in Cheshire supplies rock salt for winter use on the roads. All told, Britain is self-sufficient in salt.

Beds of gypsum and anhydrite are more restricted than salt, and anhydrite has largely been produced from one mine on Teesside and a lesser one in west Cumbria. Gypsum is more widely extracted and has multiple uses, especially for plaster and plaster board and in the making of Portland Cement. The most important sources are in the Trias near Nottingham, where there are both mines and quarries; this region also supplied the ornamental variant, alabaster. Lesser amounts have been derived from the Permian of the Vale of Eden, north-east Ireland and the Kingscourt outlier (p. 252). Purbeck gypsum beds are also worked in the central Weald (p. 308). At the time of writing there is only one commercial extraction of potash salts (sylvine, KCl), at Boulby on the coast of Cleveland; reserves appear to be large.

Evaporitic minerals introduce another theme for this chapter – the extent to which geological research has followed commercial interests, or perhaps that here the microscope follows the drill. Although alabaster has been worked for centuries and gypsum from the nineteenth century, very much less would have been known of the more soluble evaporites had not research been aided by deep drilling, largely by the chemical and oil industries. The detailed mineralogy, original and diagenetic (e.g. Stewart, 1949, 1951*a*, *b*; Armstrong *et al.* 1951), has depended on the borehole cores and from these stem deductions on the environment of deposition and the suggestion of sabkha conditions (Shearman, 1966; Smith, 1973). The results are of world-wide interest and the interchange between applied and pure geology has been especially productive.

Sedimentary ironstones

The Mesozoic bedded ironstones are partly analogous to evaporites in that both are syngenetic in a wide sense, formed originally at the same time as the accompanying sediments, though liable to later diagenesis. The iron-bearing minerals are chiefly siderite (ferrous carbonate), chamosite (hydrous aluminium silicate) and limonite (hydrous oxides). They may be formed in marine or freshwater conditions and all are low grade ores with 24–30% iron; pure siderite has a metal content of only 48.2% whereas hematite approaches 70%.

Ironstones of widely different ages have been worked at different times. Over the long pre-industrial period the main source was the clay ironstones (siderite nodules in clays) of the Wealden, especially the Wadhurst Clay (p. 324); the peak output – very small of course by any later standards – was in the sixteenth and seventeenth centuries, the time of the Sussex ironmasters. The very extensive Westphalian ores, which occur in all coalfields, only became valuable when smelting techniques developed to use coke from the neighbouring coal as a fuel; thenceforward the industry was a major contributor to the Industrial Revolution. Westphalian ores consist mainly of nodular clay ironstone (siderite-rich mudstone) but in Scotland and north Staffordshire there are also blackband ironstones; in these there is little clay or silt but a high organic content so that they are economical in smelting. The Scottish ores from the Limestone Coal Group and Productive (Westphalian) Measures had a substantial output in the latter part of the nineteenth century.

The rise of the Jurassic ores (Chapter 9) was even more pronounced, outstripping those of the Coal Measures in the middle of the nineteenth century (Hemingway, 1974*b*). In origin they are marine and in texture commonly oolitic. Siderite, chamosite and calcite are important in the matrix and the ooliths were originally of chamosite, locally altered to siderite or limonite. The workable formations occur along the main Jurassic outcrop from Cleveland to Oxfordshire, but with a good deal of variation (cf. pp. 283, 291); they are found in all three divisions of the Lias and also in the Aalenian Stage above. In recent years, in spite of cheap opencast working, the low grade of many seams has rendered even these substantial deposits seriously susceptible to competition from higher grade ore from abroad. Production, which reached 20 million tonnes per year in World War II, has now declined to 3.5 million tonnes.

Ironstones are not so inaccessible as the more soluble evaporite salts so that they have been analysed and assessed for a long time. Even here, however, economic stimulus and co-operation has enhanced research and more has been discovered about the complex mineralogy and diagenesis of the oolitic ores and the conditions under which they were formed (Hallimond, 1925; Taylor, 1949; Dunham, 1960; Hemingway, 1974*a*, *b*).

Epigenetic metalliferous ores

With the exception of these Mesozoic ironstones almost all the major metallic ore deposits in Britain are epigenetic, or discordant – that is, they were emplaced after the lithification or crystallization of the host rocks. Veins are the most common form but there are also more irregular or stratiform bodies. The ultimate sources are still much debated but deposition from hot (hydrothermal) mineralizing fluids, usually hypersaline brines, rising from deeper levels in the crust is generally the accepted vehicle. Some structural control of veins and oreshoots is seen in all orefields, in response to the contemporaneous stresses, and the orebodies are usually associated with faults or occasionally with dykes. The deposits also demonstrate varying degrees of stratigraphical control. References are numerous but recent compilations will serve as a starting point (Bowie *et al.* 1978; Dunham, 1978).

Metallogenic provinces are restricted in time and space and though in the world as a whole Precambrian rocks, especially of the major shield regions, have provided very large metallic resources this does not apply to the British Isles. The Caledonides have also been productive only locally while the great majority of Caledonian granites lack associated hydrothermal deposits. So far the metamorphic, or orthotectonic, belt has produced only minor ores, though recently a more promising barytes deposit has been found in Middle Dalradian pelites and turbidites in Perthshire. On the other hand the southern, or paratectonic, belt includes some major orebodies, particularly of copper.

Of these, two are outstanding, on either side of the Irish Sea. The copper–pyrite deposit of Avoca in Wicklow has long been known, but is the only one in current production. The lenticular mineralized zones lie within Ordovician (possibly Caradoc) slates and acid lavas and tuffs, parallel to the cleavage and all are affected by Caledonian isoclinal folding. The very rich copper ore on Parys Mountain, Anglesey, is somewhat similar but though there has been extensive prospecting it has not been worked for many years. It is associated with cleaved and sheared Ordovician and Silurian sediments, probably forming a recumbent syncline, and also a highly silicified igneous rock, previously called 'felsite' but which is probably an altered tuff. An additional check on the age here is provided by pebbles in the basal Carboniferous conglomerate.

There are other minor emplacements of copper, gold, lead and zinc in Central and North Wales and a more varied suite in the Lake District, for instance at Carrock Fell and around Keswick. Although the country rock of the Leadhills lead–zinc field in the Southern Uplands is Ordovician, structural and isotopic evidence suggest a Carboniferous or Permian age; it may be analogous to the major Pennine orefields but in a different stratigraphical setting.

Since only trivial mineralization accompanied the Tertiary vulcanism there remains the largest metallogenic province in the British Isles, which might be called Hercynian but with the proviso that isotopic data show that in many regions mineralization continued into the Permian and may have been rejuvenated in a small way later. The only important post-Triassic ores are the epigenetic hematites.

In geological setting and range of minerals there is a marked distinction in this 'Hercynian' group between the Cornubian metalliferous field and the various lead–zinc fields in the more stable blocks farther north. The latter are linked together not only in their dominant ores but also in their host rocks, which are commonly at some level in Carboniferous limestones or sandstones. In a stratigraphical context the Dinantian is thus the most widespread ore-bearing division in the country. The associated minerals include barite, fluorite and more rarely chalcopyrite. The first has been worked sporadically in many areas and several substantial mines continue. The British Isles are unique in the world for their output of witherite; fluorspar is a major product in Derbyshire and County Durham, and most of the country's lead production at present is a by-product of working for this mineral.

The individual orefields (cf. Dunham, 1978, pp. 142–3) include Derbyshire, the northern Pennines (Alston and Askrigg blocks), Halkyn–Minera of North Wales and smaller areas such as the limestones of the Mendips and South Wales. In many there is physical and chemical evidence of stratigraphical control in the distribution of oreshoots. Thus in the northern Pennines a high proportion occur in the Great or Main Limestone, some in other Viséan or Namurian limestones, and in a restricted area the top of the Great Scar Limestone and the base of the coarser Millstone Grit – all these in contrast to the much more impermeable shales in between. In Derbyshire the mineralization is dominantly in the Viséan limestones; very few veins penetrate the Namurian shales above while the several layers of lavas and tuffs are also relatively impermeable. However on the Alston Block the Whin Sill forms an effective wallrock.

As a result of these controls the oreshoots tend to be laterally extensive but shallow and they exhibit a lateral zoning, under temperature control.

The highest temperature zones (in the approximate range 100–200°C) carry fluorite, centrally on the Alston Block and on the eastern margin of the Askrigg Block and in Derbyshire; the other dominant minerals – barite, witherite and calcite – are ranged around or to the west.

The great spread of Carboniferous Limestone in central Ireland has also been the site of important lead–zinc mineralization but the thick drift cover is more hampering here than in the limestone uplands of England and Wales. Accordingly while some deposits have long been known there have also been some new major discoveries. Three examples are selected (cf. Williams and McArdle, 1978).

Silvermines Mountain, Tipperary, is formed of Old Red Sandstone with a Lower Palaeozoic inlier, and is faulted on the north side against Lower Carboniferous limestones (Tournaisian). The several orebodies include veins (Taylor and Andrew, 1978) but the most substantial are stratiform 'zones' in dolomitic breccias. The combination is especially interesting because here there is persuasive evidence of synsedimentary deposition; the hydrothermal fluids discharged into the unconsolidated bottom sediments to form major stratiform sulphides, the epigenetic oreshoots also being formed in feeder-channels. Major deposits of barytes are found nearby in a similar stratigraphical situation and have been worked opencast for several years.

Tynagh (Galway) and Navan (Meath, Fig. 31) have some similar features and some differences. In the former mineralization is also connected with faulting between Old Red Sandstone and Tournaisian limestones. The ores are mainly in reef limestones adjacent to the fault and bounded above and below by more shaly beds. The Navan mine is the newest, and very large, discovery and the accounts are only preliminary. Again the orebodies are mainly in reef facies, especially dolomitized micrite, not far above the base of the Tournaisian limestones. Sphalerite is dominant over galena and the zinc output likely to be large. Both deposits display synsedimentary and discordant features; mineralization may possibly have preceded the Viséan.

Before leaving the epigenetic ores in limestones, mention should be made of the iron ore, hematite. This has been worked in small quantities in Wales and, in hydrated form, in the Mendips and Forest of Dean. The main province is the Dinantian rim to the Lake District, in south and west Cumbria, where the limestones are overlain by Permian breccias or 'Bunter Sandstone'. Although only one mine remains large quantities of ore, valuable for its high grade, were obtained in the past. The source of the mineralizing brines is not very clear, but was possibly in the adjacent or overlying Permo-Triassic sandstones, where these are downfaulted in the Irish Sea basin (Rose and Dunham, 1977, p. 91).

In Chapters 6 and 7 it was seen how the Cornubian peninsula of South-West England stands apart from the rest of the country, in its stratigraphy, tectonics and major granite batholith. There is a similar distinction in mineralization, to make it by far the greatest metallogenic province in the British Isles (Dunham *et al.* 1978, p. 292). The host of igneous–hydrothermal minerals are directly related to the batholith, spatially and probably in origin, especially the ores of tin, copper, lead and iron, with tungsten and arsenic as accessories. Although there is a history ranging back to pre-Roman times, only a few mines operate at present, all based on tin.

The ores were emplaced in successive zones, mainly during the late stages of granite crystallization and controlled by temperature gradients. The details of mineral distribution and ore-bearing veins (the 'lodes' of Cornwall) are extremely complex, but the intensity of mineralization is related not so much to the exposed bosses, or cupolas, of the granite but to the form of the batholith itself. There is thus the greatest concentration over the centre, while veins in the Scilly Isles are thin and poor and Dartmoor is not highly mineralized. In some veins changes are found in the dominant ores according to the wallrock – granite, metamorphic aureole, slate or sandstone – but there is no regular pattern of lithological control.

An important secondary product in the region is china clay, or kaolin, derived from the feldspars of the granite by late hydrothermal alteration. Economically it is by far the most important product in the south-west at present and has a substantial export value. Isotopic dates from uranium minerals range from about 300 million years (? the age of the intrusion) to 45 million years – perhaps relating to several later pulses of mineralization or the remobilization of earlier deposits.

Superficial deposits, aggregate, groundwater

Compared with metalliferous mining or North Sea oil the quarrying of bulk constructional resources may appear prosaic, but these materials are of increasing importance and value: the tonnage extracted of sand and gravel is second only to coal. In addition to the other building components already mentioned, aggregate, for roadstone and concrete, is outstanding. As with limestone, transport costs from quarry to market are an important factor so that the most profitable sites of extraction are near centres of population and industry. In a geological context there is a very large range of source rocks.

Crushed rock aggregate may be made from almost any hard rock that is suitably placed, but over Britain as a whole limestone, igneous rocks and sandstone predominate. Only a few examples can be quoted here. Scotland is well supplied with igneous rocks and also with gravel; from northern England to Derbyshire there are limestones, the harder Carboniferous sandstones and lesser but useful igneous rocks such as the Whin Sill and Cumbrian intrusions. Similarly the Pennant Sandstone (p. 220) is employed in South Wales. In the Midlands, where soft Triassic rocks are dominant, even small outcrops have been used, such as the Lower Cambrian Hartshill Quartzite and the Mountsorrel Granite. Silurian greywackes are not ideal but are abundant and thus convenient in Wales and the Southern Uplands. In Northern Ireland the Antrim Basalts are a major source.

In the east and south of England there have been very large extractions of sand and gravel, for aggregate and other constructional uses. They are mainly in the Thames valley and tributaries but some also in the Trent gravels. Comparable but more variable resources are worked in the fluvio-glacial deposits of northern England, lowland Scotland and Ireland. Little sand and gravel is derived from older formations but some comes from the Bunter Pebble Beds in the Midlands.

As industrial and domestic demands on water supplies have increased groundwater has risen to be a major geological resource. Among the aquifer formations the Chalk ranks first, the water being stored not in the rock pores, which are minute, but in fissures; the Upper Chalk is the most productive but all three divisions will hold great quantities of water. In second place come the lower Triassic sandstones and pebble beds (p. 255) including the Waterstones – formerly Keuper Sandstones – a valuable resource in the Midlands and as far afield as Belfast. Smaller or more irregular groundwater supplies are derived from the Cretaceous greensands, Carboniferous sandstones and limestones and various sandy and calcareous beds along the Jurassic outcrop. Superficial sands and gravels yield water widely in restricted amounts, but they are more liable to pollution.

In the wetter northern and western parts of the British Isles surface supplies and reservoirs are normally adequate, but it is a very fortunate geological accident that the greatest of our aquifers crops out in the driest south-eastern region.

Borehole data

In simple stratigraphical terms the most voluminous results of economic exploration have come from the deep boreholes, and that at a remarkable rate. Allowing for short periods when results are confidential, the broad outline of sea-floor geology around the British Isles has taken only about 25 years to compile – and in several crucial regions there is much detail as well. The much smaller land areas took perhaps 150 years, though admittedly parts of them are more complex. In addition to offshore geology virtually all that we know accurately of the deeper structure of eastern and southern Britain has come from boreholes. These have been mainly the work of industry but some have been drilled by government departments or university research teams.

Logging borehole cores brings in other specialists, in petrography and particularly micropalaeontology; conversely the sequence of microfossils so established can help in working out the stratigraphy of surface exposures, especially where the standard zonal indices are sparse. The study of spores, for instance, has been applied to the stratigraphy of continental Devonian and Triassic beds, though the science was first developed for economic purposes in the Coal Measures. Tertiary micropalaeontology does not feature very widely in this country but is an essential tool in many parts of the world, originating largely from the needs of oil geology.

It is a pity that the older and more deformed rocks of northern and western Britain are unlikely to be drilled to such depths but two research projects may be quoted because of the celebrated results. The first is a classic case of drilling that proved deductions already made by geophysical surveys – the discovery of Caledonian granite beneath the northern Pennines (Dunham *et al.* 1965) where marked negative gravity anomalies had been mapped. The other is in some degree the converse. It was thought that the Mochras borehole of North Wales (p. 310) might penetrate Ordovician, Carboniferous or Trias, but what turned up was in fact Lower Jurassic – the thickest in the British Isles (Woodland, 1971), so that maps were redrawn and palaeogeographic ideas revised.

In the first edition of this book I picked out a few themes that seemed likely to advance in succeeding years. They included more knowledge of modern sediments and their processes, which could then be applied to sediments of the past; much more known of Precambrian rocks and history and the age-relations of very ancient assemblages; more discovered about rocks at depth, in the continents and below the seas. Then although the term plate tectonics was only just emerging, some form of continental drift appeared to be a promising hypothesis and made better sense of many stratigraphical relationships than permanent immovable continents and oceans. Some 15 years later these modest speculations may well appear rather obvious, though perhaps the surveys of the ocean floor and continental shelf might have been more emphasized, with their various aspects of international co-operation and economic stimulus.

It is on a similar note that I bring this edition to a close. The patient reader who has followed the story to its end will need no popular cry of an energy crisis (or any 'crisis' in raw materials) to remind him that this is a small planet and its terrestrial resources are finite. The British Isles are only a small area of that planet's surface but, partly through good fortune and the hard work of earlier generations, they can illustrate many stratigraphical and geological principles. One may hope that these will lead not only into the fascinating – and quite necessary – fields of theoretical research but also into practical and economic applications.

REFERENCES

Those quoted in the text appear in lists A or B. The following give statistics of mineral resources, published annually: *Annual Mining Review* (international in scope) and *United Kingdom Mineral Statistics*, prepared by the Institute of Geological Sciences (H.M.S.O., London).

There are also two maps published annually: *Maps showing oil and gas activities and concessions*: (1) *North Sea and European continental shelf*, (2) *Ireland and Western Approaches*; Offshore Promotional Services Ltd, Maidenhead, Berkshire.

Appendix

STRATIGRAPHICAL AND ZONAL TABLES

Stratigraphical tables are given for those systems where a representative fossiliferous sequence is available in the British Isles, with short notes and references on the remainder. In compiling the tables, I have followed principally the *Special Reports* on correlation published by the Geological Society of London, listed on p. 430. However parts of them are highly specialized, especially where modern micropalaeontology is employed, and some abridgements have been made. There are also minor differences of policy among the authors, for instance concerning the retention of many traditional terms or a more radical institution of new nomenclature. *Report* No. 11 (Holland *et al.* 1978) is a guide to the procedure and illustrates problems of conformity and principles.

An earlier, and rather simpler, set of tables was issued in *British Palaeozoic Fossils* and *British Mesozoic Fossils* by the British Museum (Natural History). They were followed in the first edition of this book and still have a use for the non-specialist reader or when comparing works published before the *Reports* were available.

Some of the succeeding tables include traditional lithological terms as well as the formal series and stages; however such terms were usually based on different criteria and it must be emphasized that in many cases the equivalence is only approximate.

QUARTERNARY

In the context of British stratigraphy Quaternary correlation is illustrated by Table 53 (pp. 384–5, cf. Mitchell *et al.* 1973) and the methods outlined in the text; normal biostratigraphy is rarely possible. Apart from the Devensian there is some difficulty in correlating the European stages even between Britain and the Netherlands. In a wider view more is likely to depend on the cold and warm periods reflected in deep-sea cores or in sequences of non-glacial sediments. In the deep-sea cores planktonic foraminifera can give direct information of surface temperatures while the ratios of the oxygen isotopes ^{16}O and ^{18}O in the tests of benthonic foraminifera are known to reflect the growth and melting of polar ice-caps. Bowen (1978) also deals with Quaternary classification.

TERTIARY

With the exception of parts of the Palaeogene (cf. Table 52), Tertiary successions are not well represented in the British Isles or fully known from the sea areas. There are also problems in the definition of an agreed table of European stages and in their application to Britain. The most complete series of zones are based on the microfossils, particularly the planktonic foraminifera and nannoplankton. The reader is referred to the tables and discussion in Curry *et al.* (1978), which also includes the Tertiary Volcanic Province and its dating. In other publications the following stages, in upward order, have been applied to British strata: Palaeocene – Danian or Montian, Thanetian, Sparnacian; Eocene – Ypresian, Lutetian, Bartonian; Oligocene (lower part) – Lattorfian.

CRETACEOUS SYSTEM

Stages	Zones	Some English lithological divisions
	UPPER CRETACEOUS	
(UPPER MAASTRICHTIAN)		
LOWER MAASTRICHTIAN	*Belemnella occidentalis* *Belemnella lanceolata*	
SENONIAN — CAMPANIAN	*Belemnitella mucronata* *Gonioteuthis quadrata* *Offaster pilula*	UPPER CHALK
SENONIAN — SANTONIAN	*Marsupites testudinarius* *Uintracrinus socialis* *Micraster coranguinum*	
SENONIAN — CONIACIAN	*Micraster cortestudinarium*	
TURONIAN	*Holaster planus* *Terebratulina lata* *Inoceramus labiatus*	MIDDLE CHALK
CENOMANIAN	*Sciponoceras gracile* *Calycoceras naviculare* *Acanthoceras rhotomagense* *Mantelliceras mantelli*	LOWER CHALK
	LOWER CRETACEOUS	
ALBIAN	*Stoliczkaia dispar* *Mortoniceras inflatum* *Euhoplites lautus* *Euhoplites loricatus* *Hoplites dentatus*	GAULT
APTIAN	*Douvilleiceras mammillatum* *Leymeriella tardefurcata* *Hypacanthoplites jacobi* *Parahoplites nutfieldiensis* *Cheloniceras martinioides* *Tropaeum bowerbanki* *Deshayesites deshayesi* *Deshayesites forbesi* *Prodeshayesites fissicostatus*	LOWER GREENSAND [of southern England]

Stages	Zones	Some English lithological divisions
BARREMIAN	*Parancyloceras bidentatum* *Hemicrioceras rude* *Hoplocrioceras fissicostatum* *Paracrioceras rarocinctum* *Simbirskites variabilis*	
HAUTERIVIAN	*Simbirskites marginatus* *Simbirskites gottschei* *Simbirskites speetonensis* *Simbirskites inversus* *Endemoceras regale* *Endemoceras noricum* *Endemoceras amblygonium*	[Stages present in marine facies of eastern England, see text]
VALANGINIAN	[unnamed zone] *Dicostella pitrei* *Dichotomites* spp. *Polyptychites* spp. *Paratollia* spp.	
RYAZANIAN	*Peregrinoceras albidum* *Surites stenomphalus* *Surites icenii* *Hectoroceras kochi* *Runctonia runctoni*	

Largely from Rawson *et al.* (1978). The zonal indices of the Lower Cretaceous are all ammonites; those of the Upper Cretaceous are drawn from several phyla. None of the stages is based on a British locality (or stratotype), several being French, and there are problems of definition between different countries and also between the Boreal and Tethyan provinces. The Coniacian Stage is particularly controversial; the base of the *M. coranguinum* Zone may well lie within it.

Appendix

JURASSIC SYSTEM

Stages	Zones	Some English lithological divisions
PORTLANDIAN	*Subcraspedites lamplughi* *Subcraspedites preplicomphalus* *Subcraspedites primitivus* *? Titanites oppressus*	[Marine beds in Eastern England]
	Titanites anguiformis *Galbanites kerberus* *Galbanites okusensis* *Glaucolithites glaucolithus* *Progalbanites albani*	PORTLAND BEDS
KIMMERIDGIAN	*Virgatopavlovia fittoni* *Pavlovia rotunda* *Pavlovia pallasioides* *Pectinatites pectinatus* *Pectinatites hudlestoni* *Pectinatites wheatleyensis* *Pectinatites scitulus* *Pectinatites elegans* *Aulacostephanus autissiodorensis* *Aulacostephanus eudoxus* *Aulacostephanus mutabilis* *Rasenia cymodoce* *Pictonia baylei*	KIMMERIDGE CLAY
OXFORDIAN	*Amoeboceras rosenkrantzi* *Amoeboceras regulare* *Amoeboceras serratum* *Amoeboceras glosense* *Cardioceras tenuiserratum* *Cardioceras densiplicatum* *Cardioceras cordatum* *Quenstedtoceras mariae*	CORALLIAN BEDS AMPTHILL CLAY, etc.
CALLOVIAN	*Quenstedtoceras lamberti* *Peltoceras athleta* *Erymnoceras coronatum* *Kosmoceras jason*	OXFORD CLAY, etc.
	Sigaloceras calloviense	KELLAWAYS BEDS
	Macrocephalites macrocephalus	UPPER CORNBRASH

Stages	Zones	Some English lithological divisions
BATHONIAN	*Clydoniceras discus*	LOWER CORNBRASH
	Oppelia aspidoides	GREAT OOLITE
	Procerites hodsoni	
	Morrisiceras morrisi	
	Tulites subcontractus	
	Procerites progracilis	
	Asphinctites tenuiplicatus	
	Zigzagiceras zigzag	
BAJOCIAN	*Parkinsonia parkinsoni*	INFERIOR OOLITE
	Strenoceras garantiana	
	Strenoceras subfurcatum	
	Stephanoceras humphriesianum	
	Emileia sauzei	
	Witchellia laeviuscula	
	Hyperlioceras discites	
AALENIAN	*Graphoceras concavum*	
	Ludwigia murchisonae	
	Leioceras opalinum	
TOARCIAN — YEOVILIAN	*Dumortieria levesquei*	UPPER LIAS
	Grammoceras thouarsense	
	Haugia variabilis	
TOARCIAN — WHITBIAN	*Hildoceras bifrons*	
	Harpoceras falciferum	
	Dactylioceras tenuicostatum	
PLIENSBACHIAN	*Pleuroceras spinatum*	MIDDLE LIAS
	Amaltheus margaritatus	
	Prodactylioceras davoei	LOWER LIAS
	Tragophylloceras ibex	
	Uptonia jamesoni	
SINEMURIAN	*Echioceras raricostatum*	
	Oxynoticeras oxynotum	
	Asteroceras obtusum	
	Caenisites turneri	
	Arnioceras 'semicostatum'	
	Arietites bucklandi	
HETTANGIAN	*Schlotheimia angulatum*	
	Alsatites liasicus	
	Psiloceras planorbis	

From Cope *et al.* (1980*a, b*). All the zonal indices are ammonites. The uppermost four zones of the Portlandian Stage are based on the sequence in eastern England and hence belong to the Boreal Province. Earlier tables included the Pre-planorbis Beds which are now relegated to the Trias and in some the Callovian Stage was included in the Upper Jurassic. The Volgian Stage of the Russian sequence is equivalent to the Portlandian and the upper part of the Kimmeridgian Stage.

PERMIAN AND TRIAS

The Permian standard succession and stages are derived from the Russian platform and Ural Mountains and those of the Trias chiefly from the eastern Alps. In the continental facies of Britain fossils are sporadic, rare or absent and though some have stratigraphical value (especially the plant spores of the Trias, allowing some correlation with Germany), no standard zonal scheme can be applied in Britain and no table is included here. The references are: Permian – Smith *et al.* (1974); Trias – Warrington *et al.* (1980).

SILESIAN SUBSYSTEM

Series	Stages*	Indices and zones		Bivalves
			Goniatites	
WESTPHALIAN	WESTPHALIAN D	'A' *'Anthracoceras'*		*Anthraconaia prolifera* *Anthraconauta tenuis*
	WESTPHALIAN C			*Anthraconauta phillipsii*
	WESTPHALIAN B			*Anthracosia similis* and *Anthraconaia pulchra* {Upper / Lower}
				Anthraconaia modiolaris
	WESTPHALIAN A			*Carbonicola communis*
		G_2	*Gastrioceras listeri* *Gastrioceras subcrenatum*	*Carbonicola lenisulcata*
NAMURIAN	YEADONIAN	G_1	*Gastrioceras cumbriense* *Gastrioceras cancellatum*	
	MARSDENIAN	R_2	*Reticuloceras superbilingue* *Reticuloceras bilingue* *Reticuloceras gracile*	
	KINDERSCOUTIAN	R_1	*Reticuloceras reticulatum* *Reticuloceras nodosum* *Reticuloceras circumplicatile*	
	ALPORTIAN	H_2	*Homoceratoides prereticulatus* *Homoceras undulatum* *Hudsonoceras proteus*	
	CHOKIERIAN	H_1	*Homoceras beyrichianum* *Homoceras subglobosum*	
	ARNSBERGIAN	E_2	*Nuculoceras nuculum* *Cravenoceratoides nitidus* *Eumorphoceras bisulcatum*	
	PENDLEIAN	E_1	*Cravenoceras malhamense* *Eumorphoceras pseudobilingue* *Cravenoceras leion*	

DINANTIAN SUBSYSTEM

Series	Stages	Corals and Brachiopods		Goniatites Indices and zones
VISÉAN	BRIGANTIAN	D_2	P_2	*Lyrogoniatites georgiensis* *Lusitanites subcirculare* *Goniatites granosus*
			P_1	*Goniatites koboldi* *Goniatites elegans* *Goniatites falcatus* *Goniatites crenistria*
	ASBIAN	D_1	B_2 B_1	*Beyrichoceras*
	HOLKERIAN	S_2		
	ARUNDIAN	(S_1C_2)		
	CHADIAN	(S_1C_2) (C_1)		*Pericyclus*
TOURNAISIAN	IVORIAN	(C_1)		
	HASTARIAN	(Z) (K)		*Gattendorfia*

Adapted from George *et al.* (1976), Ramsbottom and Mitchell (1980) and other sources. Below the *Beyrichoceras* zones the goniatite sequence comes mainly from Ireland. The coral–brachiopod indices refer to the genera *Cleistopora, Zaphrentis, Caninia, Seminula, Dibunophyllum*; below S_2 this zonal sequence has been variously interpreted and is not really satisfactory. Courceyan was an earlier British Stage, approximately equivalent to Tournaisian.

SILESIAN SUBSYSTEM

It is questionable whether the uppermost stage, Stephanian, is represented in Great Britain.

The table is abridged from Ramsbottom *et al.* (1978, Pl. 1); additional zonal groups are given there (spores, plants and conodonts), also additional stratigraphical divisions and marker bands.

DEVONIAN SYSTEM

Stages, marine		Ammonoids	Correlation	Stages recognized in the Welsh Borders	Old Red Sandstone	Continental facies with vertebrates
UPPER	FAMENNIAN	*Wocklumeria* *Clymenia* *Platyclymenia* *Cheiloceras*	Some broad correlation	FARLOVIAN	UPPER OLD RED SANDSTONE	Nairn region: 4 faunas recognized of psammosteids and other fishes and placoderms
	FRASNIAN	*Manticoceras*				
MIDDLE	GIVETIAN	*Maenioceras*		(no faunas recorded)	MIDDLE OLD RED SANDSTONE	Orcadian region: 5 faunas based mainly on coccosteids with other fishes
	EIFELIAN	*Anarcestes*				
LOWER	EMSIAN	*Anetoceras*		BRECONIAN	LOWER OLD RED SANDSTONE	*Rhinopteraspis cornubica*
	SIEGENIAN	(no ammonoids)	Little agreed correlation			*Althaspis leachi*
				DITTONIAN		*Pteraspis crouchi*
	GEDINNIAN	(*Monograptus uniformis*)		DOWNTONIAN		*Pteraspis leathensis* *Traquairaspis* zones *Hemicyclaspis* *Cyathaspis*

The marine stages and ammonoids are taken from House (1979, p. 268) with slight abridgement. The 'zones' of the Lower Old Red Sandstone can be used only in parts of the Anglo-Welsh region; the Downtonian–Dittonian boundary is debated. The Nairn faunas are adapted from Miles (1968, pp. 4–11), the Orcadian from Mykura (1976, p. 54). For general correlation see House *et al.* (1977), for vertebrate ranges Westoll (1979).

SILURIAN SYSTEM

Series	Stages	Zones
DOWNTON	(No stages yet defined)	
LUDLOW	WHITCLIFFIAN	[Higher graptolite zones not yet recorded in the British Isles] *Bohemograptus*
	LEINTWARDINIAN	*Saetograptus leintwardinensis*
	BRINGEWOODIAN	*Pristiograptus tumescens*
	ELTONIAN	*Cucullograptus scanicus* *Neodiversograptus nilssoni*
WENLOCK	HOMERIAN	*Pristiograptus ludensis* *Gothograptus nassa* *Cyrtograptus lundgreni*
	SHEINWOODIAN	*Cyrtograptus ellesae* *Cyrtograptus linnarssoni* *Cyrtograptus rigidus* *Monograptus riccartonensis* *Cyrtograptus murchisoni* *Cyrtograptus centrifugus*
LLANDOVERY	TELYCHIAN	*Monoclimacis crenulata* *Monoclimacis griestoniensis* *Monograptus crispus*
	FRONIAN	*Monograptus turriculatus* *Monograptus sedgwickii*
	IDWIDIAN	*Monograptus convolutus* *Pribylograptus leptotheca* *Diplograptus magnus* *Monograptus triangulatus*
	RHUDDANIAN	*Coronograptus cyphus* *Lagarograptus acinaces* *Atavograptus atavus* *Orthograptus acuminatus* *Glyptograptus persculptus*

All the zonal indices are graptolites, compiled largely from Bulman (1970, p. V101), Cocks *et al.* (1971) and Rickards (1976). However in the Ludlow Series a '*Bohemograptus* Zone' is added on the recommendation of Dr R. B. Rickards, and the four stages are defined principally on shelly faunas; their correspondence with the zones is therefore only approximate. The Downton Series is mainly in an Old Red Sandstone facies.

ORDOVICIAN SYSTEM

Series	Stages	Zones
ASHGILL	HIRNANTIAN	
	RAWTHEYAN	*Dicellograptus anceps*
	CAUTLEYAN	
	PUSGILLIAN	*Dicellograptus complanatus*
		Pleurograptus linearis
CARADOC	ONNIAN	
	ACTONIAN	*Dicranograptus clingani*
	MARSHBROOKIAN	
	LONGVILLIAN	
	SOUDLEYAN	*Climacograptus wilsoni*
	HARNAGIAN	*Climacograptus peltifer*
	COSTONIAN	
LLANDEILO	(upper)	*Nemagraptus gracilis*
	(middle)	
	(lower)	*Glyptograptus teretiusculus*
LLANVIRN	(upper)	*Didymograptus murchisoni*
	(lower)	*Didymograptus bifidus*
ARENIG	(upper)	*Didymograptus hirundo*
	(lower)	*Didymograptus extensus*

The zonal indices are graptolites but the Caradoc and Ashgill stages are based primarily on shelly faunas. From Williams *et al.* (1972).

CAMBRIAN SYSTEM

	Series	Zones
UPPER CAMBRIAN	Tremadoc Series	*Angelina sedgwickii* *Shumardia pusilla* *Clonograptus tenellus* *Dictyonema flabelliforme*
	Merioneth Series	*Acerocare* *Peltura*, etc. *Leptoplastus* *Parabolina spinulosa* *Olenus* *Agnostus pisiformis*
MIDDLE CAMBRIAN	St David's Series	*Paradoxides forchhammeri* *Paradoxides paradoxissimus* *Paradoxides oelandicus*
LOWER CAMBRIAN	Comley Series	Protolenid–Strenuellid Olenellid Non-trilobite zone

The table is based on the Anglo-Welsh-Scandinavian (or Acado-Baltic) faunas and does not apply to north-west Scotland; the Solva and Menevian groups of South Wales lie within the St David's Series. From Cowie *et al*. (1972).

BIBLIOGRAPHY

References are arranged under two headings. List A contains books, major compilations and regional publications; list B consists principally of individual works with a few more marginal in content or older, including the sources of certain figures.

LIST A

Special Reports published by the Geological Society of London:

No. 1 COCKS, L. R. M., HOLLAND, C. H., RICKARDS, R. B. and STRACHAN, I. 1971. A correlation of Silurian rocks in the British Isles. *Jl geol. Soc. Lond.* **127**, 103–36 (also issued separately).

No. 2 COWIE, J. W., RUSHTON, A. W. A. and STUBBLEFIELD, C. J. 1972. *A correlation of Cambrian rocks in the British Isles.* 42 pp.

No. 3 WILLIAMS, A., STRACHAN, I., BASSETT, D. A., DEAN, W. T., INGHAM, J. K., WRIGHT, A. D. and WHITTINGTON, H. B. 1972. *A correlation of Ordovician rocks in the British Isles.* 74 pp.

No. 4 MITCHELL, G. F., PENNY, L. F., SHOTTON, F. W. and WEST, R. G. 1973. *A correlation of Quaternary rocks in the British Isles.* 99 pp.

No. 5 SMITH, D. B., BRUNSTROM, R. G. W., MANNING, P. I., SIMPSON, S. and SHOTTON, F. W. 1974. *A correlation of Permian rocks in the British Isles.* 45 pp.

No. 6 HARRIS, A. L., SHACKLETON, R. M., WATSON, J., DOWNIE, C., HARLAND, W. B. and MOORBATH, S. (Editors). 1975. *A correlation of Precambrian rocks in the British Isles.* 136 pp.

No. 7 GEORGE, T. N., JOHNSON, G. A. L., MITCHELL, M., PRENTICE, J. E., RAMSBOTTOM, W. H. C., SEVASTOPULO, G. D. and WILSON, R. B. 1976. *A correlation of Dinantian rocks in the British Isles.* 87 pp.

No. 8 HOUSE, M. R., RICHARDSON, J. B., CHALONER, W. G., ALLEN, J. R. L., HOLLAND, C. H. and WESTOLL, T. S. 1977. *A correlation of Devonian rocks in the British Isles.* 110 pp.

No. 9 RAWSON, P. F., CURRY, D., DILLEY, F. C., HANCOCK, J. M., KENNEDY, W. J., NEALE, J. W., WOOD, C. J. and WORSSAM, B. C. 1978. *A correlation of Cretaceous rocks in the British Isles.* 70 pp.

No. 10 RAMSBOTTOM, W. H. C., CALVER, M. A., EAGAR, R. M. C., HODSON, F., HOLLIDAY, D. W., STUBBLEFIELD, C. J. and WILSON, R. B. 1978. *A correlation of Silesian rocks in the British Isles.* 81 pp.

No. 11 HOLLAND, C. H., AUDLEY-CHARLES, M. G., BASSETT, M. G., COWIE, J. W., CURRY, D., FITCH, F. J., HANCOCK, J. M., HOUSE, M. R., INGHAM,

J. K., KENT, P. E., MORTON, N., RAMSBOTTOM, W. H. C., RAWSON, P.
F., SMITH, D. B., STUBBLEFIELD, C. J., TORRENS, H. S., WALLACE, P.
and WOODLAND, A. W. 1978. *A guide to stratigraphical procedure.* 18 pp.

No. 12 CURRY, D., ADAMS, C. G., BOULTER, M. C., DILLEY, F. C., EAMES, F.
E., FUNNELL, B. M. and WELLS, M. K. 1978. *A correlation of Tertiary
rocks in the British Isles.* 72 pp.

No. 13 WARRINGTON, G., AUDLEY-CHARLES, M. G., ELLIOTT, R. E., EVANS, W.
B., IVIMEY-COOK, H. C., KENT, P., ROBINSON, P. L., SHOTTON, F. W.
and TAYLOR, F. M. 1980. *A correlation of Triassic rocks in the British Isles.*
78 pp.

No. 14 COPE, J. C. W., GETTY, T. A., HOWARTH, M. K., MORTON, N. and
TORRENS, H. S. 1980*a*. *A correlation of Jurassic rocks in the British Isles,
Part 1: Introduction and Lower Jurassic.* 73 pp.

No. 15 COPE, J. C. W., DUFF, K. L., PARSONS, C. F., TORRENS, H. S.,
WIMBLEDON, W. A. and WRIGHT, J. K. 1980*b*. *A correlation of Jurassic
rocks in the British Isles, Part 2: Middle and Upper Jurassic.* In press.

These reports, with extensive references, give useful tables of strata and
correlative methods but only touch incidentally on other aspects, such as
sedimentology, tectonics or magmatism. No. 6 on the Precambrian is rather wider
in scope.

British Regional Geology (Institute of Geological Sciences, H.M.S.O.). These
works cover the United Kingdom, the older issues being:
England and Wales: Bristol and Gloucester District, Kellaway and Welch (1948);
East Anglia, Chatwin (1960); East Yorkshire and Lincolnshire, Wilson (1948);
The Hampshire Basin, Chatwin (1960); London and the Thames Valley,
Sherlock (1960); North Wales, George (1961); Pennines and Adjacent Areas,
Edwards and Trotter (1954).
Scotland: The Midland Valley of Scotland, MacGregor and MacGregor (1948); The
Northern Highlands, Phemister (1960); The Tertiary Volcanic Districts, Mac-
Gregor and Anderson (1961).
More recent issues are:
HAINS, B. A. and HORTON, A. 1969. Central England.
TAYLOR, B. J., BURGESS, I. C., LAND, D. H., MILLS, D. A. C., SMITH, D. B. and
WARREN, P. T. 1971. Northern England.
GEORGE, T. N. 1970. South Wales.
EDMONDS, E. A., McKEOWN, M. C. and WILLIAMS, M. 1969. South-West
England.
GALLOIS, R. W. 1965. The Wealden District.
EARP, J. R. and HAINS, B. A. 1971. The Welsh Borderland.
JOHNSTONE, G. S. 1966. The Grampian Highlands.
MYKURA, W. 1976. Orkney and Shetland.
GREIG, D. C. 1971. The South of Scotland.
WILSON, H. E. 1972. Northern Ireland.

Books, symposia, etc. Some contributions in these appear in list B under
individual authors.
AGER, D. V. 1973. *The nature of the stratigraphical record.* 114 pp. Macmillan,
London.
ANDERTON, R., BRIDGES, P. H., LEEDER, M. R. and SELLWOOD, B. W. 1979. *A
dynamic stratigraphy of the British Isles.* 301 pp. Allen and Unwin, London.

ARKELL, W. J. 1933. *The Jurassic System in Great Britain.* 681 pp. Clarendon Press, Oxford.

ARKELL, W. J. 1947a. *The geology of Oxford.* 267 pp. Clarendon Press, Oxford.

BASSETT, M. G. (Editor). 1976. *The Ordovician System.* 696 pp. University of Wales Press, Cardiff.

BOWEN, D. Q. 1978. *Quaternary geology.* 221 pp. Pergamon Press, Oxford.

BOWES, D. R. and LEAKE, B. E. (Editors). 1978. *Crustal evolution in northwestern Britain and adjacent regions.* 492 pp. Seel House Press, Liverpool.

BOWIE, S. H. U., KVALHEIM, A. and HASLAM, H. W. (Editors). 1978. *Mineral deposits of Europe.* Vol. 1: *Northwest Europe.* 362 pp. Institution of Mining and Metallurgy, London.

CASEY, R. and RAWSON, P. F. (Editors). 1973. *The boreal Lower Cretaceous.* 448 pp. Seel House Press, Liverpool.

CHARLESWORTH, J. K. 1963. *Historical geology of Ireland.* 565 pp. Oliver and Boyd, Edinburgh.

CLIFTON-TAYLOR, A. 1972. *The pattern of English building.* 466 pp. Faber and Faber, London.

COE, K. (Editor). 1962. *Some aspects of the Variscan fold belt.* 163 pp. University Press, Manchester.

COHEE, G. V., GLAESSNER, M. F. and HEDBERG, H. D. 1978. *Contributions to the geologic time scale.* Studies in geology No. 6, 388 pp. American Association of Petroleum Geologists, Tulsa.

CRAIG, G. Y. (Editor). 1965. *The geology of Scotland.* 556 pp. Oliver and Boyd, Edinburgh.

DONOVAN, D. T. 1966. *Stratigraphy, an introduction to principles.* 287 pp. Murby, London.

EVANS, J. W. and STUBBLEFIELD, C. J. (Editors). 1929. *Handbook of the geology of Great Britain.* 556 pp. Murby, London.

FAURE, G. 1977. *Principles of isotope geology.* 464 pp. Wiley, New York.

GASS, I. G., SMITH, P. J. and WILSON, R. C. L. (Editors). 1972. *Understanding the Earth,* 2nd Edn. 383 pp. Open University (Artemis Press, Sussex).

HALLAM, A. 1973a. *A revolution in the earth sciences.* 127 pp. Clarendon Press, Oxford.

HALLAM, A. 1975. *Jurassic environments.* 269 pp. University Press, Cambridge.

HARLAND, W. B. and FRANCIS, E. H. (Editors). 1971. Supplementary papers and items, Part 1. Pp. 1–120 in *The Phanerozoic time-scale, a supplement.* Spec. Publ. No. 5, Geological Society, London.

HARLAND, W. B., GILBERT SMITH, A. and WILCOCK, B. (Editors). 1964. *The Phanerozoic time-scale.* 458 pp. Geological Society of London (also as *Q. Jl geol. Soc. Lond.* **120S**).

HARRIS, A. L., HOLLAND, C. H. and LEAKE, B. E. (Editors). 1979. *The Caledonides of Britain – reviewed.* 768 pp. Scottish Academic Press, Edinburgh.

HOLLAND, C. H. (Editor). 1974. *The Cambrian of the British Isles, Norden and Spitsbergen.* 300 pp. Wiley, London.

HOUSE, M. R., SCRUTTON, C. T. and BASSETT, M. G. 1979. *The Devonian System.* Spec. Pap. Palaeontology No. 23. 353 pp. Palaeontological Association, London.

HUGHES, N. F. (Editor). 1973. *Organisms and continents through time.* Spec. Pap. Palaeontology No. 12, 334 pp. Palaeontological Association, London.

KAY, M. (Editor). 1969. *North Atlantic – geology and continental drift.* Mem. 12, 1082 pp. American Association of Petroleum Geologists, Tulsa.

KNILL, J. L. (Editor). 1978. *Industrial geology.* 344 pp. University Press, Oxford.

LE PICHON, X., FRANCHETEAU, J. and BONNIN, J. 1973. *Plate tectonics.* 300 pp. Elsevier, Amsterdam.

MIDDLEMISS, F. A. and RAWSON, P. F. (Editors). 1971. *Faunal provinces in space and time.* 236 pp. Seel House Press, Liverpool.

MOSELEY, F. (Editor). 1978a. *The geology of the Lake District.* 284 pp. Yorkshire Geological Society, Leeds.

NAIRN, A. E. M. and STEHLI, F. G. (Editors). 1974. *The ocean basins and margins.* Vol. 2. *The North Atlantic.* 598 pp. Plenum Press, New York.

NAYLOR, D. and MOUNTENEY, S. N. 1975. *Geology of the north-west European continental shelf,* Vol. 1. 162 pp. G. T. Dudley Publishers Ltd., London.

NAYLOR, D., PHILLIPS, W. E. A., SEVASTOPULO, G. D. and SYNGE, F. M. 1980. *An introduction to the geology of Ireland.* 49 pp. Royal Irish Academy. (Excursion 058 C: G 08, 26th International Geological Congress, Paris.)

OWEN, T. R. 1976. *The geological evolution of the British Isles.* 161 pp. Pergamon Press, Oxford.

OWEN, T. R. (Editor). 1974a. *The Upper Palaeozoic and post-Palaeozoic rocks of Wales.* 426 pp. University of Wales Press, Cardiff.

PEGRUM, R. M., REES, G. and NAYLOR, D. 1975. *Geology of the north-west European continental shelf,* Vol. 2. 225 pp. G. T. Dudley Publishers Ltd., London.

PRESS, F. and SIEVER, R. 1974. *Earth.* 945 pp. W. H. Freeman, San Francisco.

RAYNER, D. H. and HEMINGWAY, J. E. (Editors). 1974. *The geology and mineral resources of Yorkshire.* 405 pp. Yorkshire Geological Society, Leeds.

READ, H. H. and WATSON, J. 1962. *Introduction to geology.* Vol. 1. *Principles.* 693 pp. Macmillan, London.

READ, H. H. and WATSON, J. 1975. *Introduction to geology.* Vol. 2. *Earth history,* parts I and II. 221 and 371 pp. Macmillan, London.

READING, H. G. (Editor). 1978. *Sedimentary environments and facies.* 557 pp. Blackwell, Oxford.

SELLEY, R. C. 1978. *Ancient sedimentary environments,* 2nd Edn. 287 pp. Chapman and Hall, London.

SHOTTON, F. W. (Editor). 1977. *British Quaternary studies.* 298 pp. Clarendon Press, Oxford.

SMITH, A. G. and BRIDEN, J. C. 1977. *Mesozoic and Cenozoic paleocontinental maps.* 63 pp. University Press, Cambridge.

SYLVESTER-BRADLEY, P. C. and FORD, T. D. (Editors). 1968. *The geology of the East Midlands.* 400 pp. University Press, Leicester.

TARLING, D. H. (Editor). 1978. *Evolution of the Earth's crust.* 443 pp. Academic Press, London.

TARLING, D. H. and RUNCORN, S. K. (Editors). 1973. *Implications of continental drift to the earth sciences,* 2 Vols. 1184 pp. Academic Press, London.

TARLING, D. H. and TARLING, M. P. 1972. *Continental drift.* 142 pp. Penguin Books, Harmondsworth.

WEST, R. G. 1968. *Pleistocene geology and biology.* 379 pp. Longmans, London.

WILLS, L. J. 1951. *A palaeogeographical atlas.* 64 pp. Blackie, London.

WINDLEY, B. F. 1977. *The evolving continents.* 385 pp. Wiley, London.

WOOD, A. (Editor). 1969. *The Pre-cambrian and Palaeozoic rocks of Wales.* 461 pp. University of Wales Press, Cardiff.

WOODLAND, A. W. (Editor). 1975. *Petroleum and the continental shelf of north-west Europe.* Vol. 1. *Geology.* 501 pp. Applied Science Publishers, Barking, Essex.

YORK, D. and FARQUHAR, R. M. 1972. *The earth's age and geochronology.* 178 pp. Pergamon Press, Oxford.

LIST B

AGER, D. V. 1975. The geological evolution of Europe. *Proc. Geol. Ass.* **86**, 127–54.

AGER, D. V. 1976. The Jurassic world ocean. Pp. JNNSS/o 1–18 in Finstad and Selley (1976). See this list, below.

ALLEN, J. R. L. 1960. The Mam Tor Sandstones: a 'turbidite' facies in the Namurian deltas of northern England. *J. sedim. Petrol.* **30**, 193–208.

ALLEN, J. R. L. 1964. Studies in fluviatile sedimentation: six cyclothems from the Lower Old Red Sandstone, Anglo-Welsh basin. *Sedimentology*, **3**, 163–98.

ALLEN, J. R. L. 1965. Upper Old Red Sandstone (Farlovian) palaeogeography in South Wales and the Welsh Borderland. *J. sedim. Petrol.* **35**, 167–95.

ALLEN, J. R. L. 1968. The Cambrian and Ordovician systems. Pp. 20–40 in Sylvester-Bradley and Ford (1968). See list A.

ALLEN, J. R. L. 1974. The Devonian rocks of Wales and the Welsh Borderland. Pp. 47–84 in Owen (1974a). See list A.

ALLEN, J. R. L. 1979. Old Red Sandstone facies in external basins, with particular reference to southern Britain. Pp. 65–80 in House et al. (1979). See list A.

ALLEN, J. R. L. and KAYE, P. 1973. Sedimentary facies of the Forest Marble (Bathonian), Shipton-on-Cherwell quarry, Oxfordshire. *Geol. Mag.* **110**, 153–63.

ALLEN, J. R. L. and TARLO, L. B. 1963. The Downtonian and Dittonian facies of the Welsh Borderland. *Geol. Mag.* **100**, 129–55.

ALLEN, P. 1965. L'âge du Purbecko-Wealdien d'Angleterre. Pp. 321–6 in *Colloque sur le Crétacé inférieur* (Lyons, 1963). *Mêm. Bur. Rech. Géol. Min.* No. 34, Paris.

ALLEN, P. 1969. Lower Cretaceous sourcelands and the North Atlantic. *Nature, Lond.* **222**, 657–58.

ALLEN, P. 1972. Wealden detrital tourmaline: implications for northwestern Europe. *Jl geol. Soc. Lond.* **128**, 273–94.

ALLEN, P. 1976. Wealden of the Weald: a new model. *Proc. Geol. Ass.* **86**, 389–437.

ANDERSON, F. W. and BAZLEY, R. A. B. 1971. The Purbeck Beds of the Weald (England). *Bull. geol. Surv. Gt Br.* No. 34, 1–172.

ANDERSON, F. W. and DUNHAM, K. C. 1966. The geology of Northern Skye. *Mem. geol. Surv. Scotland.* 216 pp. H.M.S.O., Edinburgh.

ANDERTON, R. 1976. Tidal-shelf sedimentation: an example from the Scottish Dalradian. *Sedimentology*, **23**, 429–58.

ARKELL, W. J. 1947b. The geology of the country around Weymouth, Swanage, Corfe and Lulworth. *Mem. geol. Surv. U.K.* 386 pp. H.M.S.O. London.

ARMSTRONG, G., DUNHAM, K. C., HARVEY, C. O., SABINE, P. A. and WATERS, W. F. 1951. The paragenesis of sylvine, carnallite, polyhalite and kieserite in Eskdale borings Nos. 3, 4 and 6, north-east Yorkshire. *Mineralog. Mag.* **29**, 667–89.

ARMSTRONG, R. L. 1978. Pre-Cenozoic Phanerozoic time scale – computer file of critical dates and consequences of new and in-progress decay-constant revisions. Pp. 73–91 in Cohee et al. (1978). See list A.

ARTHURTON, R. S., BURGESS, I. C. and HOLLIDAY, D. W. 1978. Permian and Triassic. Pp. 189–206 in Moseley (1978a). See list A.

AUDLEY-CHARLES, M. G. 1970a. Stratigraphical correlation of the Triassic rocks of the British Isles. *Q. Jl geol. Soc. Lond.* **126**, 19–47.

AUDLEY-CHARLES, M. G. 1970b. Triassic palaeogeography of the British Isles. *Q. Jl geol. Soc. Lond.* **126**, 49–89.

BAKER, J. W. 1973. A marginal Late Proterozoic ocean basin in the Welsh region. *Geol. Mag.* **110**, 447–55.

BALL, H. W., DINELEY, D. L. and WHITE, E. I. 1961. The Old Red Sandstone of Brown Clee hill and the adjacent areas. *Bull. Br. Mus. nat. Hist. Geol.* **5**, 175–310.

BAMFORD, D., NUNN, K., PRODEHL, C. and JACOB, B. 1977. LISPB–III. Upper crustal structure of northern Britain. *Jl geol. Soc. Lond.* **133**, 481–8.

BASSETT, M. G., COCKS, L. R. M., HOLLAND, C. H., RICKARDS, R. B. and WARREN, P. T. 1975. The type Wenlock Series. *Rep. Inst. geol. Sci.* No. 75/13, 19 pp.

BECKINSALE, R. D. and THORPE, R. S. 1979. Rubidium–strontium whole-rock isochron evidence for the age of the metamorphism and magmatism in the Mona Complex of Anglesey. *Jl geol. Soc. Lond.* **136**, 433–9.

BISHOP, A. C., ROACH, R. A. and ADAMS, C. J. D. 1975. Precambrian rocks within the Hercynides. Pp. 102–7 in Harris *et al.* (1975). See list A.

BLUCK, B. L. 1978*a*. Geology of a continental margin. 1: the Ballantrae Complex. Pp. 151–62 in Bowes and Leake (1978). See list A.

BLUCK, B. L. 1978*b*. Sedimentation in a late orogenic basin: the Old Red Sandstone of the Midland Valley of Scotland. Pp. 249–78 in Bowes and Leake (1978). See list A.

BOWEN, D. Q. 1974. The Quaternary of Wales. Pp. 373–426 in Owen (1974*a*). See list A.

BRADBURY, H. J., SMITH, R. A. and HARRIS, A. L. 1976. 'Older' granites as time-markers in Dalradian evolution. *Jl geol. Soc. Lond.* **132**, 677–84.

BRENCHLEY, P. J. 1969. The relationship between Caradocian volcanicity and sedimentation in North Wales. Pp. 181–202 in Wood, A. (1969). See list A.

BRIDEN, J. C., MORRIS, W. A. and PIPER, J. D. A. 1973. Palaeomagnetic studies in the British Caledonides. VI: regional and global implications. *Geophys. J. R. astr. Soc.* **34**, 107–34.

BRISTOW, C. R. and BAZLEY, R. A. 1972. Geology of the country around Royal Tunbridge Wells. *Mem. geol. Surv. Gt Br.* 161 pp. H.M.S.O., London.

BROOK, M., BREWER, M. S. and POWELL, D. 1976. Grenville age for rocks in the Moine of north-western Scotland. *Nature, Lond.* **260**, 515–17.

BROOKS, J. R. V. and CHESHER, J. A. 1976. Review of offshore Jurassic of the U.K. northern North Sea. Pp. JNNSS/2 1–24 in Finstad and Selley (1976). See this list, below.

BRÜCK, P. M., POTTER, T. L. and DOWNIE, C. 1974. The Lower Palaeozoic stratigraphy of the northern part of the Leinster Massif. *Proc. R. Irish Acad.* **74B**, 75–84.

BRÜCK, P. M. and REEVES, T. J. 1976. Stratigraphy, sedimentology and structure of the Bray Group in County Wicklow and south County Dublin. *Proc. R. Irish Acad.* **76B**, 53–77.

BULLARD, E. C., EVERETT, J. E. and GILBERT SMITH, A. 1965. The fit of the continents around the Atlantic. *Phil. Trans. R. Soc.* A, **258**, 41–51.

BULMAN, O. M. B. 1970. Graptolithina, 2nd Edn. Part V. 163V pp. in *Treatise on invertebrate paleontology* (Ed. R. C. Moore). University of Kansas Press, Lawrence, Kansas.

CALLOMON, J. H. and COPE, J. C. W. 1971. The stratigraphy and ammonite successions of the Oxford and Kimmeridge clays in the Warlingham borehole. *Bull. geol. Surv. Gt Br.* No. 36, 147–76.

CALVER, M. A. 1968. Distribution of Westphalian marine faunas in northern England and adjoining areas. *Proc. Yorks. geol. Soc.* **37**, 1–72.

CALVER, M. A. 1969. Westphalian of Britain. *C. R. Congr. Stratigr. Géol. Carb.* (6th, Sheffield), Vol. I, 233–54.

CASEY, R. 1961: The stratigraphical palaeontology of the Lower Greensand. *Palaeontology*, **3**, 487–621.

CASEY, R. 1971. Facies, faunas and tectonics in late Jurassic – early Cretaceous Britain. Pp. 153–68 in Middlemiss and Rawson (1971). See list A.

CLARKSON, C. M., CRAIG, G. Y. and WALTON, E. K. 1975. The Silurian rocks bordering Kirkcudbright Bay, south Scotland. *Trans. R. Soc. Edinb.* **69**, 313–25.

COLLINSON, J. D. 1969. The sedimentology of the Grindslow Shales and the Kinderscout Grit; a deltaic complex in the Namurian of northern England. *J. sedim. Petrol.* **39**, 194–221.

COLLINSON, J. D. 1970. Deep channels, massive beds and turbidity current genesis in the central Pennine basin. *Proc. Yorks. geol. Soc.* **37**, 495–519.

COPE, J. C. W. 1967. The palaeontology and stratigraphy of the lower part of the Upper Kimmeridge Clay of Dorset. *Bull. Br. Mus. nat. Hist. Geol.* **15**, 1–79.

COPE, J. C. W. 1978. The ammonite faunas and stratigraphy of the upper part of the Upper Kimmeridge Clay of Dorset. *Palaeontology*, **21**, 469–533.

COWIE, J. W. 1974. The Cambrian of Spitsbergen and Scotland. Pp. 123–55 in Holland (1974). See list A.

COX, A. H. 1925. The geology of the Cader Idris range. *Q. Jl geol. Soc. Lond.* **81**, 539–94.

CRAIG, G. Y. and WALTON, E. K. 1959. Sequence and structure in the Silurian rocks of Kirkcudbrightshire. *Geol. Mag.* **96**, 209–20.

CRIBB, S. J. 1975. Rubidium–strontium ages and strontium isotope ratios from the igneous rocks of Leicestershire. *Jl geol. Soc. Lond.* **131**, 203–12.

CURRY, D. 1965. The Palaeogene beds of south-east England. *Proc. Geol. Ass.* **76**, 151–73.

CURRY, D. and SMITH, A. J. 1975. New discoveries concerning the geology of the central and eastern parts of the English Channel. *Phil. Trans. R. Soc.* A, **279**, 155–67.

DALEY, B. 1972. Some problems concerning the early Tertiary climate of southern Britain. *Palaeogeogr. Palaeoclimatol. Palaeoecol.* **11**, 177–90.

DALEY, B. 1973. The palaeoenvironment of the Bembridge Marls (Oligocene) of the Isle of Wight, Hampshire. *Proc. Geol. Ass.* **84**, 83–93.

DEAN, W. T. 1964. The geology of the Ordovician and adjacent strata in the southern Caradoc district of Shropshire. *Bull. Br. Mus. nat. Hist. Geol.* **9**, 257–69.

DEAN, W. T., DONOVAN, D. T. and HOWARTH, M. K. 1961. The Liassic ammonite zones and subzones of the north-west European province. *Bull. Br. Mus. nat. Hist. Geol.* **4**, No. 10, 435–505.

DE RAAF, J. F. M., READING, H. G. and WALKER, R. G. 1965. Cyclic sedimentation in the lower Westphalian of North Devon. *Sedimentology* **4**, 1–52.

DEWEY, J. F. 1963. The Lower Palaeozoic stratigraphy of central Murrisk, County Mayo, and the evolution of the south Mayo trough. *Q. Jl geol. Soc. Lond.* **119**, 313–44.

DEWEY, J. F. 1969*a*. Evolution of the Appalachian/Caledonian orogen. *Nature, Lond.* **222**, 124–9.

DEWEY, J. F. 1969*b*. Structure and sequence in paratectonic British Caledonides. Pp. 309–35 in Kay (1969). See list A.

DEWEY, J. F. 1971. A model for the Lower Palaeozoic evolution of the southern margin of the early Caledonides of Scotland and Ireland. *Scott. J. Geol.* **7**, 219–240.

DEWEY, J. F. 1974. The geology of the southern termination of the Caledonides. Pp. 205–31 in Nairn and Stehli (1974). See list A.

DEWEY, J. F. and PANKHURST, R. J. 1970. Evolution of the Scottish Highlands and their radiometric pattern. *Trans. R. Soc. Edinb.* **68**, 361–89.

DHONAU, N. B. and HOLLAND, C. H. 1974. The Cambrian of Ireland. Pp 157–76 in Holland (1974). See list A.

DODSON, M. H. and REX, D. C. 1971. Potassium–argon ages of slates and phyllites from south-west England. *Q. Jl geol. Soc. Lond.* **126**, 465–99.

DONOVAN, D. T. and JONES, E. J. W. 1979. Causes of world-wide changes in sea level. *Jl geol. Soc. Lond.* **136**, 187–92.

DONOVAN, R. N., FOSTER, R. J. and WESTOLL, T. S. 1974. A stratigraphical revision of the Old Red Sandstone of north-eastern Caithness. *Trans. R. Soc. Edinb.* **69**, 171–205.

DOWNIE, C. 1975. Precambrian of the British Isles – palaeontology. Pp. 113–15 in Harris *et al.* (1975). See list A.

DOWNIE, C., LISTER, T. R., HARRIS, A. L. and FETTES, D. J. A. 1971. A palynological investigation of the Dalradian rocks of Scotland. *Rep. Inst. geol. Sci.* No. 71/9, 30 pp.

DUFF, K. L. 1975. Palaeoecology of a bituminous shale – the Lower Oxford Clay of central England. *Palaeontology* **18**, 443–82.

DUFF, P. McL. D. 1965. Economic geology. Pp. 505–33 in Craig (1965). See list A.

DUFF, P. McL. D., HALLAM, A. and WALTON, E. K. 1967. Cyclic sedimentation. *Devs Sedimentology*, **10**, 280 pp, Elsevier, Amsterdam.

DUNHAM, K. C. 1960. Syngenetic and diagenetic mineralization in Yorkshire. *Proc. Yorks. geol. Soc.* **32**, 229–84.

DUNHAM, K. C. 1978. The metallogeny of Britain. Pp. 137–65 in Knill (1978). See list A.

DUNHAM, K. C., BEER, K. E., ELLIS, R. A., GALLAGHER, M. J., NUTT, M. J. C. and WEBB, B. C. 1978. United Kingdom. Pp. 263–317 in Bowie *et al.* (1978). See list A.

DUNHAM, K. C., DUNHAM, A. C., HODGE, B. L. and JOHNSON, G. A. L. 1965. Granite beneath Viséan sediments with mineralization at Rookhope, northern Pennines. *Q. Jl geol. Soc. Lond.* **121**, 383–417.

DUNHAM, K. C. and POOLE, E. G. 1974. The Oxfordshire Coalfield. *Jl geol. Soc. Lond.* **130**, 387–91.

DUNNING, F. W. 1975. Precambrian craton of central England and the Welsh Borders. Pp. 83–94 in Harris *et al.* (1975). See list A.

EDWARDS, W. 1951. The concealed coalfield of Yorkshire and Nottinghamshire, 3rd Edn. *Mem. geol. Surv. U.K.* 285 pp. H.M.S.O., London.

ELLIOTT, R. E. 1961. The stratigraphy of the Keuper Series in southern Nottinghamshire. *Proc. Yorks. geol. Soc.* **33**, 197–234.

EVANS, A. M., FORD, T. D. and ALLEN, J. R. L. 1968. Precambrian rocks. Pp. 1–19 in Sylvester-Bradley and Ford (1968). See list A.

EVANS, D. J. and THOMPSON, M. S. 1979. The geology of the central Bristol Channel and the Lundy area, South Western Approaches, British Isles. *Proc. Geol. Ass.* **90**, 1–14.

EVANS, W. B. 1970. The Triassic salt deposits of north-western England. *Q. Jl geol. Soc. Lond.* **126**, 103–23.

FALCON, N. L. and KENT, P. E. 1960. Geological results of petroleum exploration in Britain 1945–1957. *Mem. geol. Soc. Lond.* No. 2, 56 pp.

FALLER, A. M. and BRIDEN, J. C. 1978. Palaeomagnetism of Lake District rocks. Pp. 17–24 in Moseley (1978a). See list A.

FINSTAD, K. G. and SELLEY, R. C. (Coordinators). 1976. *Jurassic northern North Sea Symposium* (Stavanger, 1975). Norwegian Petroleum Society.

FITCH, F. J., MILLER, J. A., EVANS, A. L., GRASTY, R. L. and MENEISY, M. Y. 1969. Isotopic age determinations on rocks from Wales and the Welsh Borders. Pp. 23–45 in Wood, A. (1969). See list A.

FITCH, F. J., MILLER, J. A. and WILLIAMS, S. C. 1970. Isotopic ages of British Carboniferous rocks. *C. R. Congr. Stratigr. Géol. Carb.* (6th, Sheffield). Vol. II, 771–89.

FITTON, J. G. and HUGHES, D. J. 1970. Volcanism and plate tectonics in the British Ordovician. *Earth planet. Sci. Lett.* **8**, 223–8.

FLOYD, P. A. 1972. Geochemistry, origin and tectonic environment of the basic and acidic rocks of Cornubia, England. *Proc. Geol. Ass.* **83**, 385–404.

FORD, T. D. 1958. Pre-Cambrian fossils from Charnwood Forest. *Proc. Yorks. geol. Soc.* **31**, 211–17.

FRANCIS, E. H. 1961. Thin beds of graded kaolinized tuff and tuffaceous siltstones in the Carboniferous of Fife. *Bull. geol. Surv. Gt Br.* No. 17, 191–215.

FRANCIS, E. H. 1965a. Carboniferous. Pp. 309–57 in Craig (1965). See list A.

FRANCIS, E. H. 1965b. Carboniferous–Permian igneous rocks. Pp. 359–82 in Craig (1965). See list A.

FRANCIS, E. H. 1978. Igneous activity in a fractured craton: Carboniferous volcanism in northern Britain. Pp. 279–96 in Bowes and Leake (1978). See list A.

FRANCIS, E. H. 1979. British coalfields. *Sci. Progr., Lond.* **66**, 1–23.

GALLOIS, R. W. 1965. The Wealden District, 4th Edn. *British Regional Geology*, 101 pp. H.M.S.O., London.

GALLOIS, R. W. 1976. Coccolith blooms in the Kimmeridge Clay and the origin of North Sea oil. *Nature, Lond.* **259**, 473–5.

GALLOIS, R. W. and COX, B. M. 1976. The stratigraphy of the Lower Kimmeridge Clay of eastern England. *Proc. Yorks. geol. Soc.* **41**, 13–26.

GALLOIS, R. W. and COX, B. M. 1977. The stratigraphy of the middle and upper Oxfordian sediments of Fenland. *Proc. Geol. Ass.* **88**, 207–28.

GANSSER, A. 1973. Facts and theories on the Andes. *Jl geol. Soc. Lond.* **129**, 93–131.

GEORGE, T. N. 1958. Lower Carboniferous palaeogeography of the British Isles. *Proc. Yorks. geol. Soc.* **31**, 227–318.

GEORGE, T. N. 1960. The stratigraphical evolution of the Midland Valley. *Trans. geol. Soc. Glasgow*, **24**, 32–107.

GEORGE, T. N. 1965. The geological growth of Scotland. Pp. 1–48 in Craig (1965). See list A.

GEORGE, T. N. 1969. British Dinantian stratigraphy. *C. R. Congr. Stratigr. Géol. Carb.* (6th, Sheffield). Vol. I, 193–218.

GEORGE, T. N. 1974. Lower Carboniferous rocks in Wales. Pp. 85–115 in Owen (1974a). See list A.

GOODLET, G. A. 1957. Lithological variation in the Lower Limestone Group of the Midland Valley of Scotland. *Bull. geol. Surv. Gt Br.* No. 12, 52–65.

GREEN, G. W. and DONOVAN, D. T. 1969. The Great Oolite of the Bath area. *Bull. geol. Surv. Gt Br.* No. 30, 1–63.

HALLAM, A. 1958. The concept of Jurassic axes of uplift. *Sci. Progr., Lond.* **46**, 441–9.

HALLAM, A. 1960. A sedimentary and faunal study of the Blue Lias of Dorset and Glamorgan. *Phil. Trans. R. Soc. B*, **243**, 1–94.

HALLAM, A. 1969. Faunal realms and facies in the Jurassic. *Palaeontology* **12**, 1–18.

HALLAM, A. 1971. Provinciality in Jurassic faunas in relation to facies and palaeogeography. Pp. 126–52 in Middlemiss and Rawson (1971). See list A.

HALLAM, A. (Editor). 1973*b*. *Atlas of Palaeobiogeography.* 531 pp. Elsevier, Amsterdam.

HALLAM, A. and SELLWOOD, B. W. 1976. Middle Mesozoic sedimentation in relation to tectonics in the British area. *J. Geol.* **84,** 301–21.

HALLIMOND, A. F. 1925. Iron ores: bedded ores of England and Wales. Petrography and chemistry. *Mem. geol. Surv. spec. Rep. Miner. Resour. Gt Br.* No. 29, 139 pp. H.M.S.O., London.

HANCOCK, J. M. 1961. The Cretaceous system in Northern Ireland. *Q. Jl geol. Soc. Lond.* **117,** 11–36.

HANCOCK, J. M. 1969. Transgression of the Cretaceous sea in south-west England. *Proc. Ussher Soc.* **2,** 61–83.

HANCOCK, J. M. 1975. The sequence of facies in the Upper Cretaceous of northern Europe compared with that in the Western Interior of North America. Pp. 83–118 in *Spec. Pap. geol. Ass. Can.* No. 13 (Ed. W. G. E. Caldwell). 666 pp. Geological Association of Canada, Toronto.

HANCOCK, J. M. 1976. The petrology of the Chalk. *Proc. Geol. Ass.* **86,** 499–535.

HANCOCK, J. M. and KAUFFMAN, E. G. 1979. The great transgressions of the late Cretaceous. *Jl geol. Soc. Lond.* **136,** 175–86.

HARLAND, W. B., AGER, D. V., BALL, H. W., BISHOP, W. W., BLOW, W. H., CURRY, D., DEER, W. A., GEORGE, T. N., HOLLAND, C. H., HOLMES, S. C. A., HUGHES, N. F., KENT, P. E., PITCHER, W. S., RAMSBOTTOM, W. H. C., STUBBLEFIELD, C. J., WALLACE, P. and WOODLAND, A. W. 1972. A concise guide to stratigraphical procedure. *Jl. geol. Soc. Lond.* **128,** 295–303.

HARRIS, A. L., BALDWIN, C. T., BRADBURY, H. J., JOHNSON, H. D. and SMITH, R. A. 1978. Ensialic basin sedimentation: the Dalradian Supergroup. Pp. 115–38 in Bowes and Leake (1978). See list A.

HARRIS, A. L. and PITCHER, W. S. 1975. The Dalradian Supergroup. Pp. 52–75 in Harris *et al.* (1975). See list A.

HARRISON, R. K., JEANS, C. V. and MERRIMAN, R. J. 1979. Mesozoic igneous rocks, hydrothermal mineralisation and volcanogenic sediments in Britain and adjacent regions. *Bull. geol. Surv. Gt Br.* No. 70, 57–69.

HAYS, J. D., IMBRIE, J. and SHACKLETON, N. J. 1976. Variations in the Earth's orbit: pacemaker of the ice ages. *Science,* **194,** 1121–32.

HEDBERG, H. D. 1976. *International stratigraphic guide.* 200 pp. Wiley, New York.

HEMINGWAY, J. E. 1974*a*. Jurassic. Pp. 161–223 in Rayner and Hemingway (1974). See list A.

HEMINGWAY, J. E. 1974*b*. Ironstone. Pp. 329–35 in Rayner and Hemingway (1974). See list A.

HOLLAND, C. H. 1969. Irish counterpart of Silurian of Newfoundland. Pp. 298–308 in Kay (1969). See list A.

HORNE, R. R. 1971. Aeolian cross stratification in the Devonian of the Dingle peninsula, County Kerry, Ireland. *Geol. Mag.* **108,** 151–8.

HOUSE, M. R. 1963. Devonian ammonoid successions and facies in Devon and Cornwall. *Q. Jl geol. Soc. Lond.* **119,** 1–27.

HOUSE, M. R. 1975. Facies and time in Devonian tropical areas. *Proc. Yorks. geol. Soc.* **40,** 233–88.

HOUSE, M. R. 1979. Biostratigraphy of the early ammonoidea. Pp. 263–80 in House *et al.* (1979). See List A.

HOWITT, F. 1964. Stratigraphy and structure of the Purbeck inliers of Sussex (England). *Q. Jl geol. Soc. Lond.* **120,** 77–113.

HOWITT, F., ASTON, E. L. and JACQUÉ, M. 1975. The occurrence of Jurassic volcanics in the North Sea. Pp. 379–87 in Woodland (1975). See list A.

HUDSON, J. D. 1964. The petrology of the sandstones of the Great Estuarine Series and the Jurassic palaeogeography of Scotland. *Proc. Geol. Ass.* **75**, 499–527.

HUDSON, J. D. 1977. Oxygen isotope studies on Cenozoic temperatures, oceans, and ice acumulation. *Scott. J. Geol.* **13**, 313–25.

HULL, J. H. and THOMAS, I. A. 1974. Limestones and dolomites. Pp. 345–59 in Rayner and Hemingway (1974). See list A.

HURST, J. M., HANCOCK, N. J. and McKERROW, W. S. 1978. Wenlock stratigraphy and palaeogeography of Wales and the Welsh borderland. *Proc. Geol. Ass.* **89**, 197–226.

INGHAM, J. K., McNAMARA, K. J. and RICKARDS, R. B. 1978. The Upper Ordovician and Silurian rocks. Pp. 121–45 in Moseley (1978a). See list A.

JACKSON, D. E. 1978. The Skiddaw Group. Pp. 79–98 in Moseley (1978a). See list A.

JAMES, J. H. 1956. The structure and stratigraphy of part of the Pre-Cambrian outcrop between Church Stretton and Linley, Shropshire. *Q. Jl geol. Soc. Lond.* **112**, 107–28.

JOHNSON, G. A. L. 1973. Closing of the Carboniferous sea in western Europe. Pp. 843–50 in Tarling and Runcorn (1973). See list A.

JOHNSTONE, G. S. 1975. The Moine Succession. Pp. 30–42 in Harris *et al.* (1975). See list A.

JONES, D. G. 1974. The Namurian Series in South Wales. Pp. 117–32 in Owen (1974a). See list A.

JONES, O. T. and PUGH, W. J. 1949. An early Ordovician shore-line in Radnorshire, near Builth Wells. *Q. Jl geol. Soc. Lond.* **105**, 65–99.

KELLAWAY, G. A. and WELCH, F. B. A. 1948. Bristol and Gloucester district, 2nd Edn. *British Regional Geology*, 99 pp. H.M.S.O., London.

KELLING, G. 1961. The stratigraphy and structure of the Ordovician rocks of the Rhinns of Galloway. *Q. Jl geol. Soc. Lond.* **117**, 37–75.

KELLING, G. 1962. The petrology and sedimentation of Upper Ordovician rocks in the Rhinns of Galloway, southwest Scotland. *Trans. R. Soc. Edinb.* **65**, 107–37.

KELLING, G. 1974. Upper Carboniferous sedimentation in South Wales. Pp. 185–224 in Owen (1974a). See list A.

KENNEDY, W. Q. 1958. The tectonic evolution of the Midland Valley of Scotland. *Trans. geol. Soc. Glasgow*, **23**, 106–33.

KENNEDY, W. J. and GARRISON, R. E. 1975. Morphology and genesis of nodular chalks and hardgrounds in the Upper Cretaceous of southern England. *Sedimentology* **22**, 311–86.

KENT, P. E. 1966. The structure of the concealcd Carboniferous rocks of north-eastern England. *Proc. Yorks. geol. Soc.* **35**, 323–52.

KENT, P. E. 1975a. The tectonic development of Great Britain and the surrounding seas. Pp. 3–28 in Woodland (1975). See list A.

KENT, P. E. 1975b. Review of North Sea basin development. *Jl geol. Soc. Lond.* **131**, 435–68.

KENT, P. E. 1977. The Mesozoic development of aseismic continental margins. *Jl geol. Soc. Lond.* **134**, 1–18.

KILBURN, C., PITCHER, W. S. and SHACKLETON, R. M. 1965. The stratigraphy and origin of the Portaskaig Boulder Bed Series (Dalradian). *Geol. J.* **4**, 343–60.

KNOX, R. W. O'B. and ELLISON, R. A. 1979. A lower Eocene ash sequence in S. E. England. *Jl geol. Soc. Lond.* **136**, 251–53.

KNOX, R. W. O'B. and HARLAND, R. 1979. Stratigraphical relationships of the early Palaeogene ash-series of NW Europe. *Jl geol. Soc. Lond.* **134**, 463–70.

LAMBERT, R. ST J. 1971. The pre-Pleistocene Phanerozoic time-scale – a review. Pp. 9–31 in Harland and Francis (1971). See list A.

LAMBERT, R. ST J. and MCKERROW, W. S. 1976. The Grampian orogeny, *Scott. J. Geol.* **12**, 271–92.

LAUBSCHER, H. and BERNOULI, D. 1977. Mediterranean and Tethys. Pp. 1–28 in Nairn *et al.* (1977). See this list, below.

LAWSON, D. E. 1976. Sandstone-boulder conglomerates and Torridonian cliffed shoreline between Gairloch and Stoer, northwest Scotland. *Scott. J. Geol.* **12**, 67–88.

LEAKE, B. E. 1978. Granite emplacement: the granites of Ireland and their origin. Pp. 221–48 in Bowes and Leake (1978). See list A.

LE BAS, M. J. 1972. Caledonian igneous rocks beneath central and eastern England. *Proc. Yorks. geol. Soc.* **39**, 71–86.

LEEDER, M. R. 1973. Sedimentology and palaeogeography of the Upper Old Red Sandstone in the Scottish Border basin. *Scott. J. Geol.* **9**, 117–44.

LEEDER, M. R. 1974. Lower Border Group (Tournaisian) fluvio-deltaic sedimentation and palaeogeography of the Northumberland basin. *Proc. Yorks. geol. Soc.* **40**, 129–80.

MCKERROW, W. S. 1979. Ordovician and Silurian changes in sea level. *Jl geol. Soc. Lond.* **136**, 137–45.

MCKERROW, W. S. and COCKS, L. R. M. 1976. Progressive faunal migration across Iapetus ocean. *Nature, Lond.* **263**, 304–6.

MCKERROW, W. S., JOHNSON, R. J. and JAKOBSON, M. E. 1969. Palaeoecological studies in the Great Oolite at Kirtlington, Oxfordshire. *Palaeontology*, **12**, 56–83.

MCKERROW, W. S. and ZIEGLER, A. M. 1972. Palaeozoic oceans. *Nature Phys. Sci.* **240**, 92–4.

MCLEAN, A. C. 1978. Evolution of fault-controlled ensialic basins in northwestern Britain. Pp. 325–46 in Bowes and Leake (1978). See list A.

MARTIN, A. J. 1967. Bathonian sedimentation in southern England. *Proc. Geol. Ass.* **78**, 473–88.

MATTHEWS, S. C. 1977. Carboniferous successions in Germany and south west England. *Proc. Ussher Soc.* **4**, 67–74.

MAX, D. M. 1975. Precambrian rocks of south-east Ireland. Pp. 97–101 in Harris *et al.* (1975). See list A.

MERCY, E. L. P. 1965. Caledonian igneous activity. Pp. 229–67 in Craig (1965). See list A.

MICHIE, U. McL. 1972. Further evidence of uranium mineralization in Orkney. *Trans. Instn Min. Metall.* **81B**, B53–4.

MICHIE, U. McL. and GALLAGHER, M. J. 1979. Uranium concentrations of potential economic significance in northern Scotland. *Trans. Instn Min. Metall.* **88B**, B28–9.

MILES, R. S. 1968. The Old Red Sandstone antiarchs of Scotland: family Bothriolepidae. *Palaeontogr. Soc. Monogr.* 130 pp.

MILLWARD, D., MOSELEY, F. and SOPER, N. J. 1978. The Eycott and Borrowdale Volcanic rocks. Pp. 99–120 in Moseley (1978a). See list A.

MITCHELL, A. H. G. and MCKERROW, W. S. 1975. Analogous evolution of the Burma orogen and the Scottish Caledonides. *Bull. geol. Soc. Amer.* **86**, 305–15.

MITCHELL, M. 1972. The base of the Viséan in south-west and north-west England. *Proc. Yorks. geol. Soc.* **39**, 151–60.

MITCHELL, M., TAYLOR, B. J. and RAMSBOTTOM, W. H. C. 1978. Carboniferous. Pp. 168–88 in Moseley (1978a). See list A.

MOORBATH, S. 1975a. The geological significance of early Precambrian rocks. *Proc. Geol. Ass.* **86**, 259–79.

MOORBATH, S. 1975b. Progress in isotopic dating of British Precambrian rocks. Pp. 108–12 in Harris *et al.* (1975). See list A.

MORTON, N. 1965. The Bearreraig Sandstone Series (Middle Jurassic) of Skye and Raasay. *Scott. J. Geol.* **1**, 189–216.

MOSELEY, F. 1977. Caledonian plate tectonics and the place of the English Lake District. *Bull. geol. Soc. Amer.* **88**, 764–8.

MOSELEY, F. 1978b. The geology of the Lake District, an introductory review. Pp. 1–16 in Moseley (1978a). See list A.

MUDGE, D. C. 1978. Stratigraphy and sedimentation of the Lower Inferior Oolite of the Cotswolds. *Jl geol. Soc. Lond.* **135**, 611–27.

MURCHISON, D. G. and WESTOLL, T. S. (Editors). 1968. *Coal and coal-bearing strata.* 418 pp. Oliver and Boyd, Edinburgh.

MYKURA, W. 1960. The replacement of coal by limestone and the reddening of Coal Measures in the Ayrshire Coalfield. *Bull. geol. Surv. Gt Br.* No. 16, 69–109.

NAIRN, A. E. M., KANES, W. H. and STEHLI, F. G. (Editors). 1977. *The ocean basins and margins.* Vol. 4A. *The eastern Mediterranean.* 503 pp. Plenum Press, New York.

NAYLOR, D. 1969. Facies change in Upper Devonian and Lower Carboniferous rocks of southern Ireland. *Geol. J.* **6**, 307–28.

NAYLOR, D. 1975. Upper Devonian – Lower Carboniferous stratigraphy along the south coast of Dunmanus Bay, Co. Cork. *Proc. R. Irish Acad.* **75B**, 317–37.

NAYLOR, D., JONES, P. C. and MATTHEWS, S. C. 1974. Facies relationships in the Upper Devonian – Lower Carboniferous of southwest Ireland and adjacent regions. *Geol. J.* **9**, 77–96.

NAYLOR, D. and SEVASTOPULO, G. D. 1979. The Hercynian 'Front' in Ireland. *Krystalinikum* **14**, 77–90.

NEALE, J. W. 1974. Cretaceous. Pp. 225–43 in Rayner and Hemingway (1974). See list A.

NEVES, R. and SELLEY, R. C. 1976. A review of the Jurassic rocks of north-east Scotland. Pp. JNNSS/5 1–29 in Finstad and Selley (1976). See this list, above.

NICHOLSON, R. 1974. The Scandinavian Caledonides. Pp. 161–203 in Nairn and Stehli (1974). See list A.

OWEN, T. R. 1974b. The geology of the Western Approaches. Pp. 233–72 in Nairn and Stehli (1974). See list A.

OXBURGH, E. R. 1974. The plain man's guide to plate tectonics. *Proc. Geol. Ass.* **85**, 299–357.

PANKHURST, R. J. 1970. The geochronology of the basic igneous complexes. *Scott. J. Geol.* **6**, 83–107.

PANKHURST, R. J. 1974. Rb–Sr whole-rock chronology of Caledonian events in northeast Scotland. *Bull. geol. Soc. Amer.* **85**, 345–50.

PANKHURST, R. J. and PIDGEON, R. T. 1976. Inherited isotope systems and the source region pre-history of early Caledonian granites in the Dalradian Series of Scotland. *Earth planet. Sci. Lett.* **31**, 55–8.

PATCHETT, P. J. and JOCELYN, J. 1979. U–Pb zircon ages for late Precambrian igneous rocks in South Wales. *Jl geol. Soc. Lond.* **136**, 13–19.

PATTISON, J., SMITH, D. B. and WARRINGTON, G. 1973. A review of late Permian biostratigraphy in the British Isles. Pp. 220–60 in *The Permian and Triassic systems and their mutual boundary* (Eds. A. Logan and L. V. Hills). *Mem. Can. Soc. Petrol. Geol.* No. 2. 766 pp. Canadian Society of Petroleum Geologists, Calgary.

PEACH, B. N. and HORNE, J. 1907. The geological structure of the North-West Highlands of Scotland. *Mem. geol. Surv. U.K.* 668 pp. H.M.S.O., Glasgow.

PENN, I. E. and EVANS, C. D. R. 1976. The middle Jurassic (mainly Bathonian) of Cardigan Bay and its palaeogeographical significance. *Rep. Inst. geol. Sci.* No. 76/6. 6 pp.

PENNY, L. F. 1974. Quaternary. Pp. 245–64 in Rayner and Hemingway (1974). See list A.

PHEMISTER, J. 1960. Scotland: the Northern Highlands, 3rd Edn. *British Regional Geology*, 104 pp. H.M.S.O., Edinburgh.

PHILLIPS, W. E. A., STILLMAN, C. J. and MURPHY, T. 1976. A Caledonian plate tectonic model. *Jl geol. Soc. Lond.* **132**, 579–609.

PIASECKI, M. A. J. and VAN BREEMEN, O. 1979. A Morarian age for the 'younger Moines' of central and western Scotland. *Nature, Lond.* **278**, 734–6.

PIDGEON, R. T. and AFTALION, M. 1978. Cogenetic and inherited zircon U–Pb systems in granites: Palaeozoic granites of Scotland and England. Pp. 183–220 in Bowes and Leake (1978). See list A.

PITCHER, W. S. and BERGER, A. R. 1972. *The geology of Donegal.* 435 pp. Wiley, New York.

PRENTICE, J. E. 1962. The sedimentation history of the Carboniferous in Devon. Pp. 93–108 in Coe (1962). See list A.

RAMSBOTTOM, W. H. C. 1965. A pictorial diagram of the Namurian rocks of the Pennines. *Trans. Leeds geol. Ass.* **7**, 181–4.

RAMSBOTTOM, W. H. C. 1969. The Namurian of Britain. *C. R. Congr. Stratigr. Géol. Carb.* (6th, Sheffield). Vol. I. 219–32.

RAMSBOTTOM, W. H. C. 1970. Carboniferous faunas and palaeogeography of the south-west of England region. *Proc. Ussher Soc.* **2**, 144–57.

RAMSBOTTOM, W. H. C. 1973. Transgressions and regressions in the Dinantian: a new synthesis of British Dinantian stratigraphy. *Proc. Yorks. geol. Soc.* **39**, 567–607.

RAMSBOTTOM, W. H. C. 1977. Major cycles of transgression and regression (mesothems) in the Namurian. *Proc. Yorks. geol. Soc.* **41**, 261–91.

RAMSBOTTOM, W. H. C., GOOSSENS, R. F., SMITH, E. G. and CALVER, M. A. 1974. Carboniferous. Pp. 45–114 in Rayner and Hemingway (1974). See list A.

RAMSBOTTOM, W. H. C. and MITCHELL, M. 1980. The recognition and division of the Tournaisian Series in Britain. *Jl geol. Soc. Lond.* **137**, 61–3.

RAST, N. 1969. The relationship between Ordovician structure and volcanicity in Wales. Pp. 305–35 in Wood, A. (1969). See list A.

READ, H. H. 1961. Aspects of Caledonian magmatism in Britain. *L'pool Manchr geol. J.* **2**, 653–83.

RENOUF, J. T. 1974. The Proterozoic and Palaeozoic development of the Armorican and Cornubian provinces. *Proc. Ussher Soc.* **3**, 6–43.

RICKARDS, R. B. 1976. The sequence of Silurian graptolite zones in the British Isles. *Geol. J.* **11**, 153–88.

RIDDIHOUGH, R. P. and MAX, M. D. 1976. A geological framework for the continental margin to the west of Ireland. *Geol. J.* **11**, 109–20.

RIDGWAY, J. 1975. The stratigraphy of Ordovician volcanic rocks on the southern and eastern flanks of the Harlech Dome in Merionethshire. *Geol. J.* **10**, 87–106.

ROBERTS, J. L. and TREAGUS, J. E. 1964. A reinterpretation of the Ben Lui fold. *Geol. Mag.* **101**, 512–16.

ROSE, W. C. C. and DUNHAM, K. C. 1977. Geology and hematite deposits of south Cumbria. *Econ. Mem. geol. Surv. Gt Br.* 170 pp.

RUSHTON, A. W. A. 1974. The Cambrian of Wales and England. Pp. 43–121 in Holland (1974). See list A.

SADLER, P. M. 1973. An interpretation of new stratigraphic evidence from south Cornwall. *Proc. Ussher Soc.* 2, 535–50.

SCHWARZACHER, W. 1975. Sedimentation models and quantitative stratigraphy. *Devs Sedimentology*, 19, 382 pp. Elsevier, Amsterdam.

SCOTT, A. C. 1978. Sedimentological and ecological control of Westphalian B plant assemblages from West Yorkshire. *Proc. Yorks. geol. Soc.* 41, 461–508.

SCOTT, A. C. 1979. The ecology of Coal Measure floras from northern Britain. *Proc. Geol. Ass.* 90, 97–116.

SELLWOOD, B. W. 1972. Regional environmental changes across a Lower Jurassic stage-boundary. *Palaeontology* 15, 125–57.

SELLWOOD, B. W. and McKERROW, W. S. 1974. Depositional environments in the lower part of the Great Oolite Group of Oxfordshire and north Gloucestershire. *Proc. Geol. Ass.* 85, 189–210.

SHACKLETON, R. M. 1954. The structural evolution of North Wales. *L'pool Manchr geol. J.* 1, 261–96.

SHACKLETON, R. M. 1969. The Pre-Cambrian of North Wales. Pp. 1–22 in Wood, A. (1969). See list A.

SHACKLETON, R. M. 1975. Precambrian rocks of Wales. Pp. 76–82 in Harris *et al.* (1975). See list A.

SHEARMAN, D. J. 1966. Origin of marine evaporites by diagenesis. *Trans. Instn Min. Metall.* 75B, B208–15.

SKEVINGTON, D. 1971. Palaeontological evidence bearing on the age of the Dalradian deformation and metamorphism in Ireland and Scotland. *Scott. J. Geol.* 7, 285–8.

SMITH, A. G. 1976. Plate tectonics and orogeny: a review. *Tectonophysics* 33, 215–85.

SMITH, A. J. and CURRY, D. 1975. The structure and geological evolution of the English Channel. *Phil. Trans. R. Soc.* A, 279, 3–20.

SMITH, D. B. 1973. The origin of Permian middle and upper potash deposits of Yorkshire: an alternative hypothesis. *Proc. Yorks. geol. Soc.* 39, 327–46.

SMITH, D. B. 1974. Permian. Pp. 115–44 in Rayner and Hemingway (1974). See list A.

SMITH, D. B. 1979. Rapid marine transgressions of the Upper Permian Zechstein Sea. *Jl geol. Soc. Lond.* 136, 155–6.

SPENCER, A. M. 1971. Late Precambrian glaciation in Scotland. *Mem. geol. Soc. Lond.* No. 6, 100 pp.

STEEL, R. J. 1974. New Red Sandstone floodplain and piedmont sedimentation in the Hebridean Province, Scotland. *J. sedim. Petrol.* 44, 336–57.

STEEL, R. J. 1978. Triassic rift basins of northwest Scotland – their configuration, infilling and development. Pp. MNNSS/7 1–18 in *Mesozoic northern North Sea Symposium* (Oslo, 1977). Norwegian Petroleum Society.

STEIGER, R. H. and JÄGER, E. 1977. Subcommission on geochronology: convention on the use of decay constants in geo- and cosmochronology. *Earth planet. Sci. Lett.* 36, 359–62.

STEWART, A. D. 1969. Torridonian rocks of Scotland reviewed. Pp. 595–608 in Kay (1969). See list A.

STEWART, A. D. 1975. 'Torridonian' rocks of western Scotland. Pp. 43–51 in Harris *et al.* (1975). See list A.

STEWART, F. H. 1949. The petrology of the evaporites of the Eskdale No. 2 boring, East Yorkshire. Part I, The Lower Evaporite Bed. *Mineralog. Mag.* **28**, 621–75.

STEWART, F. H. 1951*a*. Part II, The Middle Evaporite Bed. *Mineralog. Mag.* **29**, 445–75.

STEWART, F. H. 1951*b*. Part III, The Upper Evaporite Bed. *Mineralog. Mag.* **29**, 557–72.

STILLMAN, C. J., DOWNES, K. and SCHIENER, E. J. 1974. Caradocian volcanicity in east and south-east Ireland. *Sci. Proc. R. Dublin Soc.* A, **5**, 87–98.

STROGEN, P. 1974. The sub-Palaeozoic basement in central Ireland. *Nature, Lond.* **250**, 562–3.

SYKES, R. M. 1975. The stratigraphy of the Callovian and Oxfordian stages (Middle–Upper Jurassic) in northern Scotland. *Scott. J. Geol.* **11**, 51–78.

TALBOT, M. R. 1973. Major sedimentary cycles in the Corallian Beds (Oxfordian) of southern England. *Palaeogeogr. Palaeoclimatol. Palaeoecol.* **14**, 293–317.

TAYLOR, J. C. M. 1978. Sandstone diagenesis (Introduction). *Jl geol. Soc. Lond.* **135**, 3–5.

TAYLOR, J. H. 1949. Petrology of the Northampton Sand Ironstone formation. *Mem. geol. Surv. U.K.* 111 pp. H.M.S.O., London.

TAYLOR, S. and ANDREW, C. J. 1978. Silvermines orebodies, County Tipperary, Ireland. *Trans. Instn Min. Metall.* **B87**, B111–24.

THOMAS, L. P. 1974. The Westphalian (Coal Measures) in South Wales. Pp. 133–60 in Owen (1974*a*). See list A.

THOMAS, T. M. 1972. The mineral industry in Wales – a review of production trends, resources and exploitation trends. *Proc. Geol. Ass.* **83**, 365–83.

THORPE, R. S. 1974. Aspects of magmatism and plate tectonics in the Precambrian of England and Wales. *Geol. J.* **9**, 115–36.

TOWNSON, W. G. 1975. Lithostratigraphy and deposition of the type Portlandian. *Jl geol. Soc. Lond.* **131**, 619–38.

TOWNSON, W. G. and WIMBLEDON, W. A. 1979. The Portlandian strata of the Bas Boulonnais, France. *Proc. Geol. Ass.* **90**, 81–91.

TRUEMAN, A. E. (Editor). 1954. *The coalfields of Great Britain*. 396 pp. Edward Arnold, London.

UPTON, B. G. J., ASPEN, P., GRAHAM, A. and CHAPMAN, N. A. 1976. Pre-Palaeozoic basement of the Scottish Midland Valley. *Nature, Lond.* **260**, 517–18.

VAN EYSINGER, F. W. B. 1975. *Geological time table*, 3rd Edn. Elsevier, Amsterdam. [Chart showing systems, series, stages, etc.]

WADGE, A. J. 1978. Classification and stratigraphical relationships of the Lower Ordovician rocks. Pp. 68–78 in Moseley (1978*a*). See list A.

WALKER, R. G. 1966. Shale Grit and Grindslow Shales: transition from turbidite to shallow water sediments in the Upper Carboniferous of northern England. *J. sedim. Petrol.* **36**, 90–114.

WALMSLEY, P. J. 1975. The Forties field. Pp. 477–85 in Woodland (1975). See list A.

WALTON, E. K. 1963. Sedimentation and structure in the Southern Uplands. Pp. 71–97 in *The British Caledonides* (Eds. Johnson and Stewart). 280 pp. Oliver and Boyd, Edinburgh.

WALTON, E. K. 1965. Lower Palaeozoic rocks – stratigraphy. Pp. 161–200 in Craig (1965). See list A.

WARMAN, H. R. 1978. Oil and natural gas. Pp. 33–53 in Knill (1978). See list A.

WARRINGTON, G. 1970. The stratigraphy and palaeontology of the 'Keuper' Series of the Central Midlands of England. *Q. Jl geol. Soc. Lond.* **126**, 183–223.

WARRINGTON, G. 1974. Trias. Pp. 145–60 in Rayner and Hemingway (1974). See list A.

WATERSTON, C. D. 1965. Old Red Sandstone. Pp. 269–308 in Craig (1965). See list A.

WATSON, J. V. 1975a. The Precambrian rocks of the British Isles – a preliminary review. Pp. 1–10 in Harris *et al.* (1975). See list A.

WATSON, J. V. 1975b. The Lewisian complex. Pp. 15–29 in Harris *et al.* (1975). See list A.

WEST, I. M. 1964. Evaporite diagenesis in the lower Purbeck beds of Dorset. *Proc. Yorks. geol. Soc.* **34**, 315–30.

WEST, I. M. 1975. Evaporites and associated sediments of the basal Purbeck formation (Upper Jurassic) of Dorset. *Proc. Geol. Ass.* **86**, 205–55.

WEST, R. G. 1963. Problems of the British Quaternary. *Proc. Geol. Ass.* **74**, 147–86.

WESTOLL, T. S. 1979. Devonian fish biostratigraphy. Pp. 341–53 in House *et al.* (1979). See list A.

WILLIAMS, A. 1962. The Barr and Lower Ardmillan Series (Caradoc) of the Girvan district, south-west Ayrshire, with a description of the brachiopods. *Mem. geol. Soc. Lond.* No. 3, 267 pp.

WILLIAMS, A. 1976. Plate tectonics and biofacies evolution as factors in Ordovician correlation. Pp. 29–66 in Bassett (1976). See list A.

WILLIAMS, C. E. and McARDLE, P. 1978. Ireland. Pp. 319–45 in Bowie *et al.* (1978). See list A.

WILLIAMS, G. E. 1969. Petrography and origin of pebbles from Torridonian strata (late Precambrian), northwest Scotland. Pp. 609–29 in Kay (1969). See list A.

WILLIAMSON, I. A. 1978. Coal, geology applied to subsurface mining. Pp. 54–65 in Knill (1978). See list A.

WILLS, L. J. 1929. *The physiographical evolution of Britain.* 376 pp. Edward Arnold, London.

WILLS, L. J. 1973. A palaeogeological map of the Palaeozoic floor beneath the Permian and Mesozoic formations in England and Wales. *Mem. geol. Soc. Lond.* No. 7, 23 pp. [Map.]

WILLS, L. J. 1978. A palaeogeological map of the Lower Palaeozoic floor below the cover of Upper Devonian, Carboniferous and later formations. *Mem. geol. Soc. Lond.* No. 8, 36 pp. [Map.]

WIMBLEDON, W. A. and COPE, J. C. W. 1978. The ammonite faunas of the English Portland Beds and the zones of the Portlandian Stage. *Jl geol. Soc. Lond.* **135**, 183–90.

WOOD, A. and SMITH, A. J. 1959. The sedimentation and sedimentary history of the Aberystwyth Grits (Upper Llandoverian) *Q. Jl geol. Soc. Lond.* **114**, 163–95.

WOOD, D. S. 1969. The base and correlation of the Cambrian rocks of North Wales. Pp. 47–66 in Wood, A. (1969). See list A.

WOODHALL, D. and KNOX, R. W. O'B. 1979. Mesozoic volcanism in the northern North Sea and adjacent areas. *Bull. geol. Surv. Gt Br.* No. 70, 34–56.

WOODLAND, A. W. (Editor). 1971. The Llanbedr (Mochras Farm) borehole. *Rep. Inst. geol. Sci.* No. 71/18. 115 pp.

WORSSAM, B. C. and IVIMEY-COOK, H. C. 1971. The stratigraphy of the Geological Survey borehole at Warlingham, Surrey. *Bull. geol. Surv. Gt Br.* No. 36, 1–57.

ZIEGLER, A. M. 1970. Geosynclinal development of the British Isles during the Silurian period. *J. Geol.* **78**, 445–79.

ZIEGLER, A. M., COCKS, L. R. M. and McKERROW, W. S. 1968. The Llandovery transgression of the Welsh Borderland. *Palaeontology* **11**, 736–82.

ZIEGLER, A. M., RICKARDS, R. B. and McKERROW, W. S. 1974. Correlation of the Silurian rocks of the British Isles. *Spec. Pap. geol. Soc. Amer.* No. 154, 154 pp.

ZIEGLER, P. A. 1978. North-western Europe: tectonics and basin development. *Geol. Mijnbouw*, 57, 589–626.

ZIEGLER, W. L. 1977. Summing up on seismically active margins (in Conference Report on 'Structural style of continental margins'). *Jl geol. Soc. Lond.* 134, 84–6.

INDEX

Authors and zonal index fossils are given in the list of References and the Appendix respectively, and do not appear in the Index. Page numbers in *italics* refer to illustrations. Subsidiary items under the main entries are arranged in page, not alphabetical, order.

Index